Bingchuan Bian

Topological Optimization of Buckling

Also of Interest

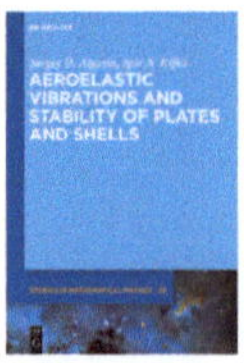

Aeroelastic Vibrations and Stability of Plates and Shells
S. Algazin, I Kijko, *2014*
ISBN 978–3-11–033836-2, e-ISBN (PDF) 978–3-11–033837-9,
e-ISBN (EPUB) 978–3-11–038945-6

Fracture Mechanics in Layered and Graded Solids
T. Xiaohong, Q. Yue, 2014
ISBN 978–3-11–029787-4, e-ISBN (PDF) 978–3-11–029797-3,
e-ISBN (EPUB) 978–3-11–036955-7

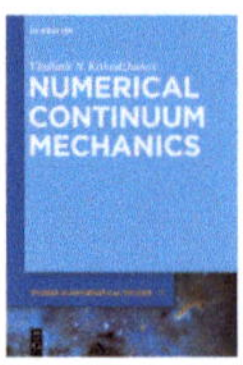

Numerical Continuum Mechanics
V. Kukudzhanov, 2012
ISBN 978–3-11–027322-9, e-ISBN (PDF) 978–3-11–027338-0

International Journal of Applied Mechanics and Engineering
4 Issues/Year
ISSN 2353–9003

Bingchuan Bian

Topological Optimization of Buckling

DE GRUYTER

華中科技大學出版社
http://www.hustp.com

Author
Prof. Bingchuan Bian
Taishan University
Dept. of Mechanical and architectural Engineering
Taian
271021 SHANDONG
China

Bingchuan Bian is associate professor at the Mechanical and Architectural engineering College of the Taishan University. Before that, he worked in Luneng Taishan Electric Power Equipment Co., Ltd in Shandong Province, China. He mainly engages in teaching and scientific research for the optimization design of mechanical structures and in the research and development of electric power equipment.

He published more than 30 papers in academic journals and international conferences.

ISBN 978-3-11-046116-9
e-ISBN (PDF) 978-3-11-046227-2
e-ISBN (EPUB) 978-3-11-046127-5

Library of Congress Cataloging-in-Publication Data
Names: Bian, Bingchuan, author.
Title: Topological optimization of buckling / Bingchuan Bian.
Description: Berlin ; Boston : de Gruyter, [2018] | Includes bibliographical references and index.
Identifiers: LCCN 2018007857 | ISBN 9783110461169 (hardcover)
Subjects: LCSH: Buckling (Mechanics) | Strength of materials. | Engineering mathematics.
Classification: LCC TA656.2 .B53 2018 | DDC 624.1/76–dc23 LC record available at https://lccn.loc.gov/2018007857

Bibliographic information published by the Deutsche Nationalbibliothek
The Deutsche Nationalbibliothek lists this publication in the Deutsche Nationalbibliografie; detailed bibliographic data are available on the Internet at http://dnb.dnb.de.

Typesetting: Integra Software Services Pvt. Ltd.
Printing and binding: CPI books GmbH, Leck
Cover image: Eskemar/ iStock / Getty Images

www.degruyter.com

Acknowledgments

I am indebted to Professor Yunkang Sui for his guidance during the preparation of this manuscript. Mr. Sui's profound wisdom, meticulous approach, and innovative methods continue to influence and inspire me. I was fortunate enough to have spent four years at Beijing University of Technology. In addition to academic training, Mr. Sui mentored and imparted various life skills. These have immensely helped me.

I am also grateful to Associate Professor Baorui Ye who provided help and guidance during my study time.

I also thank my teachers' selfless help in my doctoral study at Engineering Numerical Simulation Center of Beijing University of Technology. I also thank all students for their cooperation.

I am grateful to all the leaders, teachers at Taishan University and my family for their patience, care and support.

This research was supported by National Natural Science Foundation (10472003), Beijing Natural Science Foundation (3042002), Shandong Province Youth Scientist Fund Project (BS2009ZZ007), the Taishan University Research Fund Project (Y07-2-08) and the United States MSC company support.

https://doi.org/10.1515/9783110462272-201

Acknowledgements

Contents

1 Introduction

1.1 Background

With the development of modern industrial technology, a large number of new light and thin structures of high strength are widely used in national defense and civilian industries. Especially with the rapid development of aviation, aerospace and atomic energy industries, complex structures of the basic rod, plate and shell are applied to a large number of industries. Thus, stability of these basic components under various loads has attracted attention of many scholars to a large extent [1].

A majority of engineering accidents in the past have been due to buckling instability of the structure. For example, the three-span cantilever Quebec Bridge was built in Canada in 1907. Owing to insufficient flexural strength and buckling instability of the batten, the entire deck instantly fell into the river, thereby killing 75 workers. Designers of Hartford state's stadium in Connecticut constructed a roof, the grid structure of which considered only the bending buckling strut; however, torsional buckling, strut support and eccentric torsional buckling were not considered, and the stadium was made available for public use. On January 18, 1978 – a heavy snowy night – the space grid structure crashed down due to buckling instability of the pressure bar, causing a huge loss. The facts stated here completely explain the importance of buckling in structural design [2].

Engineering structural design works can be divided into traditional design and optimal design. Design engineers can choose the type of structure and form, including the necessary materials, in accordance with their own theoretical knowledge and practical experience in traditional design. The structures are analyzed after preparing the initial design scheme. The strength, stiffness, stability (including regional stability and global stability) and the vibration frequency of the structures are checked in accordance with the results of the analysis. When the design does not meet the economic and security requirements, then the initial design is modified, reanalyzed and rechecked until the structure meets the necessary requirements. Finding a reasonable distribution of material is the main shortcoming of traditional design, and thus it is difficult to compare the ideals of both economic and safe design. Optimal design is all about providing the optimal mathematical model, solving the optimal model based on structural optimization and design theories and methods in accordance with the established structure of the type, form, condition, material and norms under various constraints. The optimal design is in convergence after several iterations. Through optimal design, the structure achieves economic unity and security. Optimal design encompasses the structure's new development and achievements, and it has great significance and broad prospects [3–5].

https://doi.org/10.1515/9783110462272-001

Structural optimization began in the 1960s. With high-speed development of the finite element analysis theory, numerical calculation method and mathematical programming theory, and computer technology with large capacity, the structure optimizing subject has developed rapidly.

Structural optimization design can be classified into different forms conforming to different standards. On the basis of design variables, it can be divided into continuous variable design and discrete variable design. In accordance with the ease of design of the structural optimization, it can be divided into three levels: size optimization (cross section), shape optimization and topology optimization. Size optimization outlines ideal component parameters to determine the ideal size of the structure. Shape optimization is about identifying the optimal shape in that it minimizes certain cost function while satisfying given constraints. In this case, the outer boundary of the structure is modified in order to solve the optimization problem. Topology optimization helps in optimizing material layout within a given design space, for a given set of loads, boundary conditions and constraints with the goal of maximizing the performance of the system. Topological structure can be understood as a region of space arrangements.

Topological structure optimization can be divided into two categories: topological optimization of skeleton structure and topological optimization of continuum structure. Topology of the skeleton structure is the connectivity and location of the bar and the number of nodes; topological structure of the continuum is a row of the border, shape and whether the internal distribution of holes and cavities. While size (section) optimization and shape optimization are considered low-level optimizations, topology optimization is a high-level optimization technique. At present, topology optimization is one of the areas of research that attracts considerable interest. The topology optimization of structure has gone through the development from narrow sense to broad sense. Narrow sense refers to the skeleton structure of truss or frame class, to find out whether a node or rod exists or not. Broad sense is for the existence of subdomain. Whether or narrowly past the broad topology optimization, the design variables are linked to or in the cross-sectional shape optimization of the variables, the actual reduced to low levels of optimization problem. If we want to make breakthrough progress in topology optimization, we must step out from this low-level optimization. Therefore, the modeling of topology optimization should be the focus of topology optimization [6].

In 1904, Michell [7] took the truss structure as an example to explore topology optimization, which is the symbol of the birth of the theory of modern structural optimization design. However, for quite some time now, the research work is mainly focused on the optimization of the structure and shape optimization, and the topology optimization of the continuum structures has not made great progress. By the beginning of the 1980s, Cheng K T and Olhoff [8] published the results of structural optimization of an elastic plate, which attracted the attention of many scholars. In 1988, Bendsoe and Kikuch [9] published the uniform method for topology optimization of continuum structures, which set a new stage for topology optimization of continuum structures. A series of research results has been achieved in the future, but there are still many problems in the topology optimization of continuum structures.

At present, topology optimization problems that are considered more restraint include stress, displacement and frequency constraints, but buckling constraints are seldom considered; therefore, the research on the topology optimization of continuum structures with buckling constraints is a challenging task.

1.2 Current Development

With the development of computer technology, the finite element method and the mathematical programming method, the ability and method of the structural analysis have been improved, and the structural optimization has become one of the most active branches in computational mechanics.

The research scope of structure optimization is very wide, which can be divided into size (cross section) optimization, shape optimization, topology optimization and layout optimization. Depending on the complexity of the research objective, the design of a simple truss structure can be developed for the structural design of a variety of complex elements, such as beam structure, plate, shell and composite structure. The development of design variables has gone through the process which is from the continuous variable optimization design, discrete variable optimization design to distribution parameter optimization and discrete parameter optimization in continuous variable optimization. Constraint types are developed from the initial stress constraints, displacement constraints to the dynamic characteristics of the frequency constraints and buckling constraints and other advanced stages. Objective function optimization can be classified into single-objective function optimization and multi-objective function optimization. However, considering uncertain factors of engineering design, some new optimization models have appeared, such as the structure optimization design based on reliability and the fuzzy optimization of structure. The optimization method begins from the criterion of the traditional method, programming method to such as the relative difference quotient method, simulated annealing, genetic algorithm and Tabu search method.

In this book, we study the topology optimization of continuum structures under the buckling constraints (stability constraints).

1.2.1 Research on Buckling Theory

Buckling is also called structural instability, and it can be defined as follows: when a structure load reaches a certain value, and if there is an increase in a tiny disturbance force, then the structure of the equilibrium configuration will be in a new form, and even if the small disturbance force is removed from the structure, it cannot restore to the original state; this situation is called structural instability or buckling. The corresponding load is called the critical buckling load. In modern times, the structural stability theory focuses on the buckling mode of the structure (bifurcation buckling or extreme type buckling), the buckling mode and the post-buckling equilibrium path. The equilibrium state of any object has three types: stable equilibrium state, unstable equilibrium state and equilibrium state. When the object has a small deviation from the equilibrium position, and if the object could return to its original equilibrium position, the equilibrium state is called a stable equilibrium.

If the object could not return to its original position and if there is a tiny deviation from its equilibrium position, it will further continue to deviate, and this state is the unstable equilibrium state. The equilibrium state is the intermediate state of the transition from the equilibrium state to the unstable state, and it is often called a critical state.

In the field of solid mechanics, there is no subject like structural buckling, and the mechanism of buckling is still in the process of reaching perfection. Buckling theory is an ancient and new research field, and it is proving to be a more difficult field in recent years. The buckling focus has been from elastic to plastic, the static buckling to dynamic buckling problem, and buckling problems have made considerable development in all aspects.

The buckling problem of the structure can be divided into several different types according to different ways:

(1) According to the load-bearing of the structure, it can be divided into static buckling and dynamic buckling. Buckling under static load is called static buckling, and buckling under dynamic load is called dynamic buckling.
(2) According to the properties of the material, the buckling of the structure can be divided into elastic buckling, plastic buckling and elastoplastic buckling.
    Elastic buckling: the structure is still in the range of elastic state before and after buckling, so it is called elastic buckling.
    Plastic buckling: buckling occurs when the structure is in the plastic stress state.
    Elastoplastic buckling: a form in between elastic buckling and plastic buckling, where the structure is in elastic stress state before buckling, due to disturbance deformation, and a part of the material is in the plastic state, which occurred in the elastoplastic stress state.
(3) According to the existing experimental results and methods in the static buckling, the buckling behavior is divided into three kinds: bifurcation buckling, extreme buckling and jump buckling.

Figure 1.1(a) and (b) [10] denotes the buckling load and deflection curve diagram of the compression bar and compression plate, OA is the loss of balance line before

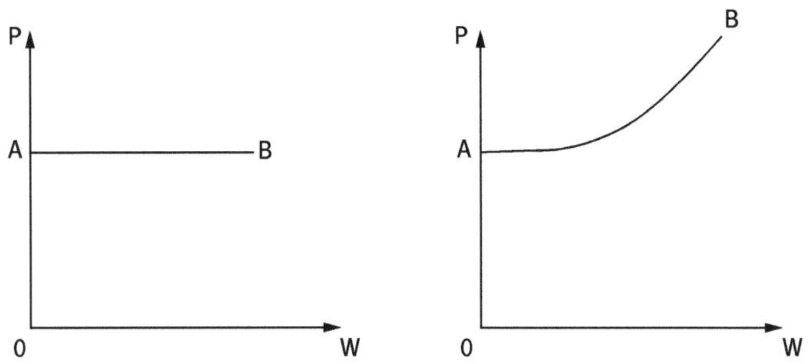

Figure 1.1: Curve of deflection and bifurcation buckling load.

stability, and it is called the main line; AB is balance line after instability, and it is called the vice line. The branch point is the main line and side line intersection. The buckling load or structure for the critical load is the load branch point.

The buckling load and deflection curve of the extreme type is shown in Figure 1.2. The characteristic of the extreme type buckling is that there is no obvious bifurcation point of the extreme point. However, when the maximum load is reached, the deformation will increase rapidly but the load will decrease.

The jump buckling curve is shown Figure 1.3; the OA section and BC section are stable, while the AB section is not stable, and when the load is increases to A, the equilibrium state has obvious jump, and suddenly it is translated to another large displacement equilibrium C.

The extremum point and the bifurcation point are two important concepts of buckling analysis. The load values $P_{\max}$ of the extremum point and the load values P_{cr} of the bifurcation point are called critical buckling loads of structural buckling. The corresponding states are called critical states. Before reaching the critical state, the state is called the buckling equilibrium state, and the state after the critical state is called the post-buckling equilibrium state [10].

In the structural design of thin plate, the load in critical state of initial buckling is the basis, that is to say that the classical linear theory is the base, and linear stability theory is limited to the discussion of the bifurcation point calculation of the initial buckling and the corresponding buckling mode. The main research direction of this book is to consider the continuum topology optimization problem with buckling constraints, so the main consideration in this book is the elastic, linear and prebuckling of the continuum structures.

The theoretical achievements of local and global static elastic buckling are mainly in the following aspects:

The theoretical analysis of the buckling of the component part and the whole elastic buckling analysis mainly have two kinds of methods: approximate method and numerical method. The approximation method is based on the effective width

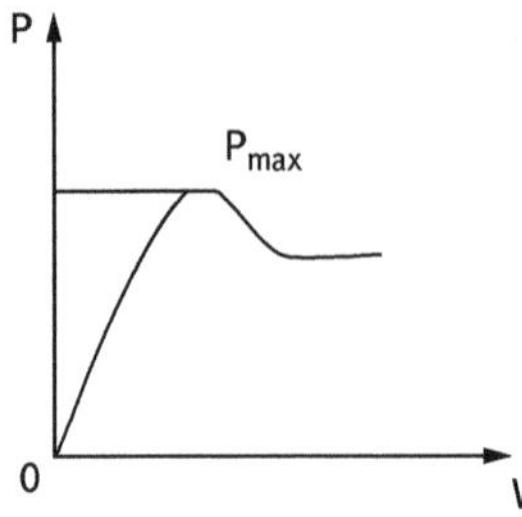

Figure 1.2: Curve of deflection and extreme type buckling load.

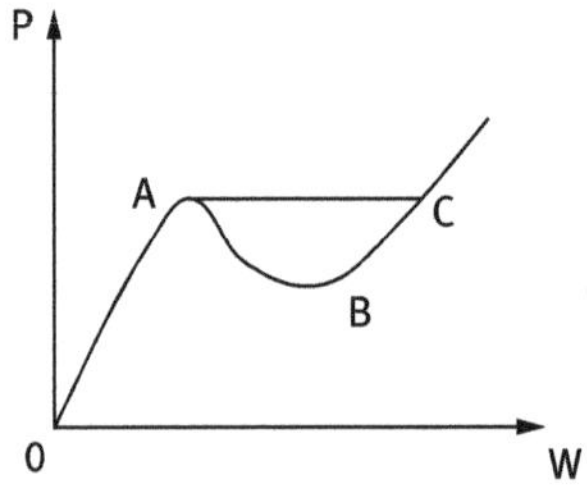

Figure 1.3: Curve of deflection and snap-through buckling load.

method [11], which is based on the concept and formula of the effective width of the short column section, and the ultimate load is determined by the stability bearing capacity formula of the column. The effective width method is used to account for the influence of the local buckling on the overall stability, which reflects the relationship between the local and the overall buckling, but not completely, because the integral component also has the opposite effect. The actual component has initial defects, and the relationship between the whole and the local buckling effects the local buckling, and at the same time the local buckling can cause the whole loss of bearing capacity early in turn.

The finite element method and finite strip method are the main methods of numerical simulation. Rajasekaran and Murray [12] studied the elastic buckling of the wide flange steel beam column by using the finite element method. The results show that the local and global correlation of the wide flange section is not obvious. Johnson and Will [13] studied the lateral torsional buckling of the beam considering the effect of cross-sectional distortion by the finite element method. The method can deal with arbitrary loads and boundary conditions. Finite strip method is used to analyze the local and global buckling of simply supported beam bending moment in plane according to Hancock [14]. The finite strip method is not as good as the finite element method, but it can reduce the calculation amount.

The thin-walled parts design of the critical load according to local buckling and global buckling is usually considered as the optimal design. However, the research of Koiter [15], VanderNeut [16], Yanlin Guo [17] and others all showed that the thin-walled component that is designed according to the "same of the local buckling and global buckling" is sensitive to the initial imperfection, so that the stability limit bearing capacity is reduced. The study of local and global elastic buckling shows that the correlation is not obvious when the initial imperfection is not considered. This conclusion lays the foundation for the further research of buckling [18].

The static buckling of thin-walled component analysis should consider the global and local buckling objectively, but the role of global and local buckling was considered, and due to the influencing factors, the analysis becomes more complex. When the initial imperfection of the structure is not considered, the thin-walled component is designed by the critical load of local buckling equal to the global buckling, and it is generally considered to be the optimal design. In view of the above reasons, and the limitation of the research content, this book only studies the static elastic buckling of the continuum.

1.2.2 Structural Topology Optimization

1.2.2.1 Topology Optimization of Skeleton Structure

Most of the results of the research on the structure of the topology optimization are focused on the truss problem, and the topology optimization of the truss was first proposed by Michel (Michell) in 1904. He first proposed the concept of topology optimization in truss, the quality as the goal function, stress as constraint, and that the optimal solution can be obtained in a single-load truss structure using the analytic method. The Michel truss topology optimization method is called the Michel criterion. The truss meet Michel criterion is called Michel truss. In this study, the classical layout theory was given by Prager and Rozvany [19], and the uniqueness of the Michell truss and the orthogonality of the pole pieces were discussed by Rozvany [20].

With the development of the finite element method and the computer technology, more attention has been paid to the numerical method for solving topology optimization problems, and many research results have been developed. The main research methods are the basic structure method and the evolutionary method.

The base structure method was proposed by Dorn, Gomory, Greenberg and others in 1964. The basic idea of this method is as follows: in the truss structure optimization, the initial structure is in some given nodes and bars, and then each node and all the other nodes are connected with a bar, and then redundant nodes and bars in base structure are deleted according to the algorithm and criterion, and the optimal structure is obtained.

There are two kinds of skeleton structure topology optimization methods: linear programming and nonlinear programming. The characteristics of the linear programming method are the internal force is the topology design variable, the nodal equilibrium equation or the force balance equation of the predetermined displacement field is the constraint condition, and the structural topology optimization problem is solved by the linear programming. The advantage of this method is that the model is simple and the computational efficiency is high. The disadvantage is that it is difficult to deal with the displacement constraints of multi-working conditions and without the compatibility conditions of deformation. For such linear programming problems, foreign research results were given by Dorn Dorn et al. [21], Ringertz [22], Kirsch [23] and so on. The domestic studies were mainly on the topology optimization design of truss with multiple working conditions by Baoyan Duan and Shanghui Ye [24] and sequences planning model with the internal force design variable and using the simplex method for solving the optimization model by Zhongfu Tan and Huanchun Sun [25].

The characteristics of the nonlinear programming method is that the most topology design variables is the cross-sectional area, and change the topology by turn the sectional area to zero. In essence the topological optimization problem is transformed into a size optimization problem(generalized section). The truss topology optimization is induced into a nonlinear programming problem, and it can deal with problems such as the multiple working conditions, including stress, displacement, frequency and stability. The typical results of this problem are such as the minimum weight optimization design of statically indeterminate truss by Sheu and Schmit [26] and the branch and bound algorithm for topology optimization design of truss by Ringertz [27]. The domestic research results are the research of topology optimization design about the discrete variables of the truss by Huanchun Sun, Yuefang Wang, Shan Chai and Shilian Shuan and so on [28, 29].

There is another method called the evolutionary method. The evolutionary method is the inverse process of basic structure method, which is based on a fixed node, load, and a simple topology structure is given first, and through certain optimization methods (criterion method, programming method) the structural elements were increased, and finally the optimal topology shape and structure were obtained. The main research achievements were: analysis and future prospects of evolutionary method by Kirsch [30]. William proposed the natural growth method in its article in 1995 [31]. Mckeown proposed the increasing node method in 1998 [32]. Liu Jingsheng proposed a variable structure growth method in the article [33, 34]. Luo Ying, Duan Baoyan proposed a triangular grid evolution method based on genetic algorithm [35]. The disadvantage of the evolutionary method is that the structural element can only increase and cannot be cut. As a result, the research on the evolutionary method is far less active than the basic structural method.

It also brings some problems when using the base structure method to optimize the topology of the truss. Zhou and Rozvany [36] pointed out the special problems encountered when considering the stress and local stability constraints in the topology optimization of truss structures. Another important issue is the "singular optimal solution problem." Singular optimal was first discovered in 1968 by Sved and Ginos [37] on the three bar truss topology optimization problem under the stress constraint and minimizing the weight. Gengdong Cheng and Jiang Zheng in 1992 [38] pointed out that the bar stress function in the zero area of discontinuity is the root cause of the existence of singular optimal solutions. So the recovery of the continuity constraint function in the bar and the vicinity of the zero section are the key solutions to the singular optimal to topological structure optimization and give a correct representation of the feasible region. Gengdong Cheng and Guo Xu [39, 40] proposed ε-relaxation method process stress constrained discrete topology optimization problems, through to rod constraint relaxation of the structure internal forces, modify the feasible domain shape, and the singular optimal solution was eliminated by this method. Zou and Zuo [41] used the biology of disuse thought, and considering the material and stability of stress constraints in the compression truss, deleting the bar that does not meet the threshold value, the optimal topology structure achieved a stable and the light weight of the target. Wei [42] studied the topology optimization of truss with stress constraint, local stability and overall stability constraints, and the solve method is proposed to the buckling length jump singularity phenomenon, and it was extended to three-dimensional large-scale problems. In order to avoid the phenomenon of "singular optimal solution," the global optimal solution was found out by Xu and Xia [43]. The simulated annealing method was proposed by May, Cai and Cheng [44, 45]. In addition, Zhao et al. [46] had studied the optimization design of stiffness and stability of beams with different cross sections. Liu et al. [47] had designed the large depressed center flat underframe structure optimization and stability.

ICM (independent, continuous, mapping) method was proposed by Sui et al. [48–56].

"Independent" means that topological variable is not depended on the design variables of lower levels. "Continuous" means that topological variable is continuum in the interval [0, 1]. "Mapping" means that there is a transformation of discrete-to-continuum topological variables and its inversing transformation of continuum-to-discrete ones. Deqing and Xin [57, 58] proposed that the new research work is also related to the truss. The truss topology optimization is extended to the framework of topology optimization by Jiazheng and Xuchun [59, 60]. Independent continuous topological variable concept did not only effectively solve the topology optimization problem of truss, but also can effectively solve the problem of continuum topology optimization. The unified framework and topology optimization model of continuum structure are realized.

1.2.2.2 Topology Optimization of Continuum Structures

Topology optimization of continuum structure is a high-level topology optimization: its mathematical model is difficult, the numerical calculation has a large amount, so for a very long period of time it is without great development.

In 1988, Bendsoe and Kikuch proposed the homogenization method for the topology optimization of continuum structures, thus creating a new stage of topology optimization of continuum structures. The basic idea of homogenization method is that the micro structure (Single cell) is added in the composition materials, the size of the micro structure cell are taken as topology variables in the process of optimization. The changing of a cell size is realized the micro structure additions and deletions, and allowing intermediate size single cells to compose the composite materials. As result the design space is expanded,and to achieve a unified and continuity of topology optimization model and size optimization model of structure.

Homogenization method is one of the most widely used methods in structural topology optimization. The basic idea of this method is the work of Gengdong Cheng and Olhoff, and their important contribution is to introduce the microstructure to the optimization design for the first time. The homogenization method is used to develop a set of Bensoussan, which is based on perturbation theory. The relationship between the size of the microstructure of the material and the macroscopic elastic constants of the material was established. So the homogenization method is a rigorous mathematical theory. Since then, many research work are carried out in the following aspects:

(1) The square hollow microstructure, two-stage arrangement layered microstructure, rectangular hollow microstructure, three-dimensional layered microstructure model, and the orthogonal microstructure hypothesis will lead to erroneous results and were put forward on the study of the microstructure of the model.

(2) The application of homogenization method was expanded in all kinds of problems, including the plane problem of multi-condition, three-dimensional continuum problem, the two-dimensional problem of vibration, thermal elastic problem, the problem of continuous buckling, three-dimensional shell, shell and composite topology optimization, etc.

The method which is to calculate micro cavity with different size in the structure at first, and the function which between density and elastic modulus in the [0,1]interval by polynomial interpolation was obtained, and then minimum the structure compliance degree in a certain volume constraints with element density as design variables and design area using finite element discrete method in the above research. However, it is more difficult to deal with the problem, and the method needs to be further improved.

Variable thickness method is a topology optimization method that was adopted earlier, and the basic idea is to take the structure element thickness as design variables, and the continuum topology optimization problem is transformed into a generalized dimension optimization problem, by removing the lower limit size thickness elements to change the topology. Variable thickness method is relatively simple to use, and it is suitable for the plane structure (film, plate, shell etc.), but it has a certain difficulty in the three-dimensional topology. Gengdong Cheng and Wang Jian research of the thin plate structure was the representative work. Tenek and Hagiwara studied on thin shell structure and Kemin Zhou and Yunchang Hu studied on the continuum plane famous research.

The variable density method is another effective method in structural topology optimization, which should be attributed to the material description method. The basic idea is to introduce a hypothetical density in [0, 1] between the variable physical parameters of the material, and assuming the material's parameter such as elastic modulus, a nonlinear relationship between stress and density, take density as the design variable, and take the least compliant material as target function, considering the equilibrium conditions and the volume constraints. Delete density less than a threshold to determine element and achieve topology changes. The main research works were by Mlejnek who established the model of variable density method; Yang designed the topology optimization of the truck body; Wang Jian and Gengdong Cheng demonstrated the equivalent conditions of the variable density method and variable thickness method; Yuan Zhen et al. studied the topology optimization based on the hybrid element and variable density.

Yunkang Sui proposed "independent, continuous, mapping method" (ICM method) by introducing a smooth function to transform the 1–0 discrete variable into the continuous variable of the [0, 1] interval, which can be used to effectively algorithm based on gradient.

After that, the smooth idea is applied to the element, and the complex element of structure topology optimization is put forward. Research on topology optimization of continuum structure with frequency constraint under ICM method and research on the global stress constraints method in topology optimization were carried out. There are 6 doctoral students Deqing Yang, Yu Xin, Jiazheng Du, Xirong Peng, Hongling Ye, Xuesheng Zhang had made some research works respectively based on the ICM method under the guidance of Professor Yunkang Sui.

The above methods are commonly used in topology optimization, and some other topology optimization methods and strategies are also proposed. Jog and Haber proposed "Equal perimeter method"; Eschenauer proposed "the Bubble method." The evolution method of Xie and the SIMP method of Zhou and Rozvany had some reference value. In evolutionary algorithm, genetic algorithm (GA) is a kind of research method. Kita and Tanie applied the boundary element method to analyze the topology and shape optimization of continuum structure. Nakanishi used the GA algorithm to optimal the topology of three-dimensional structure, and Haitao Cui used the genetic algorithm for the continuum structure topology optimization. In addition, simulated annealing algorithm (SA), particle swarm optimization (PSOA) and so on belong to the probability search method. Burezynski Tadeusz also used the GA algorithm to optimize the structure of the continuum. SA methods were used to design a parallel algorithm for topology optimization of continuum structures by S.S. Rao. The PSOA were used by Fourie et al. for topology optimization of continuum structures. These algorithms are suitable for solving small-scale problems. When the problem scale is larger, the space of the solution space is "explosive," and the optimization efficiency is low.

This is mainly due because algorithm direct was used to the discrete topological variables, and an approximate model was not established.

A large number of numerical calculations show that the topology optimization of continuum structure also has the following problems:

(1) "Chess board" phenomenon: This refers to the element interval phenomenon that occurs in the topology layout after the topology optimization, and the shape is like a chess board. The research shows that the phenomenon of "chess board" is due to the improper establishment of the numerical model.
(2) Over reliance on grid: This mainly refers that the use of different grid computing will get different results, and the encrypted grid instead of the opposite results.
(3) Local optimal solution problem: For the same topology optimization problem, different initial values are selected, and the topology optimization and optimal solutions are obtained differently.

1.2.3 Structural Optimization Considering Buckling Constraints

Buckling exists widely in the practical engineering structure. Many examples have proved that the design does not consider the buckling and will give the engineering structure the security of the great hidden danger, and in fact, whether the designer consider the buckling or not, the buckling is absolutely exist in the structure. Therefore, more and more scholars begin to realize the importance of buckling in structure optimization design, but due to the high complexity of structure optimization model and computational efficiency of buckling, the current study on structure optimization considering buckling is relatively less.

As the structural static optimization and the dynamical optimization, there are three development phases of the buckling structural optimization. They are cross-sectional, shape, and topology optimization. The cross-sectional and shape optimization considering buckling constraints were developed by many scholars. The plate's shape and frame's cross-sectional optimization were studied by D. Manickarajah et al. using the Evolutional Structural Optimization (ESO) method. The Optimum design with size and shape variables and sensitivity analysis for buckling stability of complex built-up structures and composite material were made by Gu Yuanxian et al. The studies of the optimum design and stability analysis of composite material cylinder shell were carried out by Liu Tao et al. The studies by Rong Jianhua et al. reported the cross-sectional optimum designs of the frame under the maximum bucking critical load using ESO method. Stability optimization studies of the elastic cylinder shell were reported by Liang Bing et al. The optimum design of composite material laminated wood under buckling response constraints was carried out by Zhao Guozhong et al. The topological optimization studies considering buckling are seldom due to difficulty in the process. The optimum designs reported by Zhou Ming are the shell's topological optimization under linear buckling response using variable-density method. The topology optimization model for the shell structure is constructed, in which minimized volume as the objective function is subjected to the buckling eigenvalues constraints and the maximized lowest buckling eigenvalue is subject to volume constraints. The cross section of the bar was the design variable in the model. The topological optimization of truss studies by Guo Xu reported that the model for the truss structure is constructed, in which the minimized quality as the objective function is subjected to the local buckling constraints and global buckling constraints. Esben Lindgaard and Jonas Dahl have studied the behavior of the static geometrically nonlinear structure undergoing the process of topological optimization. Philip Stanford and Beran Bret studied the topological optimization of the metal panel flutter and buckling considering the conditions of aerodynamics, inertia, elasticity, and thermal load based on the SIMP method. Katsikadelis and Babouskos studied the optimal design of thin plate thickness considering stiffness and buckling using boundary element method.

In summary, the topology optimization of continuum structures is still a hotspot in the field of structural optimization. The topology optimization considering the buckling problem adds the difficulty to the problem, but only takes into account a variety of constraints fully, and the structure of the topology is more close to the engineering practice. So it has important theoretical significance and research value to the continuum structure topology optimization with the buckling.

1.3 Software Platform

MSC.Patran/Nastran software is one of the CAE software developed by MSC. Software company. It has a good finite element analysis system, which can be directly accessed with Ansys, Abaqus, Pro/E, UG, Catia and other CAE/CAD software.

MSC.Patran/Nastran has a powerful user independent development capabilities.

Software development tool PCL language (Command Language Patran) is a high-level, modular structure of programming language and user-defined tools, and its syntax structure is similar to the C language or Fortran language. The function library contains a variety of commonly used mathematical functions, character processing functions, documents and process operation functions, MSC.Patran/Nastran access functions, common graphics functions and the user interface generation functions. Users can write applications to set the user-defined interface, display the custom graphics, read and write Patran database, and other functions. Users can write their own algorithms into the MSC.Patran/Nastran software system by PCL. Users can complete the custom function by the advanced pre-post processing system.

The study of shape optimization and topology optimization and the truss, frame, membrane, plate, shell and other research object of section optimization on the MSC. Patran/Nastran software platform was done by the leadership of Professor Sui

Yunkang in engineering numerical simulation center of Beijing University of Technology which is in collaboration with the American MSC company.

This book takes MSC.Patran/Nastran as the software platform and uses PCL language to complete the study of topology optimization of continuum structures with buckling constraints.

1.4 Key Topics in Buckling Topological Optimization Research

This book focuses on the problem of topology optimization of continuum structures with buckling constraints, displacement constraints, stress constraints and interaction each other.

The main research contents areas follows:

(1) Topology optimization of continuum structures with buckling constraints
In this chapter, according to the ICM method, the topology optimization model for the continuum structure was constructed. The model had the minimized quality as the objective function which was subjected to the buckling constraints. The continuous independent topological variables were used in this problem. Based on the Taylor expansion and the filtering function, the objective function was approximately expressed as a second-order expressions. Based on the Rayleigh quotient, the Taylor expansion and the filtering function of the buckling constraints were approximately expressed as an explicit function. The optimization model was translated into a dual programming and solved by the sequence second-order programming. The number of the variables was reduced and the model's scale was diminished. Based on the idea of homogenization, the strain energy expression of the filter function of the element stiffness and the element geometrical stiffness was derived. The concept of "quality ratio" and "buckling constraint ratio" was defined. The one-order, two-order, three-order function curve of the "quality ratio" and "quality-buckling constraints ratio" of topology optimization structure was established. The least square method is used to fit the curve of "quality-buckling constraints ratio." The method is successfully applied to the topology optimization problem of two-dimensional flat plate, cylindrical shell and three-dimensional continuum structure. The results show that the method is effective and feasible.

(2) Topology optimization of continuum structures with buckling and displacement constraints
In this chapter, according to the ICM method, the topology optimization model for the continuum structure was constructed. The model had the minimized quality as the objective function subjected to the buckling constraints and key point displacement constraints. Based on the Taylor expansion and the filtering function, the objective function was approximately expressed as a second-order expression. Based on the Rayleigh quotient, the Moore's theorem and the filtering function the constraints

were approximately expressed as an explicit function. The optimization model was translated into a dual programming and solved by the sequence second-order programming. The concept of "displacement constraint ratio" was defined. The "quality ratio" and "displacement constraints ratio" function curve of topology optimization structure was established, and the "displacement-quality constraint ratio" curve was fitted by the least squares method. The "quality-buckling constraint ratio" is compared with that of the "quality-displacement constraint ratio." Finally, the method is applied to two-dimensional and three-dimensional continuum structure topology optimization problem with single load and multi-operating conditions.

(3) Topology optimization of continuum structures with buckling and stress constraints
In this chapter, according to the ICM method, the topology optimization model for the continuum structure was constructed. The model had the minimized quality as the objective function subjected to the buckling constraints and stress constraints. The continuous independent topological variables were used in this problem. Based on the Taylor expansion and the filtering function, the objective function was approximately expressed as second-order expressions. Based on the Rayleigh quotient, the Taylor expansion and the filtering function, the buckling constraints were approximately expressed as an explicit function. Using the globalization method of stress constraints and the von Mises' yield criterion in mechanics of materials, the local stress constraints were translated into the whole structure strain energy constraint. Thus, the analysis quantity of the sensitivity was decreased. The optimization model was translated into a dual programming and solved by the sequence second-order programming. The number of the variables was reduced and the model's scale was diminished. The concept of "stress constraint ratio" was defined. The "quality ratio" and "stress constraints ratio" function curve of topology optimization structure was established, and the "stress-quality constraint ratio" curve was fitted by the least squares method. The "quality-buckling constraint ratio" is compared with that of the "quality-stress constraint ratio". By using the safety factor of the material, the correction factor of the structural strain energy constraint is determined, and the correction coefficient has the practical engineering significance.

Finally, the method is applied to two-dimensional, three-dimensional continuum structure topology optimization problem with single working condition and multi-working conditions in multiple modes. The optimal topology of Zhaozhou Bridge and the drawbridge was verified.

(4) Topology optimization of continuum structures with buckling, displacement and stress constraints
In this chapter, according to the independent continuous mapping (ICM) method, the topology optimization model for the continuum structure was constructed. The model had the minimized quality as the objective function subjected to the buckling constraints displacement constraints and stress constraints. The objective function

and three kinds of constraints are approximate and explicitly expressed, and finally the optimization model is transformed into dual programming, and solved by the sequence second-order programming. The number of the variables was reduced and the model's scale was diminished. The function curves of the "quality ratio" and the whole constraint ratio of the topology optimization structure are established, which provide a set of functional relations for the selection of the constraint values in the all static constraint topology optimization. This method is applied to two-dimensional and three-dimensional continuum structure topology optimization problem of single and multiple loading conditions.

(5) Topology optimization software user interface and instructions
Topology optimization program is based on the user development interface of the finite element analysis software MSC.Patran/Nastran, using PCL language to carry out the program. In this paper, the practical significance of each constraint parameter and optimal parameter in the input interface of the topology optimization program is introduced in detail. The final text format and graph forms output functions were added in the software, it can be exempted from the export results trouble. At last, the user manual and help document were joined in the software. They facilitate the users to use the optimization software.

(6) Application of structure optimization in engineering
The design of the optimization system is carried out based on the method of VC++ called FORTRAN mixed programming. The method combines the topology optimization design and shackle structure design for the first time in topology optimization structure design of transformer spreader. For small transformers, rectangular and oblong tank do finite element analysis, the tank plate size and strengthening section size of reinforcements under the optimal stress and displacement constraint were obtained.

2 Topological Optimization for Continuous Structure with Bulking Constrain

2.1 Introduction

The intensity, stiffness and buckling are very important factors that determine the safety of a structure. A composite structure is composed of single poles, plates and shells. Structural failure due to buckling of one part is catastrophic. So the buckling has attracted considerable attention in recent past. Structural topology optimization is one of the most challenging issues in the field of structural optimization. Depending on the buckling problem of a structure, the topology optimization problem could be more difficult

As the structural static optimization and the dynamical optimization, there are three development phases of the buckling structural optimization. They are cross-sectional, shape and topology optimization. The cross-sectional and shape optimization considering buckling constraints were developed by many scholars. The plate's shape and frame's cross-sectional optimization were studied by Manickarajah et al. using the elocutionary structural optimization method. The optimum design with size and shape variables and sensitivity analysis for buckling stability of complex built-up structures and composite material were done by Yuanxian Gu et al. The studies of the optimum design and stability analysis of composite material cylindrical shells were carried out by Liu Tao et al. The studies by Jianhua Rong et al. reported the cross-sectional optimum designs of the frame under the maximum bucking critical load using the elocutionary structural optimization method. Stability optimization studies of the elastic cylindrical shells were reported by Liang Bing et al. The optimum design of composite material-laminated wood under buckling response constraints was done by Guozhong Zhao et al. The topology optimization studies are relatively less because of its difficulty. The optimum designs reported by Zhou Ming are the shell's topological optimization under linear buckling response using the variable-density method. A topological optimization model for shell structures with minimum volume as objective function and buckling eigenvalues as constraint is established. The topological optimization of truss were studied by Guo Xu which shows the constructed model for the truss structure.The model which objective function is the minimized quality, the local buckling constraints and global buckling are constraints. The cross section of the bar was the design variable in the model.

The difficulty of structural topology optimization under the buckling constraint is that the design variables are anchored at a low level. For the continuum structure topology optimization, if the objective function is the compliance degree, which restricts the development of the field, we put the design variables

https://doi.org/10.1515/9783110462272-002

to restore its independent level, the objective function is the quality of the structure, and the application is the ICM (independent continuous mapping) method.

The ICM method has been proposed in reference [6]. "Independent" here means that the topological variable is not dependent on the design variables of lower levels. "Continuous" means that the topological variable is continuum in the interval [0, 1]. "Mapping" means that there is a transformation of discrete-to-continuum topological variables and its inverse transformation of continuum-to-discrete ones. At present, the ICM method is widely used to solve the cross-sectional, shape and topological optimization of the truss, rigid frame and continuum structures that have the stress displacement and frequency constraints. The results obtained by this method were reasonable.

In this chapter, on the basis of the ICM method, using independent continuous topological variables and a different filter function for the element mass, geometric element stiffness matrix and element stiffness matrix along with numerical examples for studying and exploring the parameters of a filter function, a reasonable filtering function is obtained. The topology optimization of continuum structures with buckling constraints was established based on the ICM method. With Taylor's expansion, the objective function is explicable by two-order approximation. Using the Rayleigh quotient, Taylor expansion and the filter function, the stability constraint will be bound to approximate explicit methods which to avoid the sensitivity of the calculation. Finally, the optimization model is transformed into dual programming, and the number of design variables is reduced. The solving scale of the model is reduced and the computational efficiency is improved by using sequential quadratic programming.

2.2 ICM Method

The ICM method [6] is an "independent, continuous, mapping" method. "Independent" refers to the established concept of independent variable topology, topological variables are extracted from low-level variables (such as the film thickness in optimization of a plane membrane structure and the section of a rod in the optimization of a truss structure) and the independent continuous topological variables concept is established. "Continuous" means that the discrete variables in 0–1 are changed into a continuous value on the [0, 1] interval. "Mapping" refers to the process of mapping and inversion. Independent, continuous structural topology variables are defined as

A real number t in a interval [0-1], which 1 is used to represent entities and 0 represent no entities and its changing process (elements or sub domains) in the structure, which 0 and 1 are respectively characterized by the corresponding entities non existence and existence, $t \in (0,1)$ is the representation of the entity in the transition state between the non - and - in the topology optimization.

From the essence of topology optimization, the ICM method reveals that the corresponding element or subdomain is defined as the filter function of 0 or 1. The discrete topological variables are turned into continuous by using the smoothing function, and then the mapping of topological variables is filtered by the filter function. Finally, a small amount of continuous value is discrete, and the transformation of topological variables "discrete–continuous–discrete" is complete .The essence of the ICM method lies in two points: one is the independent continuous topological variable, and the other is the mapping inversion process. The ICM method is different from the homogenization method: it is simple, reasonable and has a clear physical explanation.

In the traditional structure optimization, (the cross-sectional area of the rod, the thickness of the plate, the density of the material, etc.), no matter how a physical quantity is small, the value of the topology variable is 1,as long as it is not equal to zero when it is equal to zero, the topological variable will suddenly change to zero, and the relationship between physical variables and topological variables is not continuous. The relationship between the traditional physical variables and the topological variables can be described by a step function, as shown in Figure 2.1:

$$t_i = \begin{cases} 1 & v_i/v_i^0 \in (0,1] \\ 0 & v_i/v_i^0 = 0 \end{cases} \tag{2.1}$$

where t_i is topological variables, v_i is physical quantities, w_i is the quality of the element, $\boldsymbol{k}_i$ is the element stiffness matrix, $\boldsymbol{g}_i$ is the element geometric matrix and v_i^0 is the corresponding initial physical value.

Because the step function is not continuous in $v_i/v_i^0 = 0$, the optimization algorithm for continuous function theory cannot be used in topology optimization variables of the traditional structure, and so the step function is continuously different by the introduction of the smoothing function here.

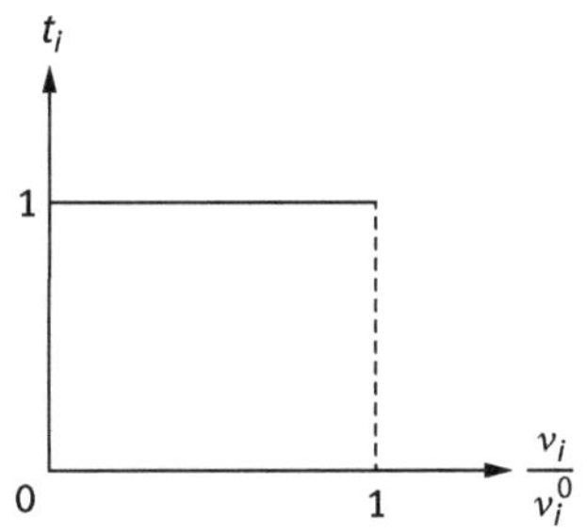

Figure 2.1: Step-up function and traditional topology variables.

The smoothing function is a power function $t_i = P(v_i/v_i^0) = (v_i/v_i^0)^{1/n}$. When n is an integer greater than 2, the smoothing function satisfies the continuity, different function and the 1–0 approximation property, and with an increase in the value of n, the approximation of the smoothing function to the step function is better, as shown in Figure 2.2. By using the smoothing function to approximate the step function, the topological variable is extended to the continuous value of the interval [0,1] by 0 or 1 of the discrete value. The function relationship between the topological variables and the physical variables is continuous and different.

When the intermediate value is taken, the close degree of the corresponding subdomain "yes" and "no" is reflected, that is, the size of each physical quantity of the subdomain structure is identified, and the inverse mapping of the step function is only 0 or 1, as shown in Figure 2.3. To get the intermediate value of the topological value, we need to recognize the size of each physical quantity, and it is also needed to approach the inverse mapping of the step function. The smoothing function approximation step function and smoothing function inverse function $v_i/v_i^0 = f(t_i) = (t_i)^n$ also approximate inverse mapping step function. The corresponding topological variables t_i and $f(t_i)$ reflect the corresponding physical quantities of subdomain "physical quantities intrinsic value" and "zero" proximity. So, the filter function is taken as a power function $f(t_i) = t_i^n$.

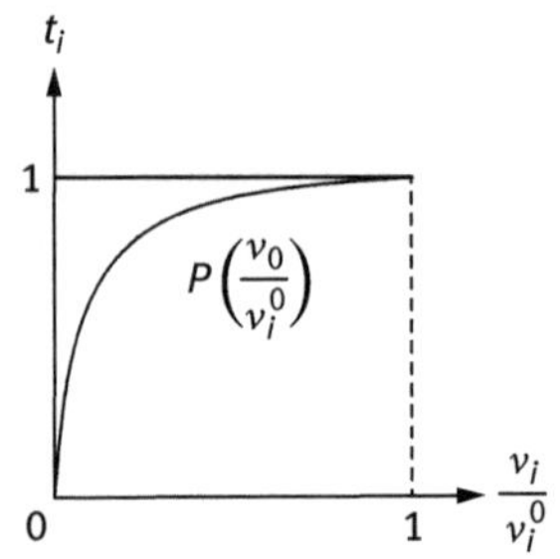

Figure 2.2: Filter function and independent continuous topology variables.

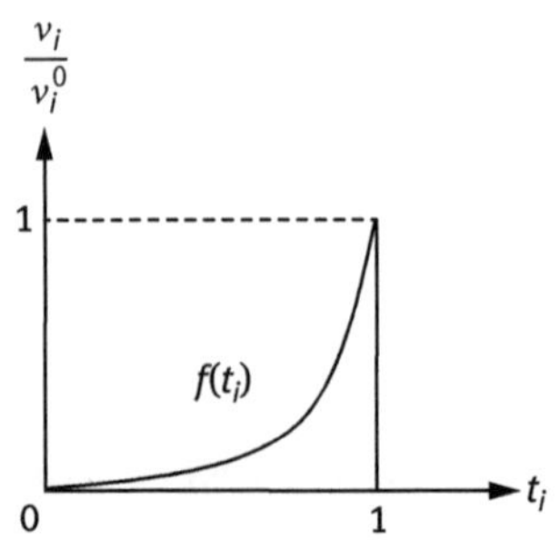

Figure 2.3: Polish function as the inverse function of filter function.

2.3 Determination of Filter Function

Element quality, element stiffness matrix and element geometric stiffness matrix and topological design variables relationship used with different filter functions can be expressed as follows:

$$w_i = f_w(t_i) w_i^0 \tag{2.2}$$

$$\boldsymbol{k}_i = f_k(t_i) \boldsymbol{k}_i^0 \tag{2.3}$$

$$\boldsymbol{g}_i = f_g(t_i) \boldsymbol{g}_i^0 \tag{2.4}$$

where $w_i^0, \boldsymbol{k}_i^0, \boldsymbol{g}_i^0$ are element quality, element stiffness matrix and element geometric stiffness matrix, respectively.

In the ICM method, the filter function is used to identify the element quality, the element stiffness and the geometric stiffness of the element. The filter function is $f_w(t_i)$, $f_k(t_i)$, $f_g(t_i)$. According to the aforementioned discussion, the corresponding filter function can be expressed as

$$f_w(t_i) = t_i^a \quad f_k(t_i) = t_i^b \quad f_g(t_i) = t_i^c \tag{2.5}$$

where a, b, c are constants, and for different types of constrained optimization, they have different values. In this chapter, a numerical test is used by the strain energy method to determine the value of the index in the filter function.

Many researchers use different geometric quantities or physical quantities to "load" topological variables to realize the transformation of an element or a subdomain, "yes" and "no." They were not only the rod cross-sectional area or thickness of a film, plate and block, but also the length of the rod piece or the density of a film and block. To make the relationship more generic, shell digging element and not digging element were used to establish the definition of the topological variables t_i by the uniformity idea.

The traditional topological variables 0–1 discrete state are changed to 0 and 1 between topological variables $t_i \in [0, 1]$. The traditional definition of topological variables and an element of the structure. where $0 \Leftrightarrow$ no and $1 \Leftrightarrow$ yes in a direct relationship, does not exist any longer; therefore, the relationship between special topological variables t_i and element or subdomain "no" or "yes" should be given.

The element topology variable is defined as the ratio of the area of the element after the hole was dug and the area of all the elements. Topological variables anchored in the low level can be expressed as $t_i = s_i / s_i^0$, where s_i is the remaining

Figure 2.4: Element without a hole.

Figure 2.5: Element with a hole.

area of the hole element and s_i^0 is the total element area. The schematic diagram of the element without digging and with digging holes is shown in Figures 2.4 and 2.5, respectively:

2.3.1 Filter Function of Element Quality

The material density is ρ, the thickness is h, the acceleration due to gravity is g and the element area is s_i^0. After digging, the area of the remaining elements is s_i, and we can draw

$$t_i = \frac{s_i}{s_i^0} = \frac{s_i \cdot h \cdot \rho \cdot g}{s_i^0 \cdot h \cdot \rho \cdot g} = \frac{w_i}{w_i^0} \tag{2.6}$$

So we can get

$$w_i = t_i w_i^0 \tag{2.7}$$

According to the formula, we can get the expression of the quality filter function:

$$f_w(t_i) = t_i^a = t_i \tag{2.8}$$

Then the exponential of the quality filter function is $a = 1$.

2.3.2 Filtering Function of Element Stiffness Matrix and Element Geometric Stiffness Matrix

For topological variables $t_i \in [0, 1]$, when $t_i = 0$ meaning fully digging the element, the element does not exist. When $t_i = 1$ meaning no hole being made in the element, the whole area is present. When $0 < t_i < 1$, meaning the element is being dug, the hole is from big to small. This is the process of the element from zero to all elements.

The finite element equation of the structure can be expressed as follows:

$$\boldsymbol{K}\boldsymbol{u} = \boldsymbol{F} \tag{2.9}$$

where $\boldsymbol{K}$ is the stiffness matrix of each nodal force applied to the system, $\boldsymbol{u}$ is the displacement matrix of each point of the system and $\boldsymbol{F}$ is the external force for the system. When the system maintains the same external force, and when all the elements in the system are not digging holes, it can be indicated that

$$\boldsymbol{K}^0\boldsymbol{u}^0 = \boldsymbol{F} \tag{2.10}$$

When all the elements in the system are digging holes, it is expressed as

$$\boldsymbol{K}\boldsymbol{u} = \boldsymbol{F} \tag{2.11}$$

The stiffness matrix is expressed as

$$\boldsymbol{K} = \sum_{i=1}^{N} \boldsymbol{k}_i = \sum_{i=1}^{N} f_k(t_i)\boldsymbol{k}_i{}^0 \tag{2.12}$$

There is

$$\boldsymbol{K} = f_k(t_i) \sum_{i=1}^{N} \boldsymbol{k}_i{}^0 = f_k(t_i)\boldsymbol{K}^0 \tag{2.13}$$

Substituting the formula into formula (2.11), we get

$$\boldsymbol{K}^0(f_k(t_i)\boldsymbol{u}) = \boldsymbol{F} \tag{2.14}$$

Comparing formulas (2.10) and (2.14), the following equation isobtained:

$$f_k(t_i)\boldsymbol{u} = \boldsymbol{u}^0 \tag{2.15}$$

When the formula is being multiplied by $\boldsymbol{F}^{\mathrm{T}}$ on both sides at the same time, we obtain

$$f_k(t_i)\boldsymbol{F}^{\mathrm{T}}\boldsymbol{u} = \boldsymbol{F}^{\mathrm{T}}\boldsymbol{u}^0 \tag{2.16}$$

In addition, because

$$\boldsymbol{F}^{\mathrm{T}} = \boldsymbol{u}^{0^{\mathrm{T}}}\boldsymbol{K}^{0^{\mathrm{T}}} = \boldsymbol{u}^{\mathrm{T}}\boldsymbol{K}^{\mathrm{T}}, \text{ and } \boldsymbol{K}^{0^{\mathrm{T}}} = \boldsymbol{K}^0,\ \boldsymbol{K}^{\mathrm{T}} = \boldsymbol{K}$$

$$f_k(t_i)\boldsymbol{u}^{\mathrm{T}}\boldsymbol{K}\boldsymbol{u} = \boldsymbol{u}^{0^{\mathrm{T}}}\boldsymbol{K}^0\boldsymbol{u}^0 \tag{2.17}$$

$$f_k(t_i) = (\boldsymbol{u}^{0^{\mathrm{T}}}\boldsymbol{K}^0\boldsymbol{u}^0)/(\boldsymbol{u}^{\mathrm{T}}\boldsymbol{K}\boldsymbol{u}) \tag{2.18}$$

Both sides of the formula are multiplied by 0.5, that is,

$$f_k(t_i) = ((1/2)\boldsymbol{u}^{0^{\mathrm{T}}}\boldsymbol{K}^0\boldsymbol{u}^0)/((1/2)\boldsymbol{u}^{\mathrm{T}}\boldsymbol{K}\boldsymbol{u}) \tag{2.19}$$

The following conclusions can be drawn from the aforementioned discussion: the ratio is the size of the filter function that is equal to the total strain energy of the structure with no hole structure and the total strain energy of the structure.

2.3.3 Numerical Test to Determine the Stiffness Filter Function

2.3.3.1 The Establishment of a Numerical Test Model

To obtain the power exponent b in the stiffness matrix filter function, the foundation structure with no hole, as shown in Figure 2.6, is obtained. In $0 < t_i \leq 1$ section, a series of corresponding $t_i^{(v)}$ $(v = 1, \ldots, N)$ value were taken in the formula (2.6), the dig hole geometry graph 2.7 was built, the strain energy was calculated, so we can get expression filtering function from the formula (2.19).

The base structure is a plane elastic body with a size of $10 \times 10 \times 1.0$ mm^3. The elastic modulus of structure material is $E = 6.889 \times 10^4$ N/mm^2, and the material density is $\rho = 1.0$ g/mm^3. The left boundary has fixed constraints. A concentrated force $P = 15{,}600$ N, which is in the negative direction of the X axis, is located at the right edge midpoint of the structure. Quadrilateral element with four nodes is used and the number of total elements is 20×20.The finite element model is shown in Figures 2.6 and 2.7.

2.3.3.2 Using the Least Square Method to Solve the Power Exponent of the Filter Function

According to the N numerical results for the b value of stiffness matrix filter function, $t_i^{(v)}$ will be used as the filter function:

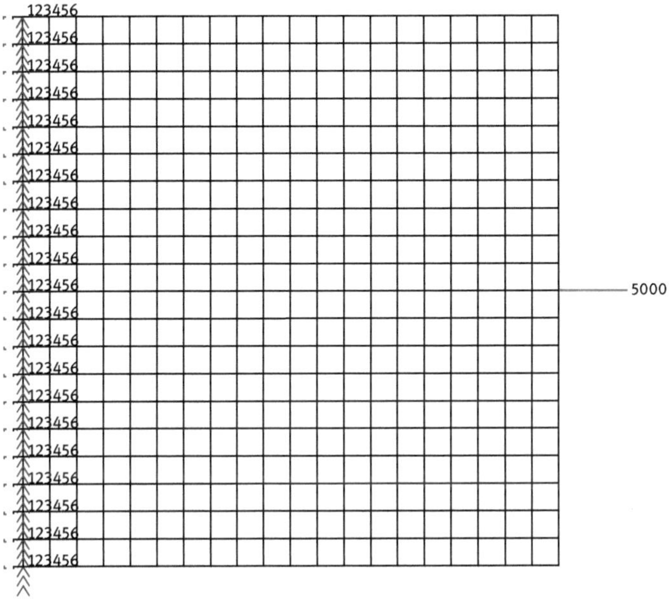

Figure 2.6: Geometry structure without holes.

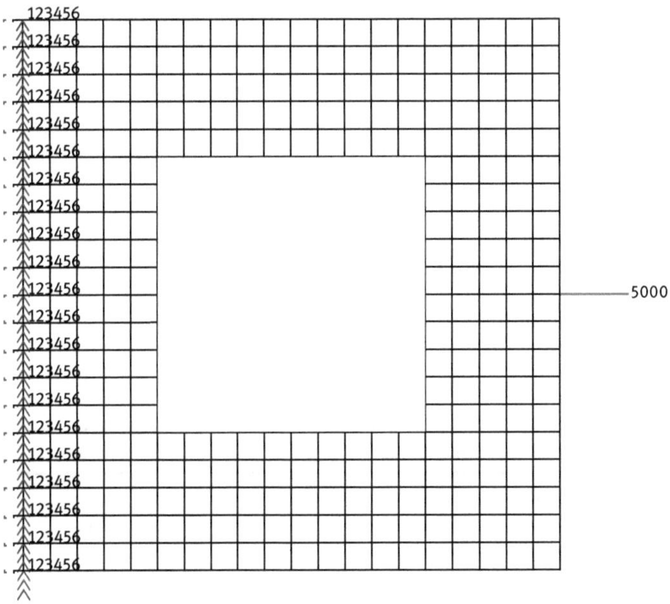

Figure 2.7: Geometry structure with holes.

$$f_k(t_i^{(v)}) = (t_i^{(v)})^b (v = 1, \ldots, N) \tag{2.20}$$

Taking log on both sides of the equation, we obtain:

$$\ln f_k(t_i^{(v)}) = b \ln t_i^{(v)} \tag{2.21}$$

According to the principle of least square method, the minimum square error is

$$\sum_{i=1}^{N} \delta_i^2 = \sum_{i=1}^{N} [\ln f_k(t_i^{(v)}) - b \ln t_i^{(v)}]^2 \to \min \tag{2.22}$$

The minimum value of the formula was made, the formula derivation by b and the value is zero

$$\sum_{i=1}^{N} \ln t_i^{(v)} \ln f_k(t_i^{(v)}) = b \sum_{i=1}^{N} (\ln t_i^{(v)})^2 \tag{2.23}$$

The b value is obtained from the aforementioned formula:

$$b = \sum_{i=1}^{N} \ln t_i^{(v)} \ln f_k(t_i^{(v)}) / \sum_{i=1}^{N} (\ln t_i^{(v)})^2 \tag{2.24}$$

Table 2.1 lists the relationship between the topological variables and hole length. Table 2.2 lists the hole structure and nonhole structure strain energy value as well as the filter function value. Table 2.3 lists the parameters of the filter function by using the least square method.

On the basis of data listed in Table 2.3, the least square method is used to solve the equation:

$$b = 15.107837/4.578186505 = 3.2999513$$

Taking b = 3.3, the filter function of the curve is shown in Figure 2.8:

Table 2.1: Corresponding connection with topology variables and edge length of the holes.

v	1	2	3	4	5	6	7	8	9	10
Hole length	0	1	2	3	4	5	6	7	8	9
Hole area	0	1	4	9	16	25	36	49	64	81
$s_i^{(v)}$	100	99	96	91	84	75	64	51	36	19
s^0_i	100	100	100	100	100	100	100	100	100	100
$t_i^{(v)}$	1	0.99	0.96	0.91	0.84	0.75	0.64	0.51	0.36	0.19

Table 2.2: Total strain energy of the structure without hole and with hole and filter functions.

Topological variable t_i	Element strain energy without holes	Element strain energy with holes	Filter function $f_k(t_i)$
1	577.4683	577.4683	1
0.99	577.4683	590.0093	0.9787693
0.96	577.4683	636.4040	0.9073926
0.91	577.4683	732.8549	0.7879708
0.84	577.4683	920.8782	0.6270843
0.75	577.4683	1307.942	0.4415091
0.64	577.4683	2220.322	0.2600831
0.51	577.4683	4937.057	0.1169661
0.36	577.4683	17044.49	0.0338800
0.19	577.4683	152476.8	0.0037872

Table 2.3: Result of numerical simulation for the filter function.

Topological variable t_i	$f_k(t_i)$	$\ln t_i$	$\ln f_k(t_i)$	$\ln t_i \times \ln f_k(t_i)$	$\ln t_i \times \ln t_i$
0.99	0.9787693	−0.0100503	−0.0214593	0.0002156	0.000101009
0.96	0.9073926	−0.0408220	−0.0971800	0.003967	0.001666435
0.91	0.7879708	−0.0943107	−0.2382942	0.0224736	0.008894504
0.84	0.6270843	−0.1743534	−0.4666743	0.0813662	0.030399104
0.75	0.4415091	−0.2876821	−0.8175566	0.2351964	0.082760975
0.64	0.2600831	−0.4462871	−1.3467541	0.6010389	0.199172178
0.51	0.1169661	−0.6733446	−2.1458711	1.4449107	0.453392887
0.36	0.0338800	−1.0216512	−3.3849304	3.4582185	1.043771272
0.19	0.0037872	−1.6607312	−5.5761283	9.2604502	2.758028141
$\sum$				15.107837	4.578186505

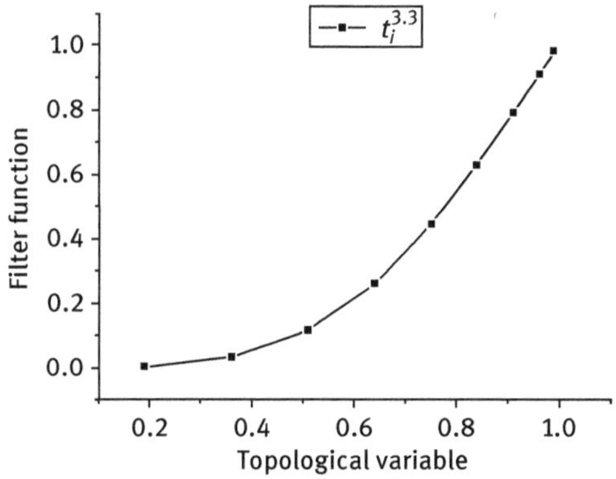

Figure 2.8: Curve of the filter function.

By the numerical simulation process, the topology variables of optimization are in the low level, and so $a = 1$, $b = 3.3$. To avoid $t_i \to 0$, the stiffness matrix and geometric stiffness matrix are not of the same order of infinitesimal, and so b and c are equal. In this chapter, the numerical experiments show that when $b = c$, the convergence effect is good.

2.4 Model Establishment and Realization

2.4.1 Continuum Topology Optimization Model

Once the form of the structure is determined, the buckling will have a variety of modes, and there will be more corresponding critical force, because the structure has not worked before buckling critical force to higher modes, so people care is usually for the minimum critical force, and it is the critical force of first buckling mode.

Eigenvalue equations for buckling analysis are

$$\left(\boldsymbol{K} + \lambda_j \boldsymbol{G}\right) \boldsymbol{u}_j = \boldsymbol{0} \quad (j = 1, \ldots J) \tag{2.25}$$

The symbols $\boldsymbol{K}$, $\boldsymbol{G}$, λ_j and $\boldsymbol{u}_j$ denote stiffness matrix, the geometric stiffness matrix, the jth eigenvalue and the corresponding eigenvector, respectively.

The relationship between critical load and outer load can be rewritten as

$$\lambda_j = \zeta_j \times P \tag{2.26}$$

where ζ_j is the buckling factor and P the is outer load.

Finally, the topology optimization model for the continuum structure is constructed, in which the minimized quality as the objective function has been subjected to the buckling constraints:

$$\begin{cases} \text{find} & \boldsymbol{t} \in E^N \\ \min & W = \sum_{i=1}^{N} f_w(t_i) w_i^0 \\ \text{s.t} & \lambda_{jl}(f_k(t_i), f_g(t_i)) \le \bar{\lambda}_{jl} \qquad (j = 1, \ldots, J, l = 1, \ldots, L) \\ & 0 \le t_i \le 1 \quad (i = 1, \ldots, N) \end{cases} \tag{2.27}$$

where $\boldsymbol{t}$ is the topology design variables, W is the total quality of the structure, w_i and N are the quality of ith element and the total number of element, $\bar{\lambda}_{jl}$ is the upper limit of the critical load, J is the total number of the buckling mode,N is the total number of elements and L is the condition number.

2.4.2 Explicit Approximate Function of the Buckling Constraint

To establish a stable topology optimization model, we need to obtain the explicit function of characteristic value to the topology design variable, and formula (2.25) can be obtained:

$$\lambda_j = -\boldsymbol{u}_j^{\mathrm{T}}\boldsymbol{K}\boldsymbol{u}_j/\boldsymbol{u}_j^{\mathrm{T}}\boldsymbol{G}\boldsymbol{u}_j \tag{2.28}$$

Unfortunately, λ eigenvalue is the implicit function of topological variables $\boldsymbol{t}$, and if you can find the first-order derivative, it can be used by the first-order Taylor expansion to approximate. Thus, the sensitivity analysis of characteristic value is to find the derivative of the design variable, which is an important part of the optimization design.

For sensitivity analysis, we have to observe the stiffness matrix and geometric matrix:

$$\boldsymbol{K} = \sum_{i=1}^{N} \boldsymbol{k}_i{}^0 t_i{}^b = \boldsymbol{K}(\boldsymbol{t}), \quad \boldsymbol{G} = \sum_{i=1}^{N} \boldsymbol{g}_i{}^0 t_i{}^b = \boldsymbol{G}(\boldsymbol{t}) \tag{2.29}$$

Substituting formula (2.29) into formula (2.25),

$$\left(\boldsymbol{K}(\boldsymbol{t}) + \lambda_j \boldsymbol{G}(\boldsymbol{t})\right)\boldsymbol{u}_j = \boldsymbol{0} \tag{2.30}$$

The solution of the eigenvalue equation group (2.30) is

$$\boldsymbol{u}_j = \boldsymbol{u}_j(\boldsymbol{t}) \tag{2.31}$$

Substituting formula (2.31) into formula (2.30),

$$\left(\boldsymbol{K}(\boldsymbol{t}) + \lambda_j \boldsymbol{G}(\boldsymbol{t})\right)\boldsymbol{u}_j(\boldsymbol{t}) \equiv \boldsymbol{0} \tag{2.32}$$

Note that

$$\lambda_j = \lambda_j(\boldsymbol{t}) \tag{2.33}$$

To facilitate, get the derivation $\partial\lambda_j/\partial t_i$. The left-hand side of formula (2.32) is multiplied by $\boldsymbol{u}_j^{\mathrm{T}}(\boldsymbol{t})$:

$$\boldsymbol{u}_j^{\mathrm{T}}(\boldsymbol{t})\boldsymbol{K}(\boldsymbol{t})\boldsymbol{u}_j(\boldsymbol{t}) + \lambda_j \boldsymbol{u}_j^{\mathrm{T}}(\boldsymbol{t})\boldsymbol{G}(\boldsymbol{t})\boldsymbol{u}_j(\boldsymbol{t}) \equiv \boldsymbol{0} \tag{2.34}$$

Our study shows that if the inverted variable is introduced,

$$x_i = 1/t_i^b \tag{2.35}$$

Both for modeling and solving, formulas (2.29) and (2.34), respectively, are favorable:

$$\boldsymbol{K} = \sum_{i=1}^{N} \boldsymbol{k}_i^0 / x_i, \quad \boldsymbol{G} = \sum_{i=1}^{N} \boldsymbol{g}_i^0 / x_i \tag{2.36}$$

$$\boldsymbol{u}_j^{\mathrm{T}}(\boldsymbol{x})\boldsymbol{K}(\boldsymbol{x})\boldsymbol{u}_j(\boldsymbol{x}) + \lambda_j(\boldsymbol{x})\boldsymbol{u}_j^{\mathrm{T}}(\boldsymbol{x})\boldsymbol{G}(\boldsymbol{x})\boldsymbol{u}_j(\boldsymbol{x}) \equiv \boldsymbol{0} \tag{2.37}$$

According to formula (2.37) on the partial derivative with respect to x_i,

$$2\left(\partial \boldsymbol{u}_j/\partial x_i\right)^{\mathrm{T}}(\boldsymbol{K} + \lambda_j \boldsymbol{G})\boldsymbol{u}_j + \left(\partial \lambda_j/\partial x_j\right)\boldsymbol{u}_j^{\mathrm{T}}\boldsymbol{G}\boldsymbol{u}_j + \boldsymbol{u}_j^{\mathrm{T}}(\partial \boldsymbol{K}/\partial x_i)\boldsymbol{u}_j + \lambda_j \boldsymbol{u}_j^{\mathrm{T}}(\partial \boldsymbol{G}/\partial x_i)\boldsymbol{u}_j = 0 \tag{2.38}$$

Substituting formula (2.28) into it:

$$\partial \lambda_j/\partial x_i = -\left[\boldsymbol{u}_j^{\mathrm{T}}(\partial \boldsymbol{K}/\partial x_i)\boldsymbol{u}_j + \lambda_j \boldsymbol{u}_j^{\mathrm{T}} + \lambda_j \boldsymbol{u}_j^{\mathrm{T}}(\partial \boldsymbol{G}/\partial x_i)\boldsymbol{u}_j\right]/\boldsymbol{u}_j^{\mathrm{T}}\boldsymbol{G}\boldsymbol{u}_j \tag{2.39}$$

$$\partial \boldsymbol{K}/\partial x_i = -\boldsymbol{k}_i^0/x_i^2 = -\boldsymbol{k}_i/x_i, \qquad \partial \boldsymbol{G}/\partial x_i = -\boldsymbol{g}_i^0/x_i^2 = -\boldsymbol{g}_i/x_i \tag{2.40}$$

Substituting formula (2.40) into formula (2.39), we obtain

$$\begin{aligned} \lambda_j/\partial x_i &= \left(\boldsymbol{u}_j^{\mathrm{T}}\boldsymbol{k}_i\boldsymbol{u}_j/2 + \lambda_j \boldsymbol{u}_j^{\mathrm{T}}\boldsymbol{g}_i\boldsymbol{u}_j/2\right)/\left(\boldsymbol{u}_j^{\mathrm{T}}\boldsymbol{G}\boldsymbol{u}_j x_i/2\right) \\ &= \left(U_{ij} + \lambda_j V_{ij}\right)/V_{\Sigma j} x_i \end{aligned} \tag{2.41}$$

where $U_{ij} = \boldsymbol{u}_j^{\mathrm{T}}\boldsymbol{k}_i\boldsymbol{u}_j/2$ is the mode strain energy for the ith element under the jth mode. $V_{ij} = \boldsymbol{u}_j^{\mathrm{T}}\boldsymbol{g}_i\boldsymbol{u}_j/2$ is the mode geometric strain energy for the ith element under the jth mode. $V_{\Sigma j} = \boldsymbol{u}_j^{\mathrm{T}}\boldsymbol{G}\boldsymbol{u}_j x_j/2$ is the total mode geometric strain energy for the structure under the jth mode.

In the finite element analysis of buckling mode, the structure element modal strain energy and the geometric modal strain energy can be put forward, and the total geometric strain energy can be calculated and the eigenvalue derivative of each design variable can also be calculated.

The constraints are approximated by first-order Taylor expansions:

$$\begin{aligned} \lambda_{jl}(\boldsymbol{t}) &= \lambda_{jl}(\boldsymbol{x}^{(k)}) + \sum_{i=1}^{N} \partial \lambda_{jl}/\partial x_i (x_i - x_i^{(k)}) \\ &= \lambda_{jl}(\boldsymbol{x}^{(k)}) + \sum_{i=1}^{N} x_i/V_{\Sigma jl} x_i^{(k)} (U_{ijl}^{(k)} + \lambda_{jl} V_{ijl}^{(k)}) - \sum_{i=1}^{N} (U_{ijl}^{(k)} + \lambda_{jl} V_{ijl}^{(k)})/V_{\Sigma jl} \end{aligned} \tag{2.42}$$

The value of the k index in the formula refers to the kth iteration time.

$\lambda_{jl} \le \bar{\lambda}_{jl}$ is:

$$\sum_{i=1}^{N} x_i (U_{ijl}^{(k)} + \lambda_{jl} V_{ijl}^{(k)}) / V_{\Sigma jl} x_i^{(k)} \le \bar{\lambda}_{jl} - \lambda_{jl}(\mathbf{x}^{(k)}) + \sum_{i=1}^{N} (U_{ijl}^{(k)} + \lambda_{jl} V_{ijl}^{(k)}) / V_{\Sigma jl} \tag{2.43}$$

Let $A_{ijl}^{(k)} = (U_{ijl}^{(k)} + \lambda_{jl} V_{ijl}^{(k)}) / V_{\Sigma jl}$. The left-hand side of inequality (2.43) is equal to:

$$\begin{aligned} &\sum_{i=1}^{N} x_i (U_{ijl}^{(k)} + \lambda_{jl} V_{ijl}^{(k)}) / V_{\Sigma jl} x_i^{(k)} \\ &= \sum_{i=1}^{N} A_{ijl}^{(k)} x_i / x_i^{(k)} = \sum_{i=1}^{N} (t_i^{(k)})^b A_{ijl}^{(k)} x_i = \sum_{i=1}^{N} d_{ijl} x_i \end{aligned} \tag{2.44}$$

$d_{ijl} = (t_i^{(k)})^b A_{ijl}^{(k)}$。 The right-hand side of inequality (2.43) is equal to:

$$\bar{\lambda}_{jl} - \lambda_{jl}(x^{(k)}) + \sum_{i=1}^{N} (U_{ijl}^{(k)} + \lambda_{jl} V_{ijl}^{(k)}) / V_{\Sigma jl} = \bar{\lambda}_{jl} - \lambda_{jl}(x^{(k)}) + \sum_{i=1}^{N} A_{ijl}^{(k)} = e_{jl} \tag{2.45}$$

Uniform expression formula can be obtained:

$$\sum_{i=1}^{N} d_{ijl} x_i \le e_{jl} \tag{2.46}$$

2.4.3 Explicit Approximate Function of the Objective Function

In formula (2.27), the objective function is expressed as

$$W = \sum_{i=1}^{N} f_w(t_i) w_i^0 \tag{2.47}$$

where the filter function is $f_w(t_i) = t_i^a$. According to relationship between topological variables and design variables, we can obtain $t_i = 1/x_i^{1/b}$ and the filter function can be rewritten as:

$$f_w(t_i) = 1/x_i^{a/b} \tag{2.48}$$

If we define $a/b = \beta$, eq. (2.48) is translated into $f_w(t_i) = 1/x_i^{\beta}$. Substituting into eq. (2.47), the object function is rewritten as

$$W = \sum_{i=1}^{N} w_i^0 / x_i^{\beta} \tag{2.49}$$

The object function is translated into explicit expression by the second-order Taylor expansion:

$$W=\sum_{i=1}^{N}\{[w_i^0\beta(\beta+1)/2x_i^{(k)\beta+2}]\times x_i^2+[-w_i^0\beta(\beta+2)/x_i^{(k)\beta+1}]\times x_i\} \tag{2.50}$$

where $x_i^{(k)}$ is the initial design variable of ith element at the kth iteration.

We define $A_i=w_i^0\beta(\beta+1)/2x_i^{(k)\beta+2}$ $B_i=-w_i^0\beta(\beta+2)/x_i^{(k)\beta+1}$,

The objective function can be written as $W=\sum_{i=1}^{N}(A_ix_i^2+B_ix_i)$. The whole topology optimization model is changed into

$$\begin{cases}\text{find} & \mathbf{x}\in E^N\\ \min & W=\sum\limits_{i=1}^{N}(A_ix_i^2+B_ix_i)\\ \text{s.t.} & \sum\limits_{i=1}^{N}d_{ijl}x_i\le e_{jl}\quad (j=1,\dots,J,l=1,\dots,L)\\ & 1\le x_i\le\overline{x_i}\qquad (i=1,\dots,N)\end{cases} \tag{2.51}$$

2.5 Solution to Optimization

According to the relationship between the topology and the design variables, the topology optimization model is translated into the model with the design variables. To reduce the number of the variable and the model's scale, the optimization model is translated into a dual programming:

$$\begin{cases}\text{find} & \boldsymbol{\mu}\\ \max & \Phi(\boldsymbol{\mu})\\ \text{s.t.} & \boldsymbol{\mu}\ge 0\end{cases} \tag{2.52}$$

In Eq. (2.52), $\Phi(\boldsymbol{\mu})=\min\limits_{1\le x_i\le x_i}(L(\mathbf{x},\boldsymbol{\mu}))$

$$L(\mathbf{x},\boldsymbol{\mu})=\sum_{i=1}^{N}(A_ix_i^2+B_ix_i)+\sum_{j=1}^{J}\sum_{l=1}^{L}\mu_{jl}(\sum_{i=1}^{N}d_{ijl}x_i-e_{jl}) \tag{2.53}$$

According to the Kutta condition, the derivative of augmented Lagrangian function with respect to the design variable can be written as follows:

$$\partial L(\mathbf{x},\boldsymbol{\mu})/\partial x_i=2A_ix_i^*+B_i+\sum_{j=1}^{J}\sum_{l=1}^{L}\boldsymbol{\mu}_{jl}d_{ijl}$$

$$\begin{cases}\le 0 & (x_i^*=\overline{x}_i)\\ =0 & (1<x_i^*<\overline{x}_i)\\ \ge 0 & (x_i^*=1)\end{cases} \tag{2.54}$$

Assuming $I_a = \{i|1 < x_i^\star < \bar{x}_i\}$ as the active variable set, the following equation is established:

$$2A_i x_i^\star + B_i + \sum_{j=1}^{J}\sum_{l=1}^{L}\mu_{jl}d_{ijl} \equiv 0$$

For μ_{kl} and μ_{jk}, respectively, we obtain

$$\partial x_i^\star/\partial \boldsymbol{\mu}_{kl} = -d_{ikl}/2A_i,\ \partial x_i^\star/\partial \mu_{jk} = -d_{ijk}/2A_i \tag{2.55}$$

Because it is difficult to obtain the explicit expression of $\Phi(\boldsymbol{\mu})$, it can be transformed into two-order Taylor expansion approximation. According to the dual theory, the objective function of formula (2.53) for the design of a variable's one- and two-order derivative is

$$\partial\Phi(\boldsymbol{\mu})/\partial\mu_{jl} = \sum_{i=1}^{N} d_{ijl}x_i^\star - e_{jl} \tag{2.56}$$

$$\begin{aligned}\partial^2\Phi(\boldsymbol{\mu})/\partial\mu_{jl}\partial\mu_{kl} &= \sum_{i=1}^{N} d_{ijl}\partial x_i^\star/\partial\mu_{kl}\\ \partial^2\Phi(\boldsymbol{\mu})/\partial\mu_{jl}\partial\mu_{jk} &= \sum_{i=1}^{N} d_{ijl}\partial x_i^\star/\partial\mu_{jk}\end{aligned} \tag{2.57}$$

Substituting Formula (2.55) into formula (2.57)

$$\begin{aligned}\partial^2\Phi(\boldsymbol{\mu})/\partial\mu_{jl}\partial\mu_{kl} &= -\sum_{i=1}^{N} d_{ijl}d_{ikl}/2A_i\\ \partial^2\Phi(\boldsymbol{\mu})/\partial\mu_{jl}\partial\mu_{jk} &= -\sum_{i=1}^{N} d_{ijl}d_{ijk}/2A_i\end{aligned} \tag{2.58}$$

where $\Phi(\boldsymbol{\mu})$ will be two-order approximation and omitting the constant term, we can get the second-order programming model:

$$\begin{cases}\text{find} & \boldsymbol{\mu}\\ \text{min} & -\Phi(\boldsymbol{\mu}) = \frac{1}{2}\boldsymbol{\mu}^{\mathrm{T}}\mathbf{D}\boldsymbol{\mu} + \mathbf{H}^{\mathrm{T}}\boldsymbol{\mu}\\ \text{s.t.} & \boldsymbol{\mu} \geq 0\end{cases} \tag{2.59}$$

where $\boldsymbol{D} = -\nabla^2\phi(\mu^0)$, $\boldsymbol{H}^{\mathrm{T}} = -[\nabla\phi(\mu^0)^{\mathrm{T}} - (\mu^0)^{\mathrm{T}}\nabla^2\phi(\mu^0)]$.

The second-order programming model was solved, and we can get $\boldsymbol{\mu}$. The active variable set was updated. Therefore, this process is repeated until

$$\|\boldsymbol{x}^{(k+1)} - \boldsymbol{x}^{(k)}\|/\|\boldsymbol{x}^{(k)}\| \leq \varepsilon$$

The optimum values of design variables $\boldsymbol{x}^*$ are solved; the optimum values of topological variables $\boldsymbol{t}^*$ are accordingly obtained. With the new topological variables $\boldsymbol{t}^{(k+1)} = \boldsymbol{t}^*$, the structure is modified. Then the structure and topological optimization are renewable begun until satisfying the following convergence condition which is

$$\Delta W = \left|(W^{(k+1)} - W^{(k)})/W^{(k+1)}\right| \le \varepsilon \tag{2.60}$$

where $W^{(k)}$ and $W^{(k+1)}$ are the structural quality of previous iteration and current iteration, respectively, and ε is the convergence precision. In this study, $\varepsilon = 0.001$ is employed.

2.6 Optimization Strategy

2.6.1 Mechanism Judgment Criterion

In the base structure of the topology optimization model, the "reservation elements" and "delete elements" bear the load together. In the process of optimization iteration, with the deleting elements number increasing, the structure more likely becomes a kind of mechanism. In the topology, if the "delete elements" load over a certain ratio, it is explained that the final topology structure is a mechanism, and not a structure. Obviously, the "delete elements" total strain energy sum is larger, which means that the "delete elements" share more load, and the "reserve elements" of the structure are more likely to be a mechanism. To prevent the topology structure from changing to a mechanism, the judgment criterion of the mechanism is constructed in the topology optimization [57].

The "delete elements" strain energy sum is U_d and the "reserve elements" strain energy sum is U_r, if

$$U_d/U_r \le 5\% \tag{2.61}$$

It is considered that the "delete elements" have no effect on the topology, and the "reserve elements" form a structure; otherwise it is a mechanism.

2.6.2 Inversion Criterion of Element Topological Variable

In the iterative process, the topological variables are continuous variables, which guarantee the continuity of the topology structure. In the optimization process, the topological variables are gradually close to 0 or 1. At the last step of the iteration, the topology optimization structure satisfies the convergence precision, and the topology structure is obtained. In the optimal topology, there are still a few nonwhite, nonred regions, namely, the topological variable values between 0 and 1 of the elements. To get a clearer topology, we have set a deletion threshold T. When the topology variable is lower than the threshold value, the topology variable is 0. When the topology is higher than the threshold value, the topology variable is 1.

$$t_i = \begin{cases} 1 & t_i \geq T \\ 0 & t_i < T \end{cases} \tag{2.62}$$

The threshold T is generally taken as 0.3–0.7, so that the continuous topological variables were inversed to 0–1 type of discrete topology, and we can get a clear topology.

2.6.3 Selection of Topology Optimization Design Region

Because of the load, the support and so on, it is needed to explicitly retain a part of region in the base finite element structure, which is not to participate in the optimization, but is only the reserved part and other parts are connected together. According to the base structure, the reserved area and the design area can be divided, which can effectively reduce the design area, which in turn can reduce the calculation scale and speed up the calculation. As shown in Figure 2.9, the whole area is divided into two: the design area and the reserved region. This chapter adopts the following strategies: the reserved region element values are set to 1 at the beginning of topology optimization, and the number involved in optimization design elements is equal to the number of elements of the whole structure minus the number of reserved regions. The number of elements in the entire model is

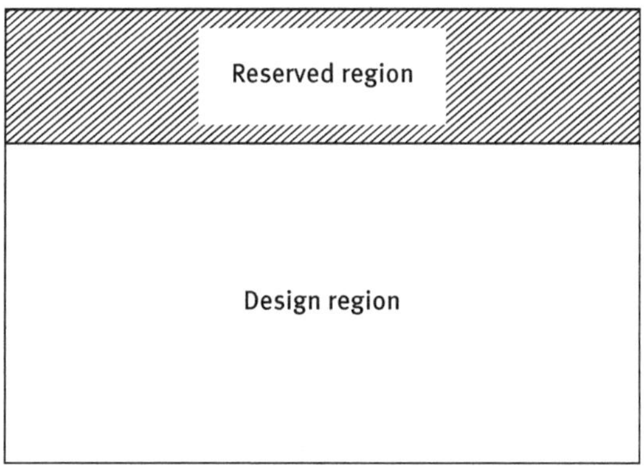

Figure 2.9: Illustrating figure for selecting a design region.

assumed to be I. The number of elements of the design region is I_d. The number of elements of the reserved region is I_s. The relationship between the three is expressed as $I_d = I - I_s$

2.7 Constrain Selection

2.7.1 Constraint Primary Select

In order to improve efficiency, the number of reanalysis of the structure, the effective constraint primaries select methods are taken in section optimization, shape optimization with multiple constraints, so as to reduce the number of constraints and save the cost of computation [59]. In topology optimization with buckling constraint, this strategy can also be considered to remove unnecessary constraints, to reduce the amount of calculation model and to improve calculation speed, and so a satisfactory result to topology optimization can be obtained quickly.

2.7.2 Selection Criterion of Buckling Critical Force Upper Limit

The upper limit of buckling critical force can also affect the optimal topology shape and the quality of the optimal topology. Accurate selection of the upper limit of the critical buckling force should be quick and effective to obtain a reasonable topology. Similarly, the size of the upper bound can be estimated by the quality of the topology. The corresponding function will give the designer more convenience to apply the topology optimization at the product conceptual design stage.

Numerical example: The base structure is a plane elastic body with a size of $520 \times 260 \times 2.0$ mm^3 (Figure 2.10). The elastic modulus of structure material is $E = 68.89 \times 10^3$ MPa, Poisson's ratio is $\gamma = 0.3$ and the material density is $\rho = 2.7 \times 10^{-6}$ kg/mm^3. The right-hand and left-hand underpart corner points of the structure are clasped. A concentrated force $P = 15{,}000$ N along with the

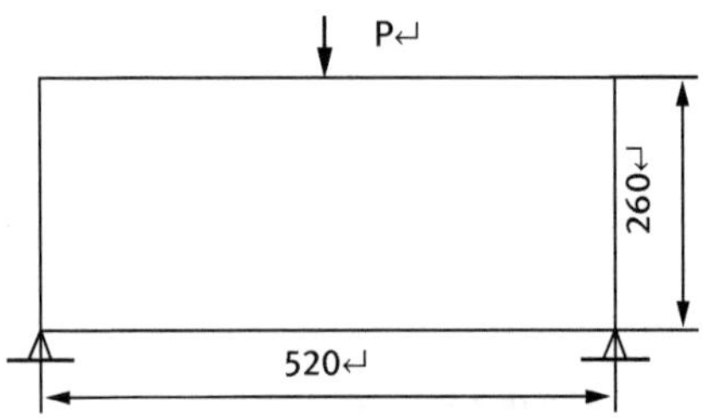

Figure 2.10: Base structure.

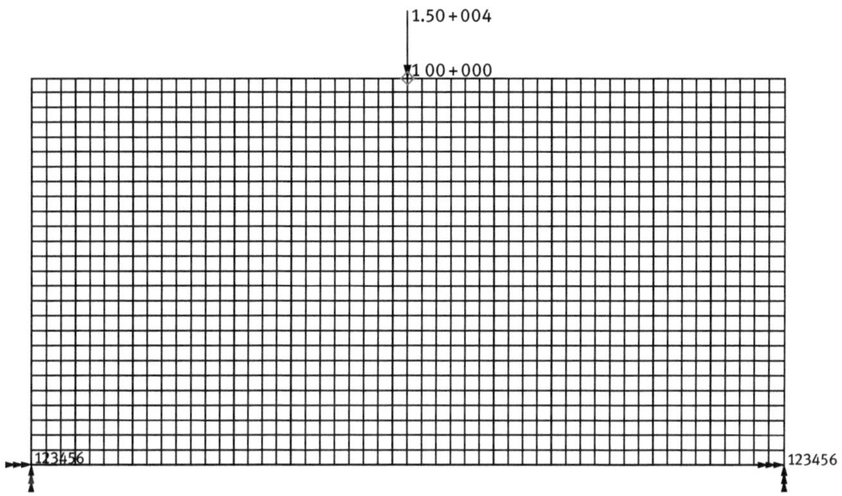

Figure 2.11: Finite element model.

up-to-down (-y) direction is located at the midpoint of upper boundary of the structure. Quadrilateral element with four nodes is used and the number of total elements is 52 × 26. The base structure's quality is 0.73 kg (Figure 2.11).

Consider the top three-order buckling critical force as the research object, the first-order buckling factor ζ_1 = 0.00794, the critical buckling force λ_1 = 119 N; second-order buckling factor ζ_2 = 0.088112, the critical buckling force λ_2 = 1,322 N and third-order buckling factor ζ_3 = 0.10186, the critical buckling force λ_3 = 1,528 N.

For doing the topology optimization design in this case, a critical force limit should be less than 119 N, and so we take the eight critical limits. The obtained results are listed in Table 2.4.

2.7.2.1 The Relationship between the First-order Critical Force Upper Limit and the Topological Structure Quality

The eight kinds of topology optimization results are listed in Table 2.4, which make us think whether the buckling critical force and its corresponding topology optimization quality have a function relation or not.

To eliminate the influence of the corresponding relationship between the buckling constraint and the quality of the topological structure, the concept of "quality ratio" and "buckling constraint ratio" is introduced here. The "quality ratio" is the ratio of the quality of the optimal structure and the quality of the base structure. The "buckling constraint ratio" is the ratio of the upper limit of the buckling critical force and the buckling critical force of the whole structure.

Table 2.4: First-order mode buckling constraint and quality of optimal structure.

Test number	Bucking constraint	Topological quality	Repeated analysis times	Topological graph
1	119	0.73	1	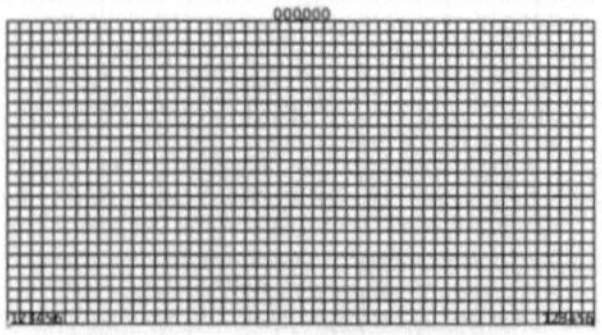
2	115	0.54567	28	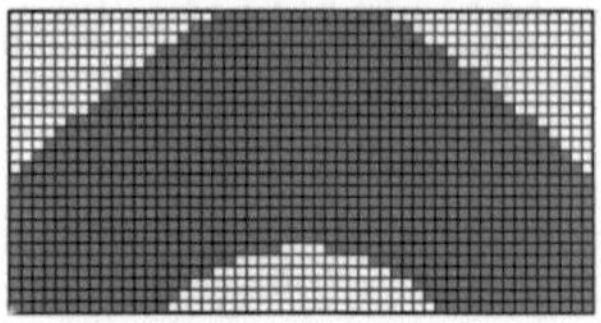
3	111	0.46953	30	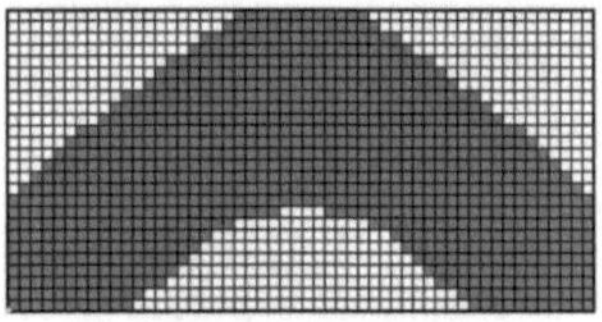
4	107	0.41472	31	
5	95	0.30672	34	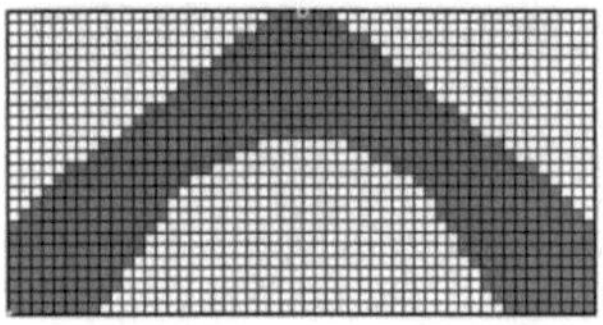
6	83	0.23976	39	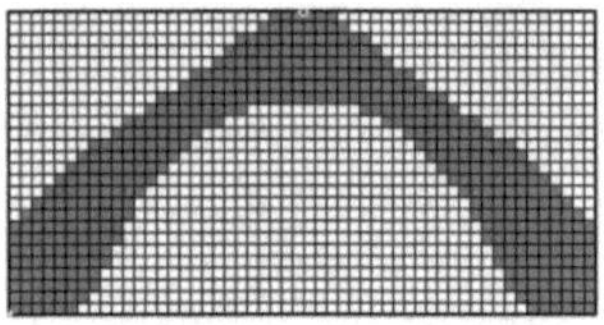
7	71	0.20358	41	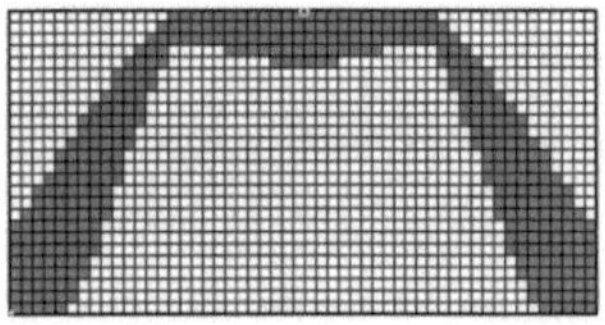

Table 2.4: (continued)

Test number	Bucking constraint	Topological quality	Repeated analysis times	Topological graph
8	59	0.16389	40	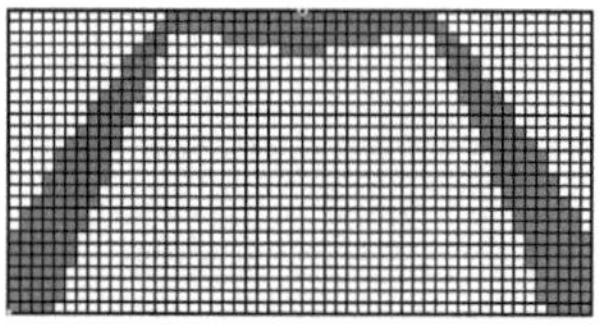
9	47	0.13365	41	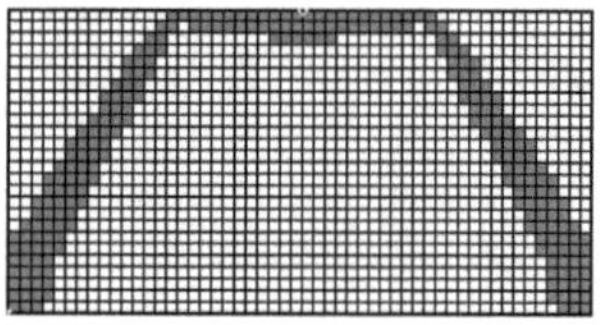

Table 2.5: The ratio of the first-order mode buckling constraint and the ratio of quality.

Test number	Buckling constraint ratio	Quality ratio of topological structure
1	1	1
2	0.966	0.74734
3	0.933	0.64324
4	0.899	0.56805
5	0.798	0.42012
6	0.697	0.3285
7	0.596	0.27895
8	0.496	0.22455
9	0.395	0.18315

Table 2.5 lists the “quality ratio” and “buckling constraint ratio” in the nine topology optimization. The “quality ratio” is the horizontal coordinate X, and the “buckling constraint ratio” is the vertical coordinate y curve as shown in Figure 2.12. It can be seen from the graph that the curve is nonlinear. To get the function relationship between the quality ratio and buckling constraint ratio, the least square method is used to fit the curve. The hyperbolic function is selected according to the distribution of the discrete points:

$$\frac{1}{y} = a + \frac{b}{x} \tag{2.63}$$

Using the data in Table 2.6, we obtain the following equations:

Table 2.6: Simulation result of the first-order mode buckling constraint curve.

Serial number	x	y	$X=1/x$	$Y=1/y$	X^2	XY
1	1	1	1	1	1	1
2	0.747337	0.966	1.3380	1.0352	1.7902	1.3851
3	0.643240	0.933	1.5546	1.0718	2.4168	1.6662
4	0.568047	0.899	1.7604	1.1123	3.0990	1.9581
5	0.420118	0.798	2.3803	1.2531	5.6658	2.9828
6	0.328513	0.697	3.0440	1.4347	9.2659	4.3672
7	0.278953	0.596	3.5848	1.6779	12.8508	6.0149
8	0.224545	0.496	4.4534	2.0161	19.8328	8.9785
9	0.183146	0.395	5.4601	2.5316	29.8127	13.8228
$\sum$			24.5756	13.1327	85.734	42.1756

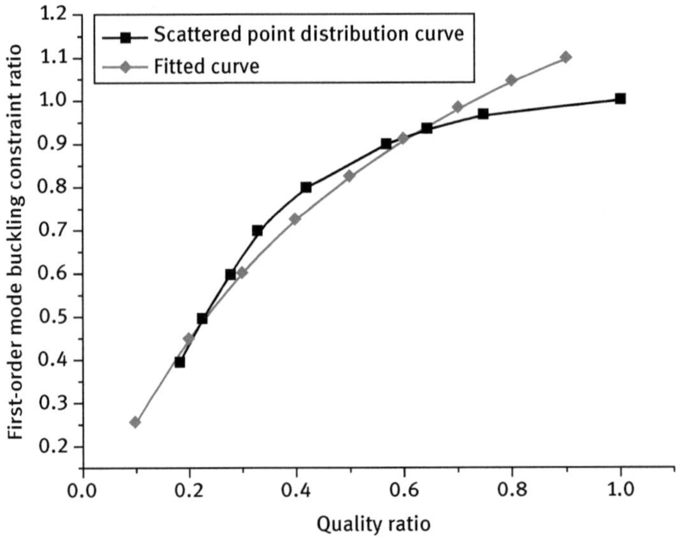

Figure 2.12: Curve of quality ratio and constraint ratio with the first-order mode.

$$\begin{bmatrix} \sum_{i=1}^{9} 1 & \sum_{i=1}^{9} X_i \\ \sum_{i=1}^{9} X_i & \sum_{i=1}^{9} X_i^2 \end{bmatrix} \begin{bmatrix} a \\ b \end{bmatrix} = \begin{bmatrix} \sum_{i=1}^{9} Y_i \\ \sum_{i=1}^{9} X_i Y_i \end{bmatrix} \tag{2.64}$$

$$\begin{cases} 9a + 24.5756b = 13.1327 \\ 24.5756a + 85.734b = 42.1756 \end{cases}$$

We obtain $a = 0.5335$ and $b = 0.339$. Substituting these into the hyperbolic equation, we get

$$y = \frac{x}{0.5335x + 0.339} \tag{2.65}$$

In the same manner, the critical force of the second-order and third-order buckling modes is, respectively, used as the single constraint to carry on the topology optimization of the model, and the curve is fitted by the least square method. According to the second-order and third-order buckling modes of the corresponding discrete point, the straight line is used to fit it.

2.7.2.2 The Relationship Between the Second-order Critical Force Upper Limit and the Topological Structure Quality

Table 2.7 lists the second-order buckling constraints and topology quality; Table 2.8 lists the second-order buckling ratio and quality ratio; Graph 2.13 shows the second-

Table 2.7: The second-order mode buckling constraint and quality of optimal structure.

Test number	Buckling constraint	Topological quality	Repeated analysis times	Topological graph
1	1322	0.2704	1	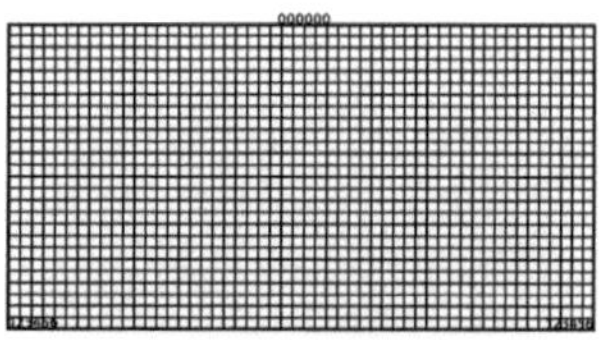
2	1260	0.2468	20	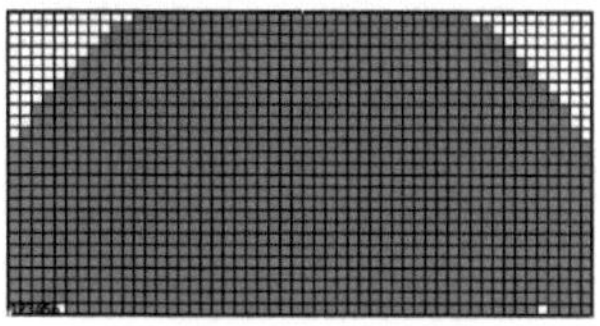
3	1200	0.2364	22	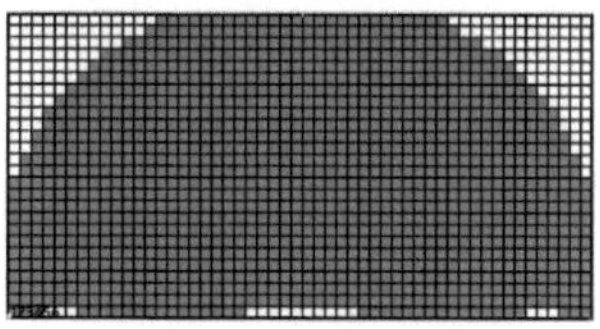
4	1140	0.2260	25	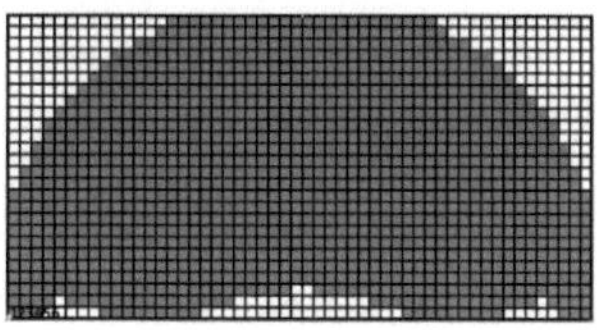

(continued)

Table 2.7: (continued)

Test number	Buckling constraint	Topological quality	Repeated analysis times	Topological graph
5	1080	0.2133	28	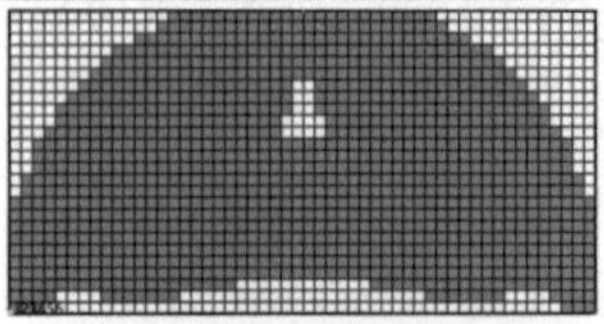
6	1020	0.2009	29	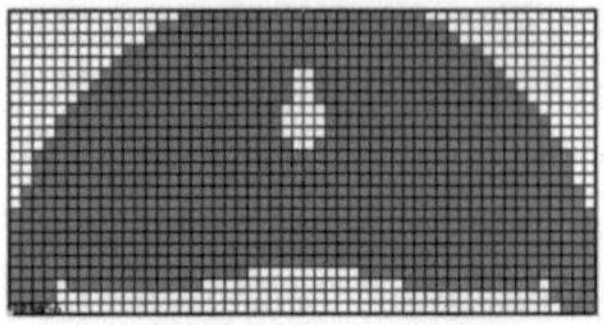
7	960	0.1897	29	
8	900	0.1789	30	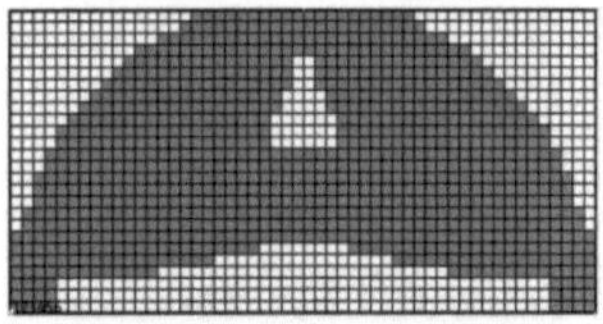
9	840	0.1669	31	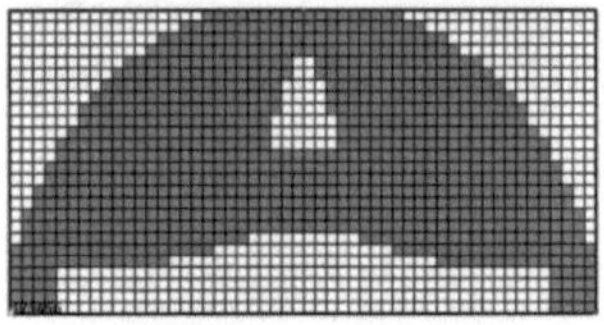
10	780	0.1577	31	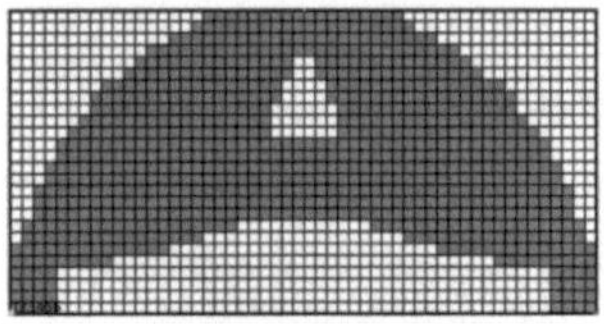
11	730	0.1493	31	

Table 2.8: The ratio of second-order mode buckling constraint and ratio of quality.

Test number	Buckling constraint ratio	Quality ratio of topological structure
1	1	1
2	0.9531	0.91272
3	0.90772	0.87426
4	0.86233	0.8358
5	0.81694	0.78883
6	0.77156	0.74297
7	0.72617	0.70155
8	0.68079	0.66161
9	0.6354	0.61723
10	0.59002	0.58321
11	0.55219	0.55214

Table 2.9: Simulation result of the first-order mode buckling constraint curve.

Serial number	x	y	x^2	xy
1	1	1	1	1
2	0.91272	0.9531	0.83306	0.86991
3	0.87426	0.90772	0.76433	0.79358
4	0.8358	0.86233	0.69856	0.72074
5	0.78883	0.81694	0.62225	0.64443
6	0.74297	0.77156	0.552	0.57325
7	0.70155	0.72617	0.49217	0.50944
8	0.66161	0.68079	0.43773	0.45042
9	0.61723	0.6354	0.38097	0.39219
10	0.58321	0.59002	0.34013	0.34411
11	0.55214	0.55219	0.30486	0.30489
$\sum$	8.27032	8.49622	6.42606	6.60296

order quality ratio and constraint ratio curve and Table 2.9 lists the fitting data of the second modes of buckling constraint curve. The least square method is used to fit the curve, and the linear fitting is used according to the distribution of the discrete points.

$$y = a + bx \tag{2.66}$$

Using the data from Table 2.9, we obtain the following equations

$$\begin{bmatrix} \sum_{i=1}^{11} 1 & \sum_{i=1}^{11} x_i \\ \sum_{i=1}^{11} x_i & \sum_{i=1}^{11} x_i^2 \end{bmatrix} \begin{bmatrix} a \\ b \end{bmatrix} = \begin{bmatrix} \sum_{i=1}^{11} y_i \\ \sum_{i=1}^{11} x_i y_i \end{bmatrix} \tag{2.67}$$

$$\begin{cases} 11a + 8.27032b = 8.49622 \\ 8.27032a + 6.42606b = 6.60296 \end{cases}$$

Solution of equations $a = -0.005$, $b = 1.034$. Substituting these into equation (2.66), we obtain:

$$y = -0.005 + 1.034x \tag{2.68}$$

2.7.2.3 The Relationship between the Third-order Critical Force Upper Limit and the Topological Structure Quality

Table 2.10 lists the third-order mode buckling constraints and topology quality; Table 2.11 lists the third-order buckling ratio and quality ratio; Graph 2.14 shows the third-order quality ratio and constraint ratio curve and Table 2.12 lists the fitting data of the third modes of the buckling constraint curve. The least square method is used to fit the curve, and the linear fitting is used according to the distribution of the discrete points.

$$y = a + bx \tag{2.69}$$

Table 2.10: The third-order mode buckling constraint and quality of optimal structure.

Test number	Buckling constraint	Topological quality	Repeated analysis times	Topological graph
1	1528	0.2704	1	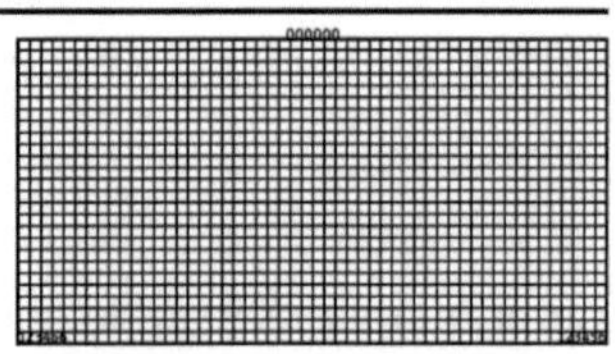
2	1400	0.2624	7	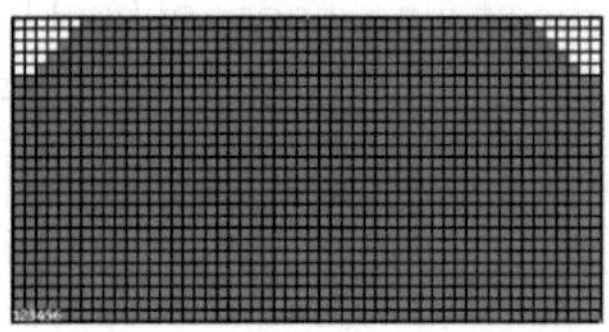

Table 2.10: (continued)

Test number	Buckling constraint	Topological quality	Repeated analysis times	Topological graph
3	1330	0.2464	8	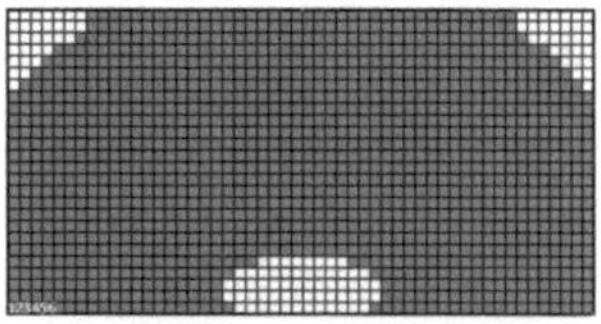
4	1250	0.2292	9	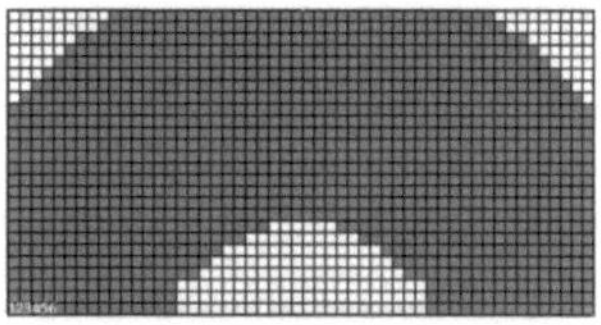
5	1200	0.2181	10	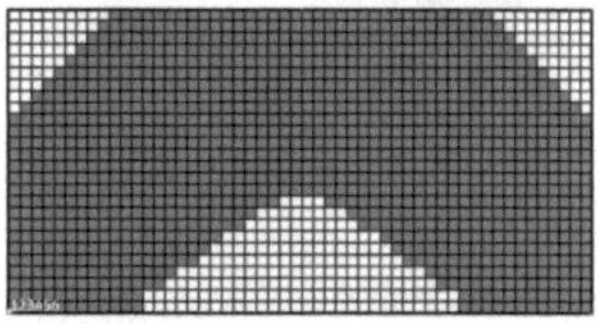
6	1150	0.2084	10	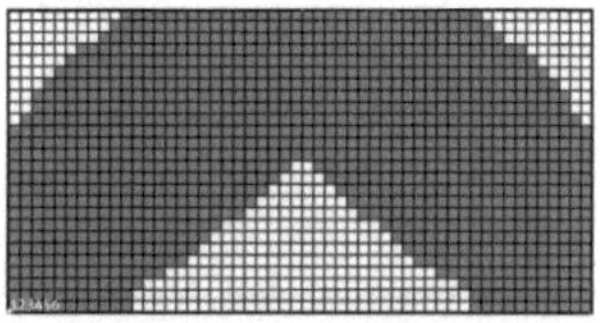
7	1100	0.2103	15	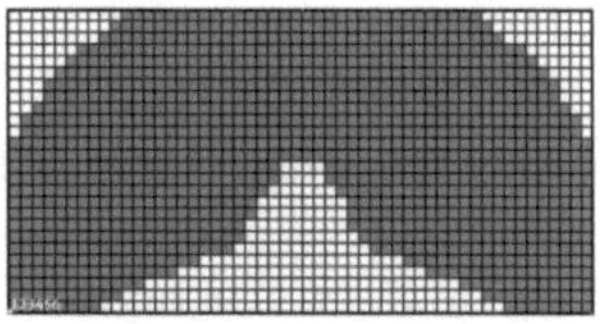
8	1050	0.2051	11	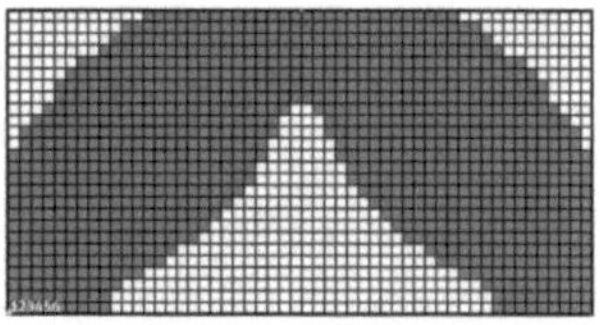
9	1030	0.1993	15	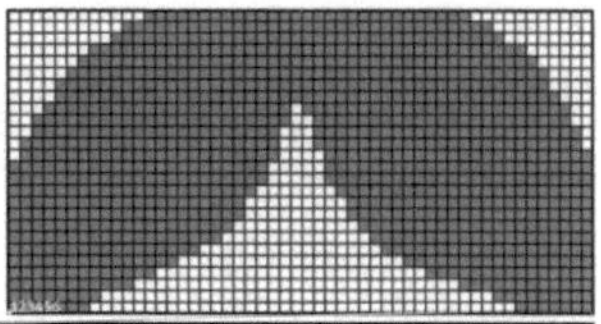

Table 2.11: The ratio of second-order mode buckling constraint and ratio of quality.

Test number	Buckling constraint ratio	Quality ratio of topological structure
1	1	1
2	0.91623	0.97041
3	0.87042	0.91124
4	0.81806	0.84763
5	0.78534	0.80658
6	0.75262	0.77071
7	0.7199	0.77774
8	0.68717	0.75851
9	0.67408	0.73706

Table 2.12: Simulation result of first-order buckling constraint curve.

Serial number	x	y	x^2	xy
1	1	1	1	1
2	0.97041	0.91623	0.9417	0.88912
3	0.91124	0.87042	0.83036	0.79316
4	0.84763	0.81806	0.71848	0.69341
5	0.80658	0.78534	0.65057	0.63344
6	0.77071	0.75262	0.59399	0.58005
7	0.77774	0.7199	0.60488	0.5599
8	0.75851	0.68717	0.57534	0.52123
9	0.73706	0.67408	0.54326	0.49684
$\sum$	7.57988	7.22382	6.45858	6.16715

Using the data in Table 2.12, we obtain the following equations:

$$\begin{bmatrix} \sum_{i=1}^{9} 1 & \sum_{i=1}^{9} x_i \\ \sum_{i=1}^{9} x_i & \sum_{i=1}^{9} x_i^2 \end{bmatrix} \begin{bmatrix} a \\ b \end{bmatrix} = \begin{bmatrix} \sum_{i=1}^{9} y_i \\ \sum_{i=1}^{9} x_i y_i \end{bmatrix} \tag{2.70}$$

$$\begin{cases} 9a + 7.57988b = 7.22382 \\ 7.57988a + 6.45858b = 6.16715 \end{cases}$$

Solution of equations is $a = -0.186$ and $b = 1.1738$. Substituting these into equation (2.69), we get

$$y = -0.186 + 1.1738x \tag{2.71}$$

The relationship between the quality of the structure and the critical force of buckling can be expressed by the approximate hyperbolic curve, which can be obtained from Graph 2.12. From the graph, we can see that the constraint ratio is in the range of 0.3–1.0, and when the upper limit of the buckling critical force is in the range of 0.3–0.7 of the whole structure, the quality of the structure changes very little, but the quality changes in the range of 0.17–0.32. This means that the optimization of the topology change is smaller, and by taking the constraint value in this interval we can get a reasonable topology graphics.

As shown in Figure 2.13, in the constrained value range of 0.55–0.7, the second-order buckling mode can be a better topological structure, and the corresponding topological structure quality is in the 0.55–0.68 range. As shown in Figure 2.14, in the constrained value interval of 0.67–0.75, the third-order buckling modes can be obtained with a better topological structure, and the corresponding topological structure quality ratio is between 0.73 and 0.78.

The aforementioned three cases can be used as a reference when the structure is used as a single constraint to carry topology optimization. The three kinds of curves drawn in the same coordinate system are shown in Figure 2.15, and we can conclude from the figure that the best range of the critical force of buckling restrained is 0.4–0.7, and the quality ratio of the topology optimization structure is 0.2–0.78.

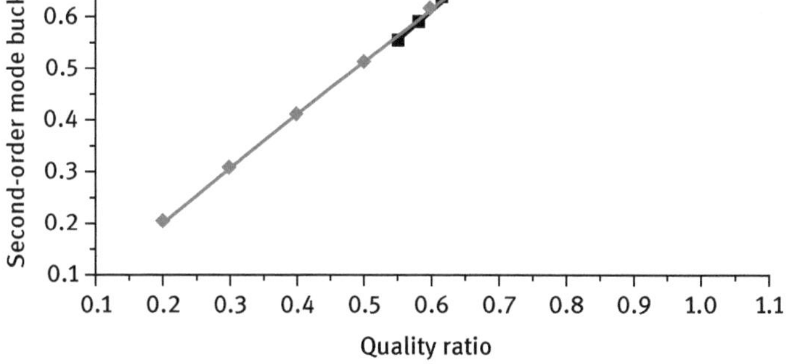

Figure 2.13: Curve of quality ratio and constraint ratio with the second-order mode.

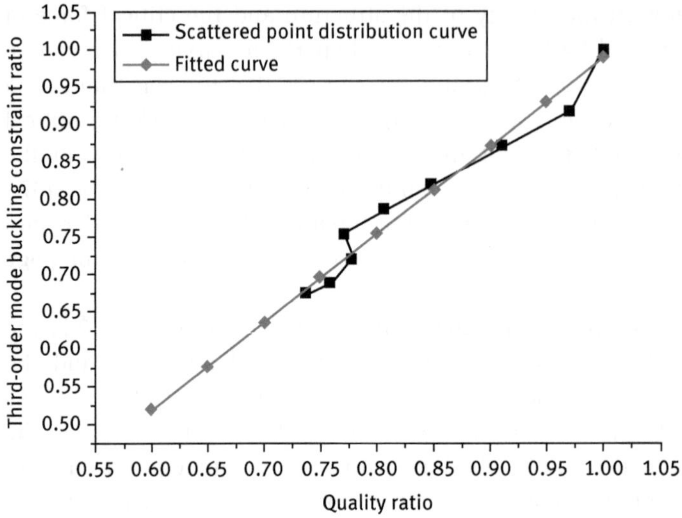

Figure 2.14: Curve of quality ratio and constraint ratio with third-order mode.

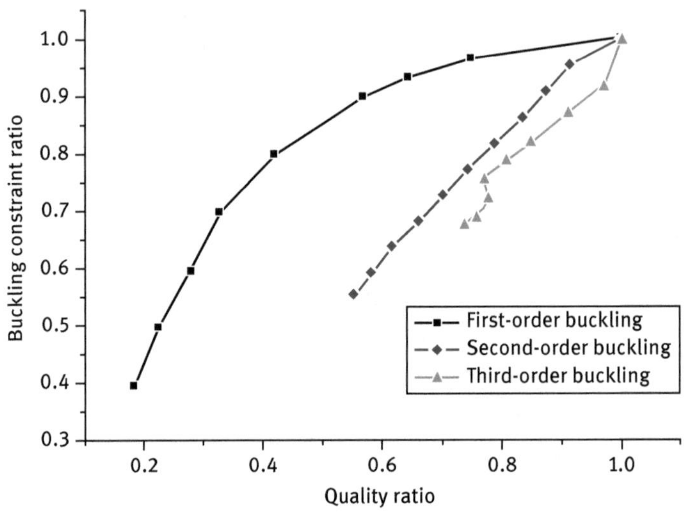

Figure 2.15: Curves of quality ratio and constraint ratio with three-order modes.

2.8 Optimization Algorithms

The optimization algorithm of the displacement and buckling constraints is based on the second development platform of MSC.Patran/Nastran software, The optimization algorithm program and user input interface were designed in PCL language. The specific steps of optimization are as follows:

(1) Finite element model
he geometric model of continuum structure is established on the MSC.Patran platform. Then the model was divided into finite element mesh, the material properties was defined, the boundary conditions, and the properties of the element were given. Because of the structural topology optimization by buckling constraints, the properties of planar elements are set to shell elements.

(2) Basic optimization parameter input: This includes the convergence accuracy and the radius of the filter, the sum of the total number of buckling modes, the name of the modal constraint, the numerical value of the buckling constraint and the number of corresponding buckling modes and condition name.

(3) Submit MSC. Nastran solution: The finite element analysis of the buckling mode of the structure is carried out, and the solution is SOL105.

(4) The buckling critical force is calculated and extracted.

(5) The effective buckling constraint is selected.

(6) Generate input files required for the dual quadratic programming.

(7) Call dual quadratic programming to solve optimization model.

(8) According to the iterative convergence criterion to determine whether the convergence conditions are met, if not satisfied, go back to step (3), and if satisfied, then stop the iteration, record the data, determine the optimal topology and inversion topology variables, get the optimal topology graph and end the optimization calculation.

The block diagram of the optimization program is shown in Figure 2.16.

2.9 Numerical Simulation and Case Study

2.9.1 Numerical Example of Two-dimensional Plane

Example 1
As illustrated in Figure 2.17, the base structure is a plane elastic body with a size of 520 mm × 260 mm × 2 mm. The elastic modulus of structure material is $E = 68.89 \times 10^3$ MPa, Poisson's ratio is $y = 0.3$ and the material density is $\rho = 2.7 \times 10^{-6}$ kg/mm^3. The left-hand and right-hand points of the structure are clasped. A concentrated force $P = 15{,}000$ N along the up-to-down (-y) direction is located in

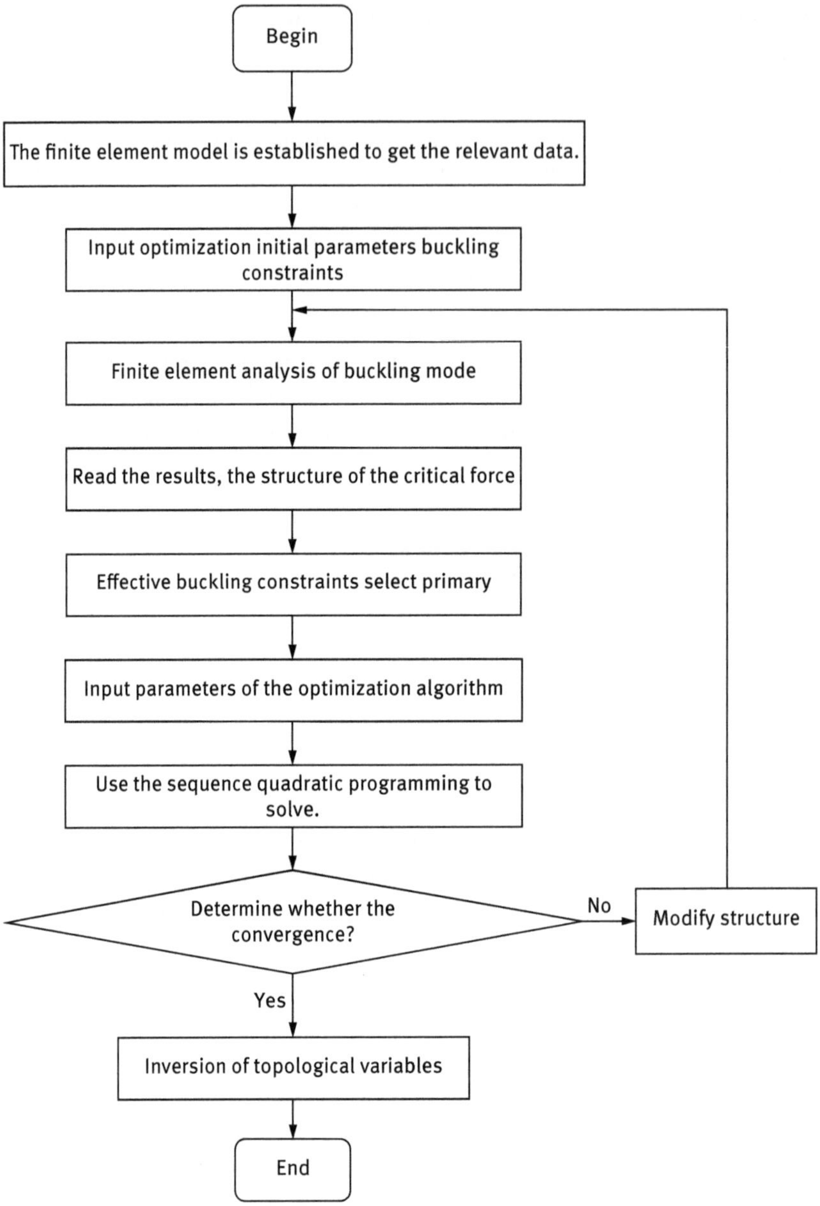

Figure 2.16: Flow chart for the algorithm of topology optimization of continuum structure subjected to buckling constraints.

the upper edge midpoint of the structure. A quadrilateral element with four nodes is used and the total number of elements is 52 × 26. The base structure's quality is 0.73 kg (Figure 2.18).

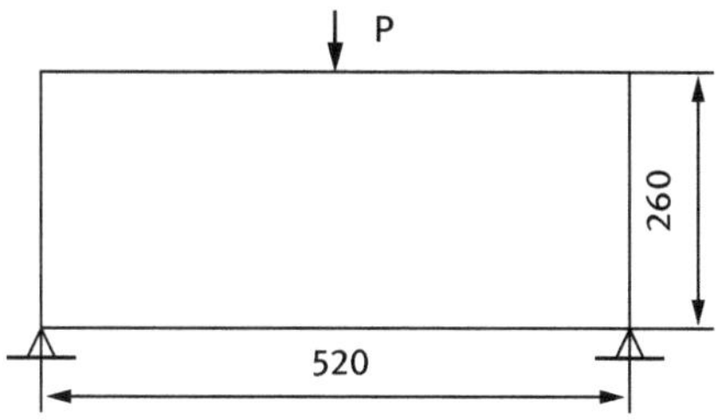

Figure 2.17: Base structure.

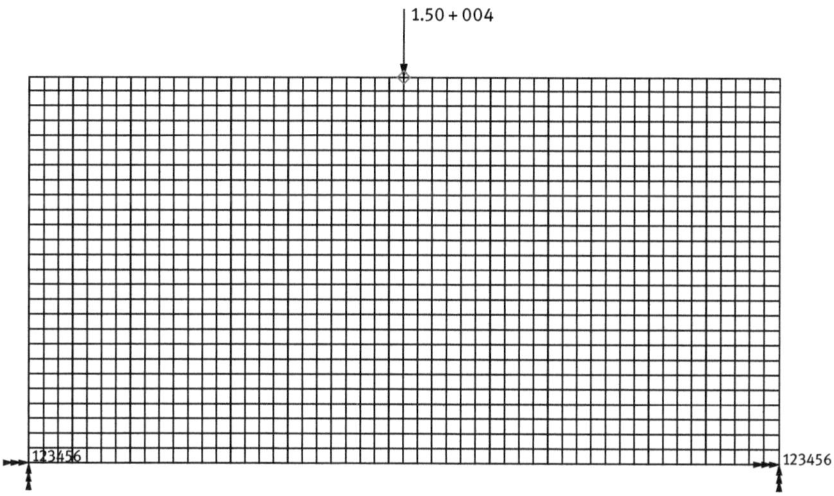

Figure 2.18: Finite element model.

Taking the top third-order buckling mode as the research object, through the finite element analysis of buckling, the first-order buckling factor ζ_1 = 0.0079, and the critical buckling force λ_1 = 119 N. The second-order buckling factor ζ_2 = 0.088112 and the critical buckling force λ_2 = 1322 N and the third-order buckling factor ζ_3 = 0.10186 and the critical buckling force λ_3 = 1528 N.

(1) Single constraint effect

The single critical force in the three-order buckling mode is used as the buckling constraint.

In the following graph, the optimal topology of the inversion of the single constraint, the iterative change curve of the buckling critical force constraint and the iteration curve of the topology structure are obtained. In this case, the threshold value T is 0.5.

In single constraint condition state the value is taken as the critical force base structure 75%, the upper limit of first-order critical force is $\bar{\lambda}_1 = 90\text{N}$, after 36 iterations of convergence, as shown in Figure 2.19. The optimized structure

Figure 2.19: Optimal topological configuration of the first-order mode.

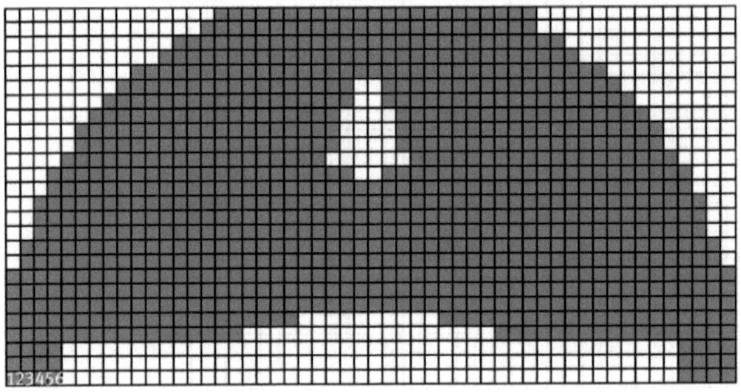

Figure 2.20: Optimal topological configuration of the second-order mode.

quality is 0.2684 kg, and it is about 37% of the quality of the base structure. The upper limit of second-order critical force is $\bar{\lambda}_2 = 960\,\text{N}$ after 29 iterations of convergence, as shown in Figure 2.20. The optimized structure quality is 0.513 kg, and it is about 70% of the quality of the base structure. The upper limit of third-order critical force is $\bar{\lambda}_3 = 1150\text{N}$ after 10 iterations of convergence, as shown in Figure 2.21. The optimized structure quality is 0.5616 kg, and it is of about 77% of the quality of the base structure.

To verify the "quality–buckling constraints ratio" function expressions, the above data were substituted into the formula. In the first order, the quality ratio is 0.37; according to the function formula, the buckling constraint ratio is 0.69. The results obtained for the function relation calculation and numerical calculation have some errors. In the second order, the quality ratio is 0.7; according to the function formula, the buckling constraint ratio is 0.72, which is close to 0.75. In the third order, the quality ratio is 0.77; according to the function formula, the buckling constraint ratio is 0.72, which is close to 0.75.

The results obtained for the function relation calculation and numerical calculation have some errors, which are mainly due to the following reasons:

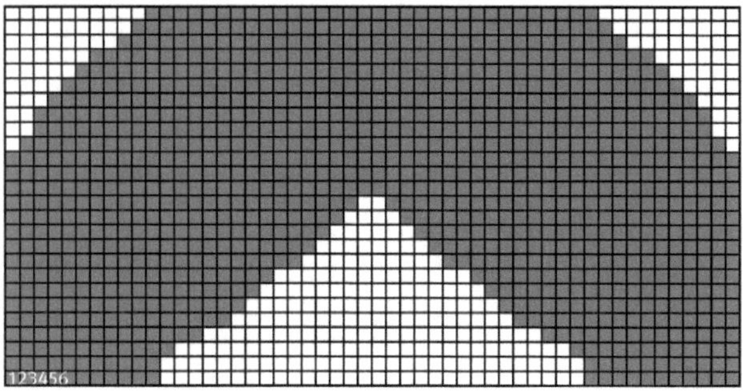

Figure 2.21: Optimal topological configuration of the third-order mode.

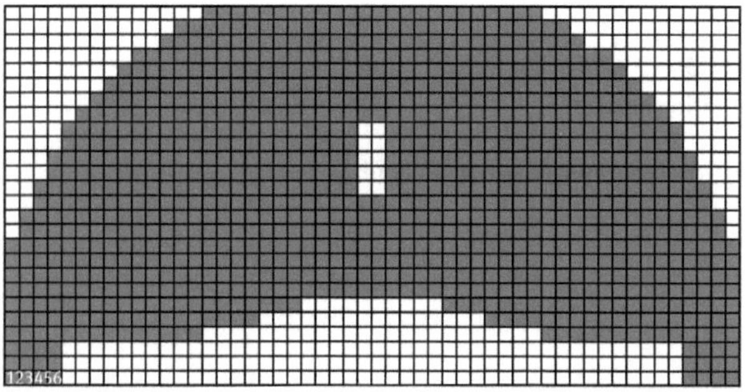

Figure 2.22: Optimal topological configuration with multiple load cases.

the calculation error, fitting error quality inversion topology and curve obtained. The results show that the function expression of the "quality–buckling constraint ratio" is reliable. The following conclusions can be drawn from the first three-order modes of the topology optimization graph and the iteration curve: when the same proportion of each order constraint based on the structure of the critical force, the high-order buckling critical force is larger than that of the low order, and so the higher-order optimal topological structure is more robust than the lower-order topology. Optimization iterations will be reduced accordingly.

(2) Multiconstraint effect

At the same time, considering the interaction of the three-order constraints, the first-, second- and third-order buckling critical force are 90 N, 960 N and 1150 N, respectively; after 29 iterations, the topological structure as in Figure 2.22 is obtained. The optimized structure quality is 0.5265 kg, and the quality of the optimal topology structure is about 72% of the base structure quality. The iteration curves of critical force and mass are given in Figures 2.23–2.30. It

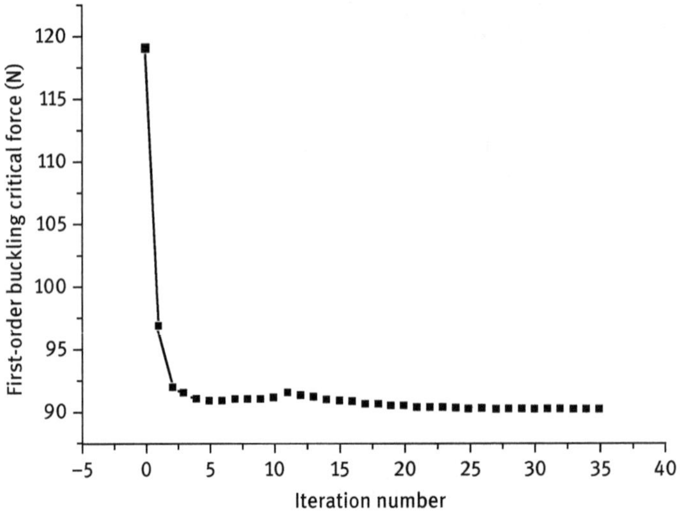

Figure 2.23: Iterative history of the first-order buckling critical load.

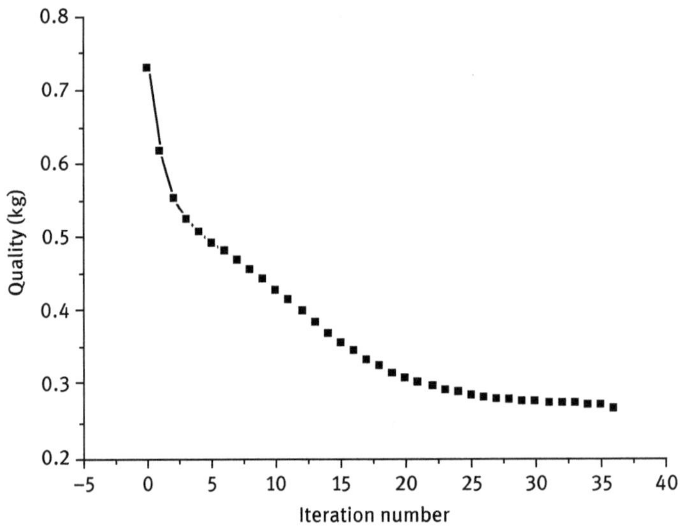

Figure 2.24: Iterative history of the first-order mode quality.

can be concluded from the topological graph and the iterative curve that the multiconstraint optimal topology graph is close to the second- and third-order topology graph, and this shows that under the multiconstrained buckling, the influence of the higher-order constraints on the topology shape is larger. It has "strong" constraints.

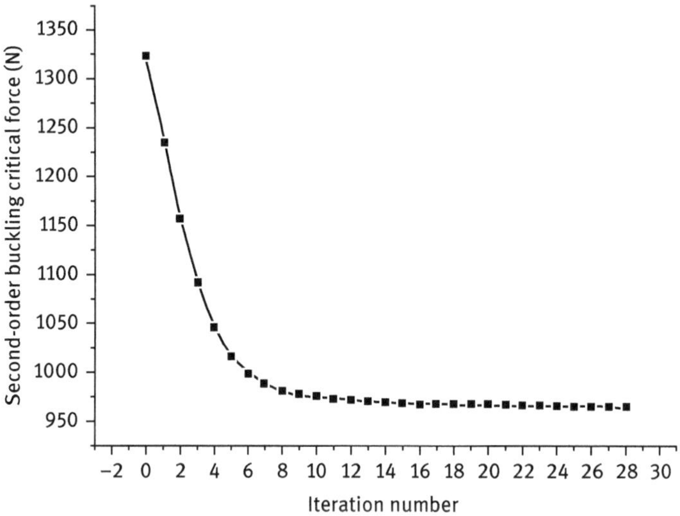

Figure 2.25: Iterative history of the second-order buckling critical load.

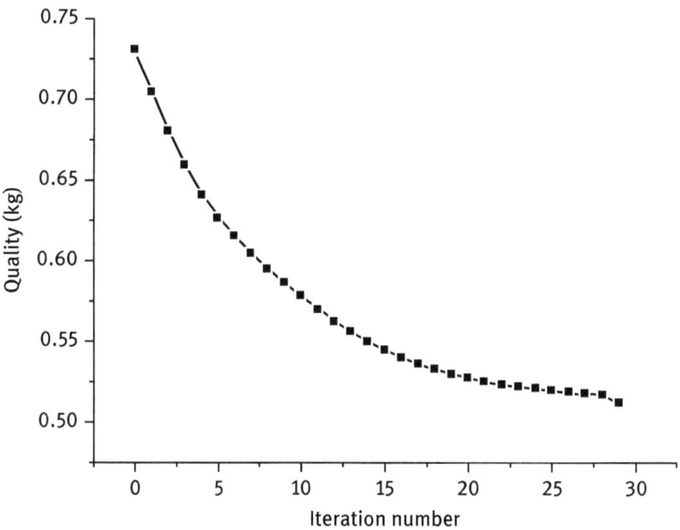

Figure 2.26: Iterative history of the second-order mode quality.

Example 2

As illustrated in Figure 2.31, the base structure is a plane elastic body with a size $520 \times 260 \times 2.0$ mm^3. The elastic modulus of structure material is $E = 68.89 \times 10^3$ MPa, Poisson's ratio is $\gamma = 0.3$ and the material density is $\rho = 2.7 \times 10^{-6}$ kg/mm^3. The left-hand

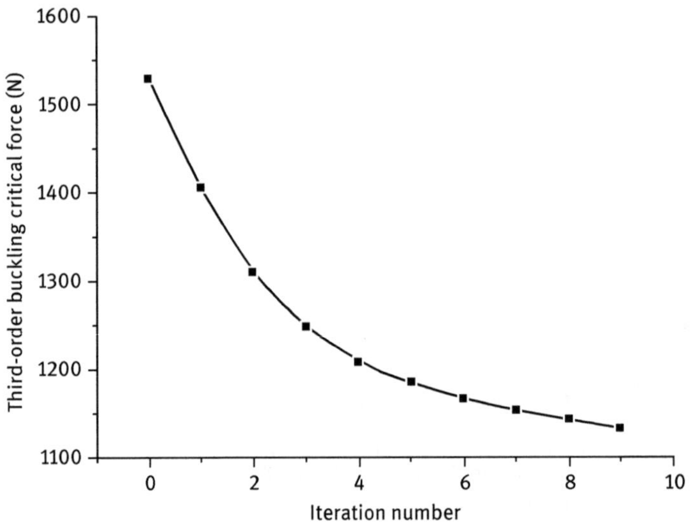

Figure 2.27: Iterative history of third-order buckling critical load.

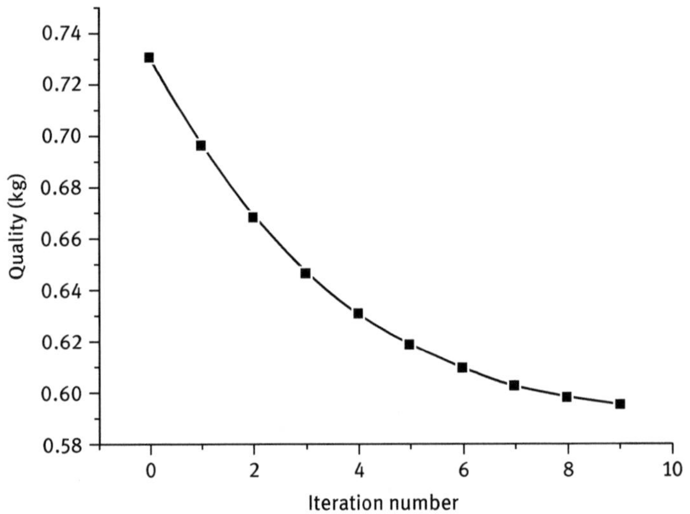

Figure 2.28: Iterative history of a third-order mode quality.

midpoint and right-hand points of the structure are clasped. A concentrated force $P = 15{,}000$ N along the up-to-down ($-y$) direction is located in upper edge midpoint of the structure. A quadrilateral element with four nodes is used and the total number of elements is 52×26. The base structure's quality is 0.73 kg (Figure 2.32).

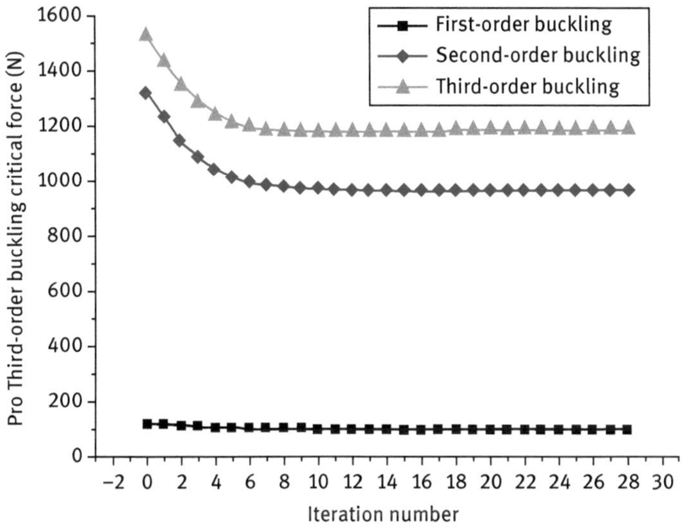

Figure 2.29: Iterative history of buckling critical load with multiple load cases.

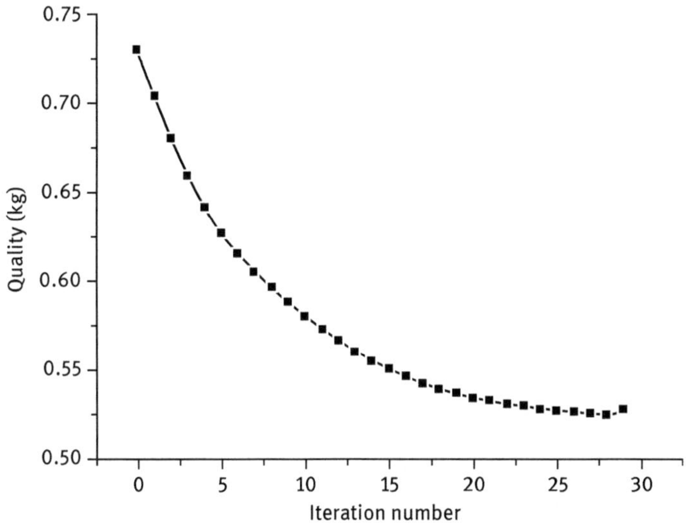

Figure 2.30: Iterative history of quality with multiple load cases.

Taking the top three-order buckling mode as the research object, by the finite element analysis of buckling, the first-order buckling factor $\zeta_1 = 0.17286$, the critical buckling force $\lambda_1 = 259$ N; the second-order buckling factor $\zeta_2 = 0.16906$, the critical buckling force $\lambda_2 = 2536$ N and the third-order buckling factor $\zeta_3 = 0.20925$, the critical buckling force $\lambda_3 = 3139$ N.

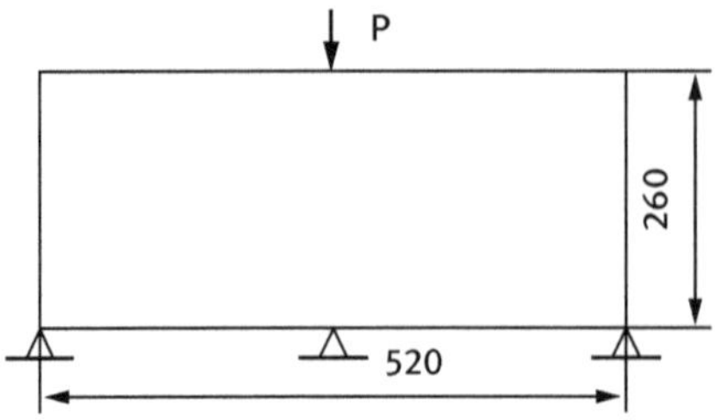

Figure 2.31: Base structure.

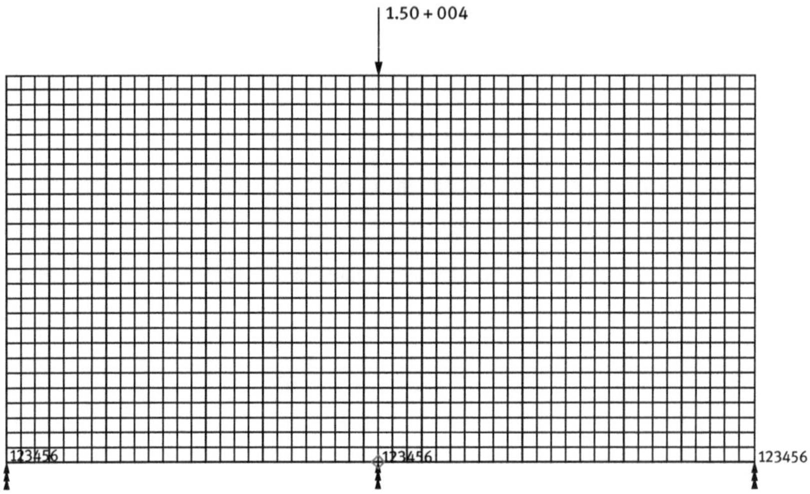

Figure 2.32: Finite element model.

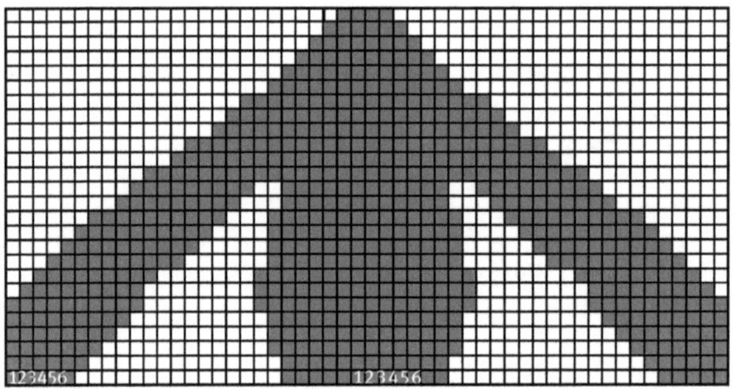

Figure 2.33: Optimal topological configuration of the first-order mode.

(1) Single constraint effect

At a single constraint condition, the value is taken as the critical force base structure of 77%, and the upper limit of first-order critical force is $\bar{\lambda}_1 = 200$ N after 34 iterations of convergence, as shown in Figure 2.33. The optimized structure

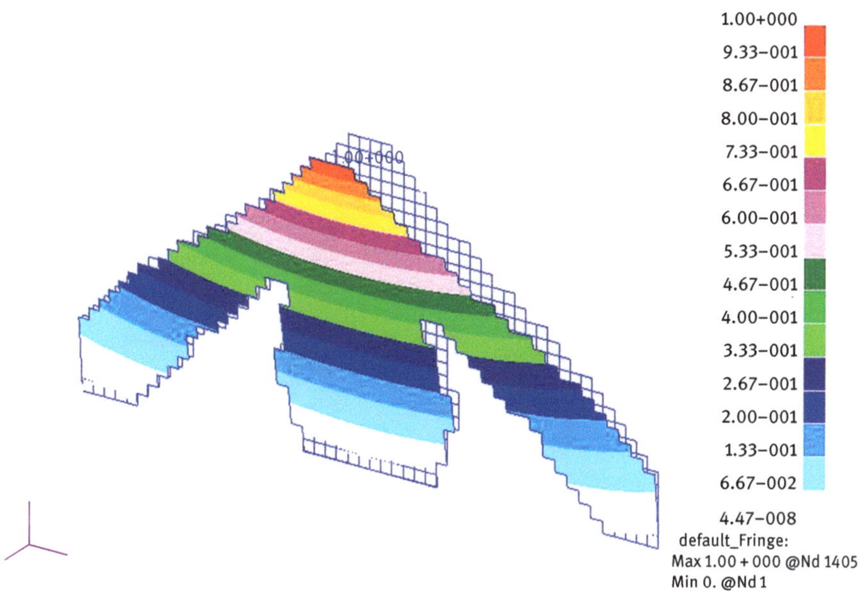

Figure 2.34: The first-order mode of topological configuration.

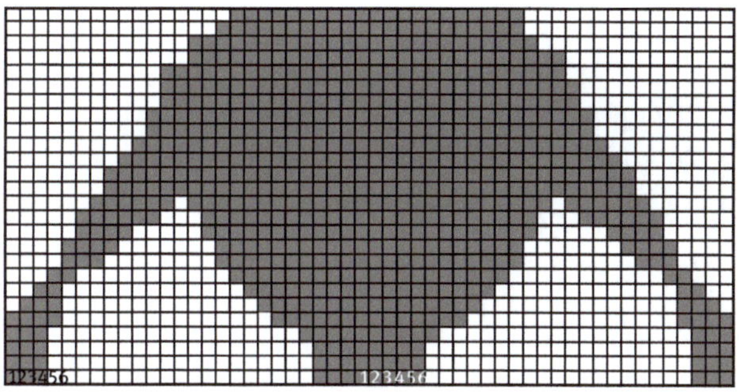

Figure 2.35: Optimal topological configuration of the second-order mode.

quality is 0.3429 kg, and it is of about 47% of the quality of the base structure. The upper limit of second-order critical force is $\bar{\lambda}_2 = 2000\text{N}$ after 35 iterations of convergence, as shown in Figure 2.35. The optimized structure quality is 0.378 kg, and it is of about 52% of the quality of the base structure. The upper limit of third-order critical force is $\bar{\lambda}_3 = 2500\,\text{N}$ after 28 iterations of convergence, as shown in Figure 2.37. The optimized structure quality is 0.5076 kg, and it is about 70% of the quality of the base structure.

To verify the "quality–buckling constraints ratio" function expressions, the aforementioned data were substituted into the formula. In the first order, the quality

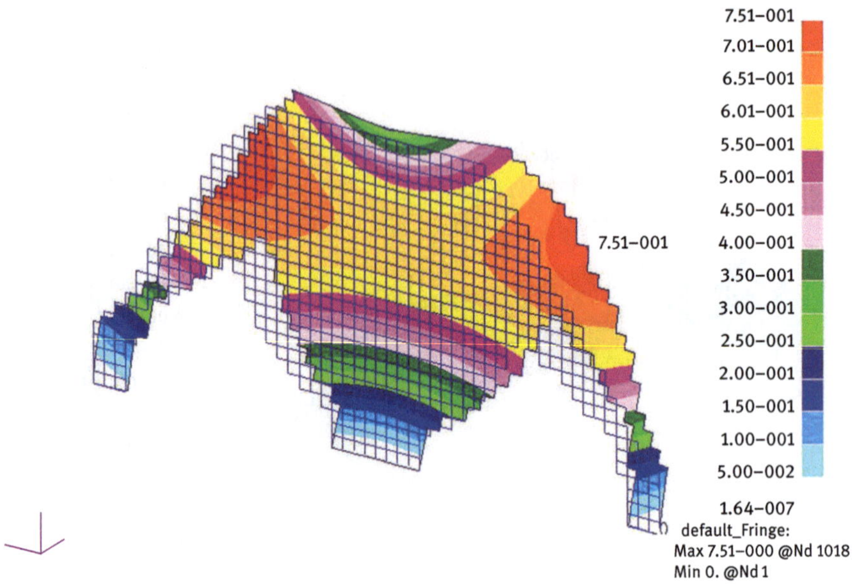

Figure 2.36: The second-order mode of topological configuration.

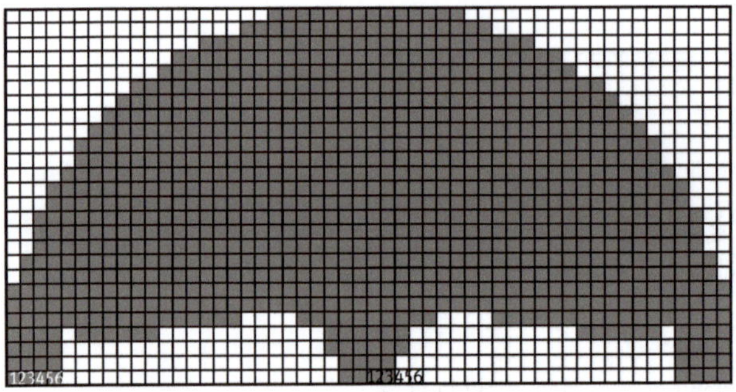

Figure 2.37: Optimal topological configuration of the third-order mode.

ratio is 0.47, and according to the function formula, the buckling constraint ratio is 0.79. In the second order, the quality ratio is 0.52, and according to the function formula, the buckling constraint ratio is 0.54, which is not close to 0.77. In the third order, the quality ratio is 0.70, and according to the function formula, the buckling constraint ratio is 0.64, which is close to 0.77.

From the three-order topology optimization graph and the iteration curve, we can get the same conclusion as in example 1. The high-order buckling mode topology optimization structure quality is larger than that of the low-order buckling mode. In Figures 2.33–2.38, the deformation of the three-order buckling mode is given, and the numerical value is the deformation. Figures 2.39–2.47, the iterative history of the three order buckling mode are given.

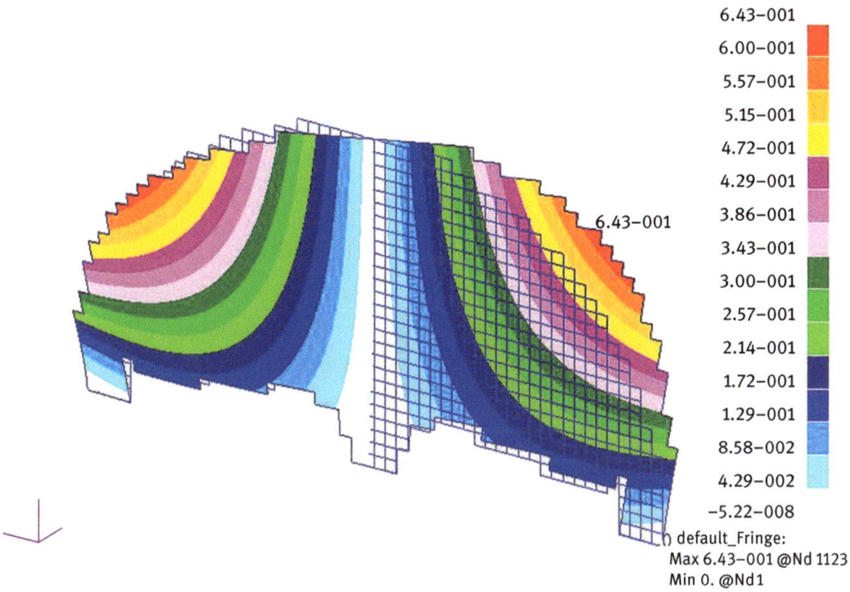

Figure 2.38: The third-order mode of topological configuration.

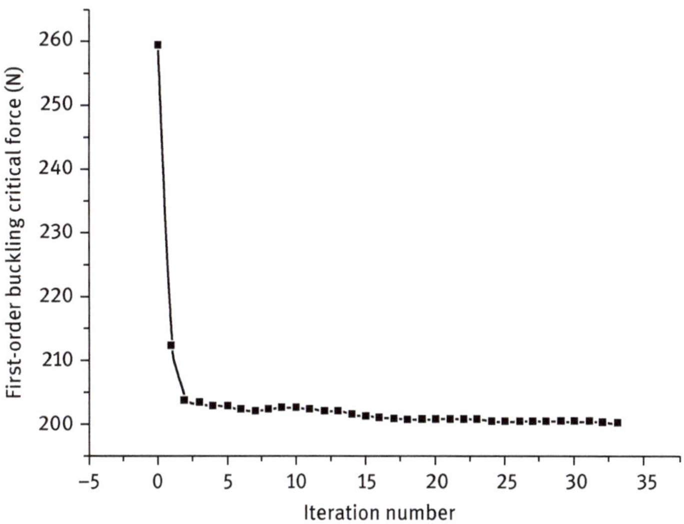

Figure 2.39: Iterative history of the first-order buckling critical load.

(2) Multiconstraint effect

At the same time, considering the interaction of the three-order constraints, the first-, second- and third-order buckling critical forces are 200 N, 2000 N and 2500 N,

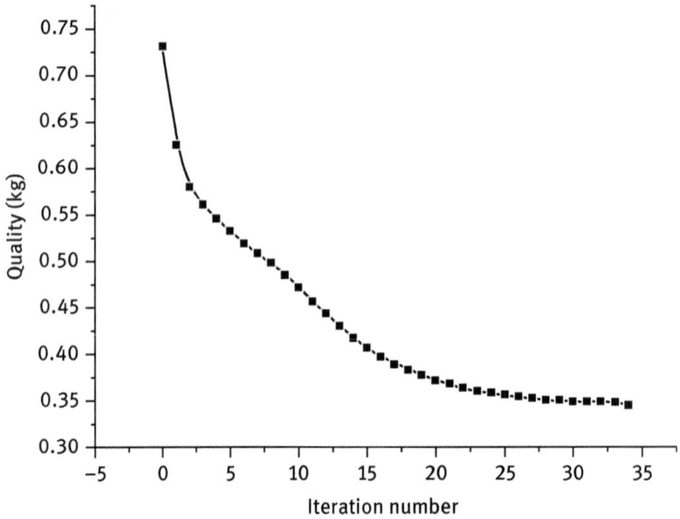

Figure 2.40: Iterative history of the first-order mode quality.

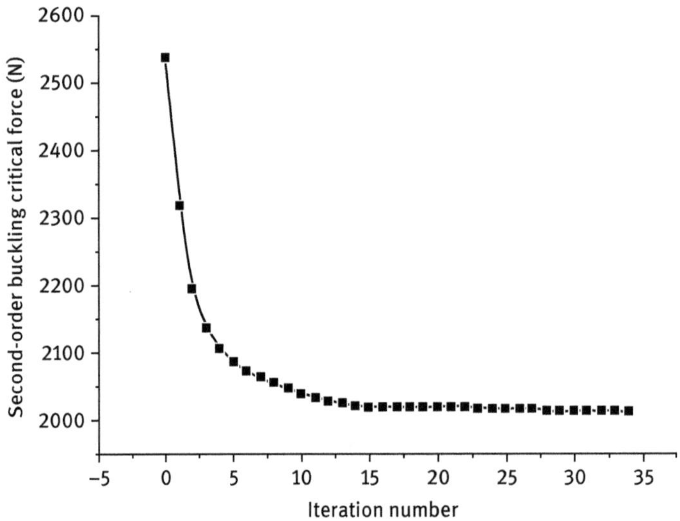

Figure 2.41: Iterative history of second-order buckling critical load.

respectively, after 39 iterations, and the topological structure as in Graph 2.45 is obtained. The optimized structure quality is 0.4806 kg, and the quality of the optimal topology structure is about 66% of the base structure quality. The iteration curves of critical force and mass are given in Figures 2.47–2.48. In this case, the final topology

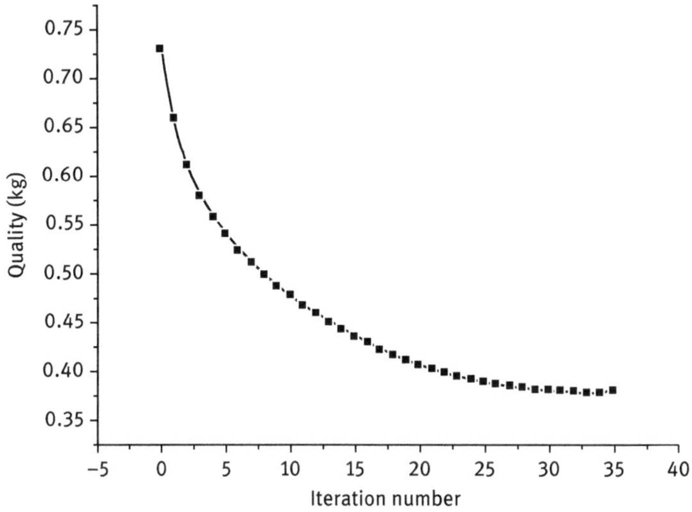

Figure 2.42: Iterative history of the second-order mode quality.

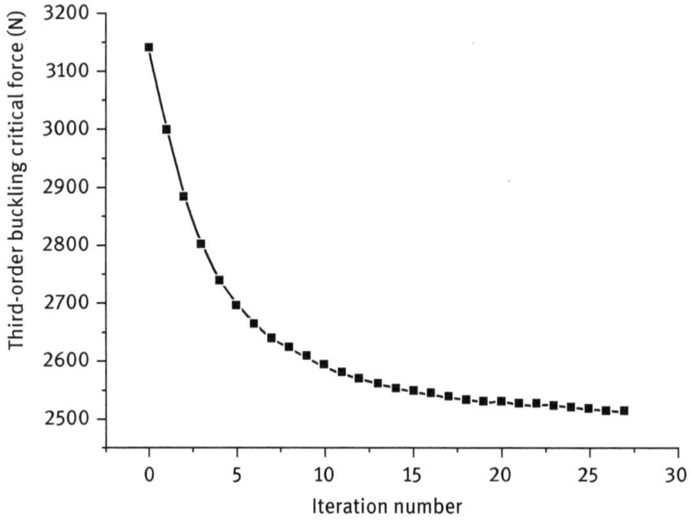

Figure 2.43: Iterative history of buckling critical load with multiple load cases.

and the high-order topology are basically the same; owing to the deletion of a few elements in the high-order topology, the topological structure of quality is larger, which can meet the low-order constraints, and so when there is the multiconstraint interaction, the topology structure will be similar to the higher-order topological structure.

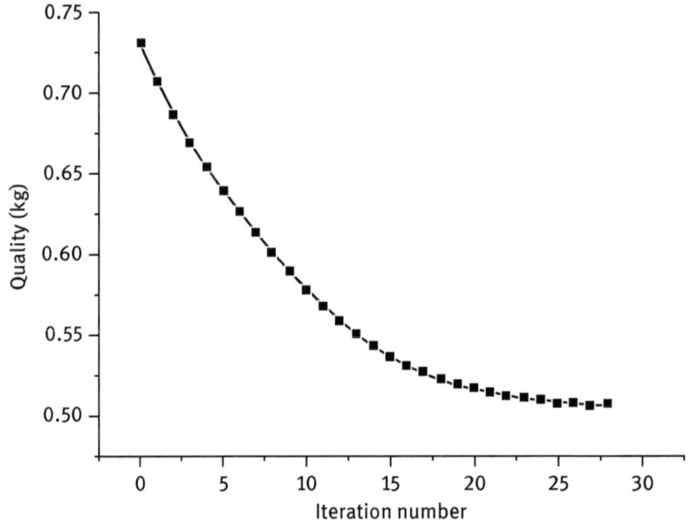

Figure 2.44: Iterative history of quality with multiple load cases.

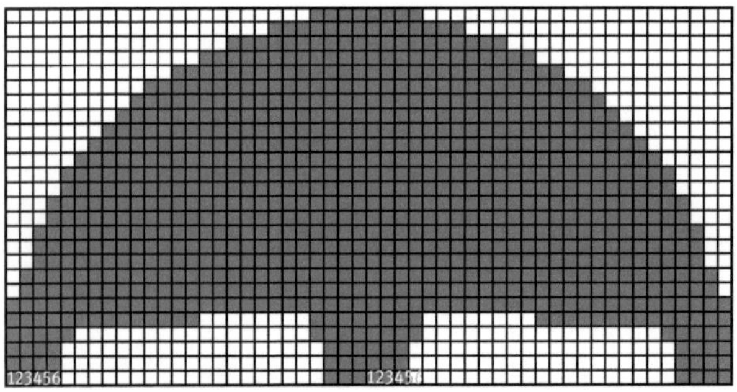

Figure 2.45: Topological configuration with multiple load cases.

Example 3

As illustrated in Figure 2.49, the base structure is a plane elastic body with a size of $520 \times 260 \times 2.0$ mm^3. The elastic modulus of the structure material is $E = 68.89 \times 10^3$ MPa, Poisson's ratio is $\gamma = 0.3$ and the material density is $\rho = 2.7 \times 10^{-6}$ kg/mm^3. The left-hand and right-hand points of the structure are clasped. A load has line pressure of 100 N/mm, and total force $P = 52{,}000$ N along the up-to-down (-y) direction is located in upper edge of the structure. A quadrilateral element with four nodes is used, and the total number of elements is 52×26. The base structure's quality is 0.73 kg.

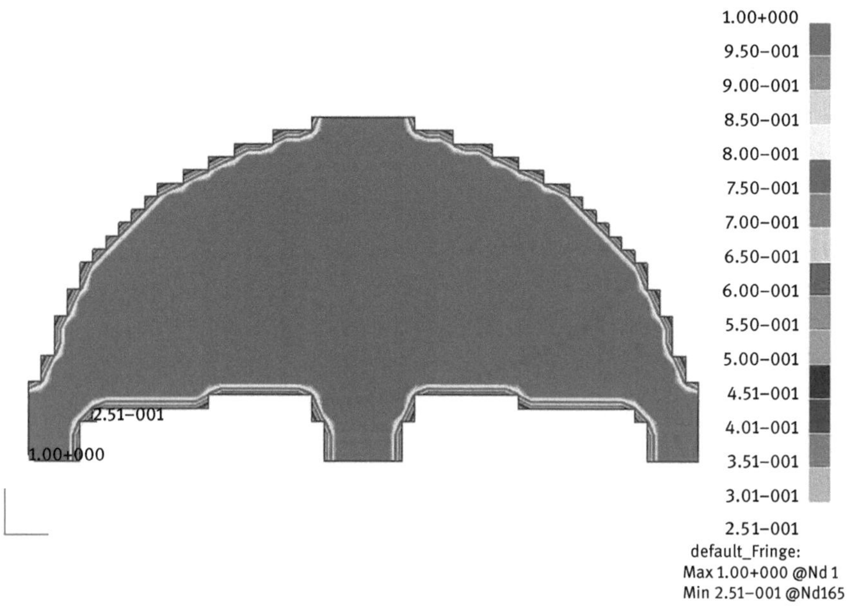

Figure 2.46: Topological variables of topological configuration.

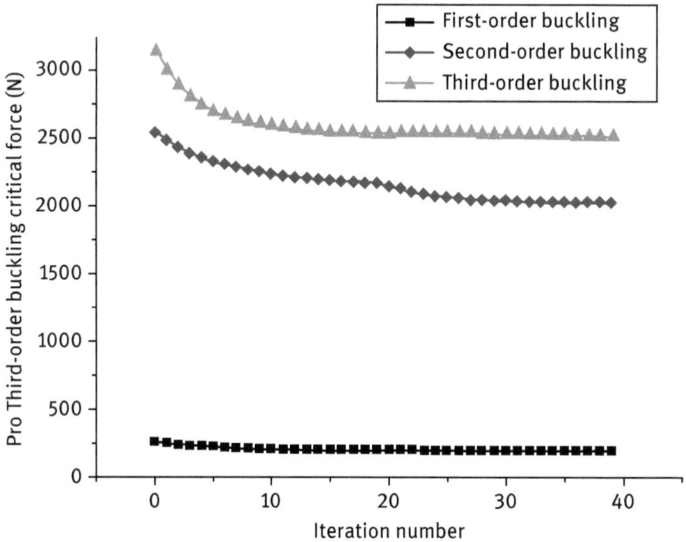

Figure 2.47: Iterative history of critical load with multiple load cases.

The base structural first-order buckling factor is $\zeta_1 = 0.002511$, and the first-order buckling critical force is 130.572 N. In this case, the critical buckling upper limit is $\bar{\lambda}_1 = 100$ N after 51 iterations of convergence, such as the topology shown in

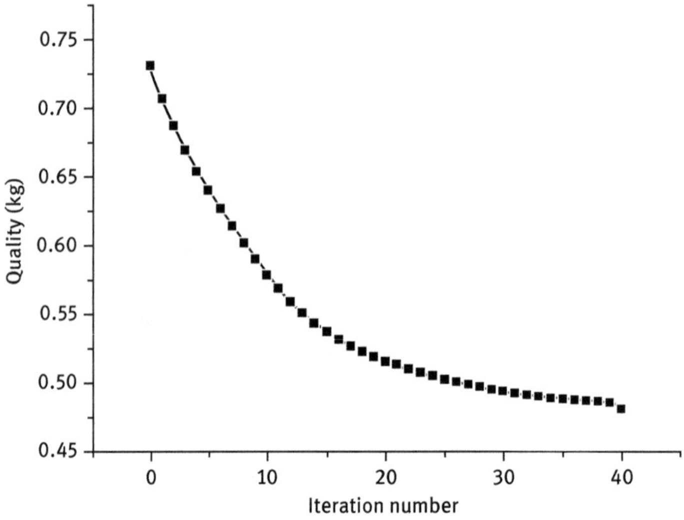

Figure 2.48: Iterative history of quality with multiple load cases.

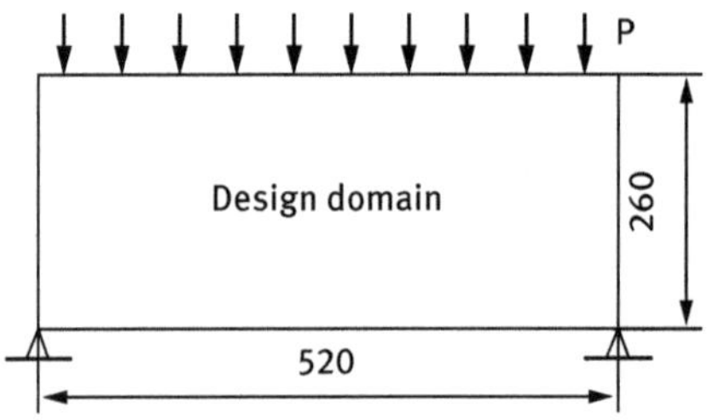

Figure 2.49: Base structure.

Figure 2.50. The optimized structure quality is 0.3132 kg, and the buckling critical force is 100.5888 N; the upper boundary of the model is applied to the nondesign area and the two layers above are retained.

According to the optimization results, the critical force of the topology structure is 100.5888 N (Figure 2.51), and the error between the critical force upper limit value and the critical force is 0.59%. The quality of the optimal topology is 0.3132 kg. It is 42.78% of the whole quality (Figure 2.52), and save 57.22% quality of the base structure.

Example 4

As illustrated in Figure 2.53, the base structure is a plane elastic body with a size of $520 \times 260 \times 2.0\ \text{mm}^3$. The elastic modulus of structure material is $E = 68.89 \times 10^3$ MPa,

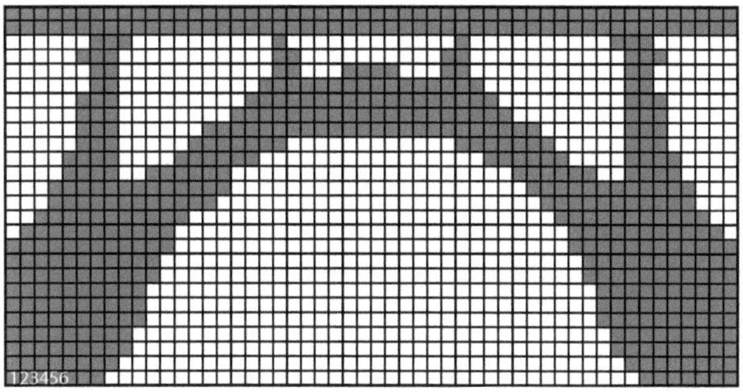

Figure 2.50: Optimal topology structure.

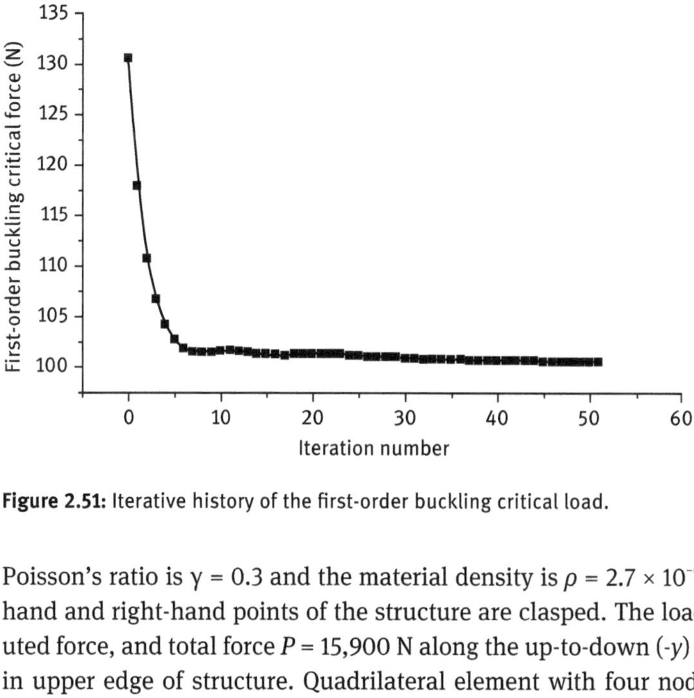

Figure 2.51: Iterative history of the first-order buckling critical load.

Poisson's ratio is γ = 0.3 and the material density is $\rho = 2.7 \times 10^{-6}$ kg/mm^3. The left-hand and right-hand points of the structure are clasped. The load is kind of distributed force, and total force P = 15,900 N along the up-to-down (-y) direction is located in upper edge of structure. Quadrilateral element with four nodes is used and the total number of elements is 52 × 26. The base structure's quality is 0.73 kg.

The base structural first-order buckling factor is $\zeta_1 = 0.0177$, and the first-order buckling critical force is 282 N. In this case, the critical buckling upper limit is $\bar{\lambda}_1 = 200$ N after 47 iterations of convergence, such as the topology shown in Figure 2.54. The optimized structure quality is 0.3375 kg (Figure 2.56), and the buckling critical force is 200.976 N (Figure 2.55); the upper boundary of the model is applied to the nondesign area, and the two layers above are retained.

Example 5

As illustrated in Figure 2.57, the base structure is a plane elastic body with a size of 300 × 100 × 1.0 mm^3. The elastic modulus of structure material is E = 70,000 MPa,

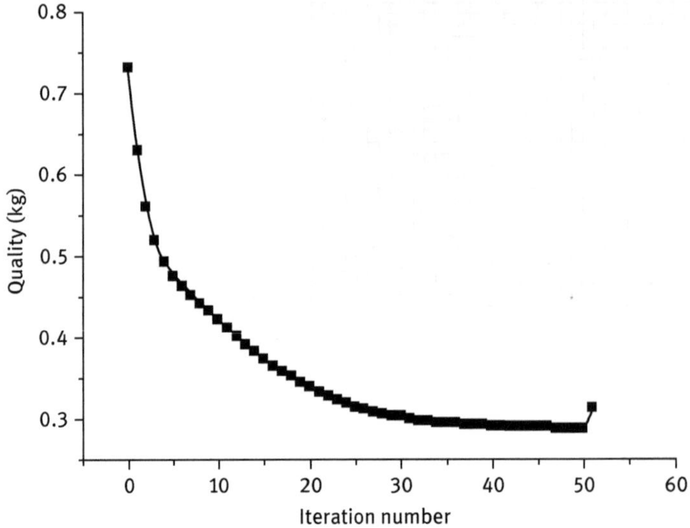

Figure 2.52: Iterative history of the first-order mode quality.

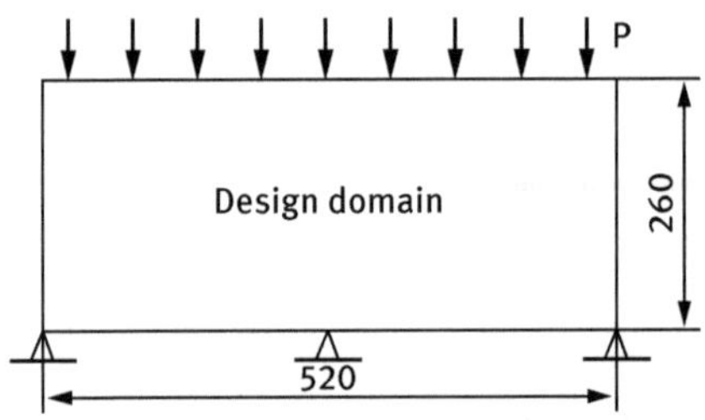

Figure 2.53: Base structure.

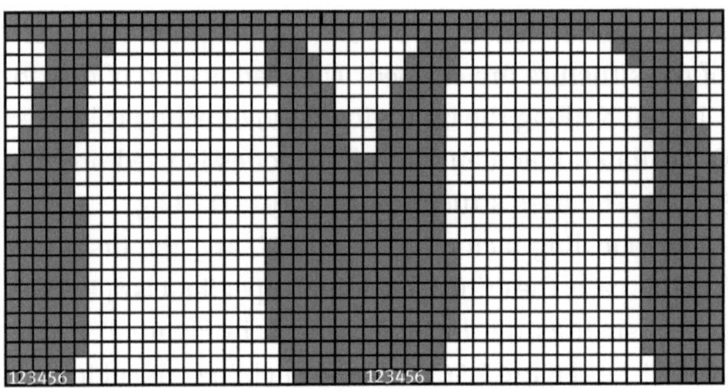

Figure 2.54: Optimal topology structure.

Poisson's ratio is γ = 0.33 and the material density is $\rho = 2.7 \times 10^{-6}$ kg/mm^3. The right-hand points of the structure are clasped. A load P = 2,000 N along the x-direction is located in left-hand edge midpoint of the structure. A quadrilateral element with four

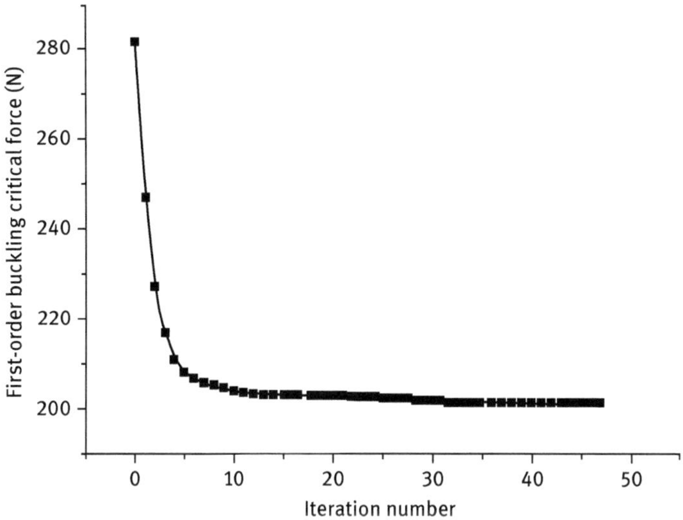

Figure 2.55: Iterative history of the first-order buckling critical load.

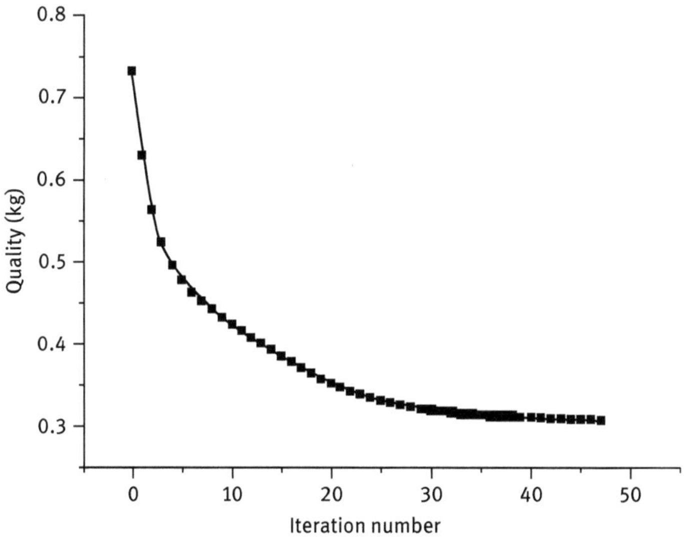

Figure 2.56: Iterative history of the first-order mode quality.

nodes is used and the total number of elements is 60 × 20. The base structure's quality is 0.081 kg.

The base structural first-order buckling factor is $\zeta_1 = 0.0082$, and the first-order buckling critical force is 16.4 N. In this case, the critical buckling upper limit

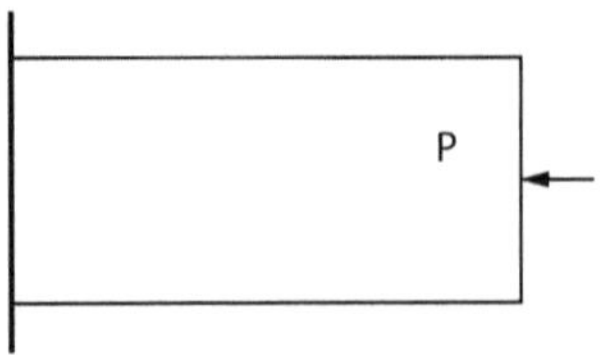

Figure 2.57: Base structure.

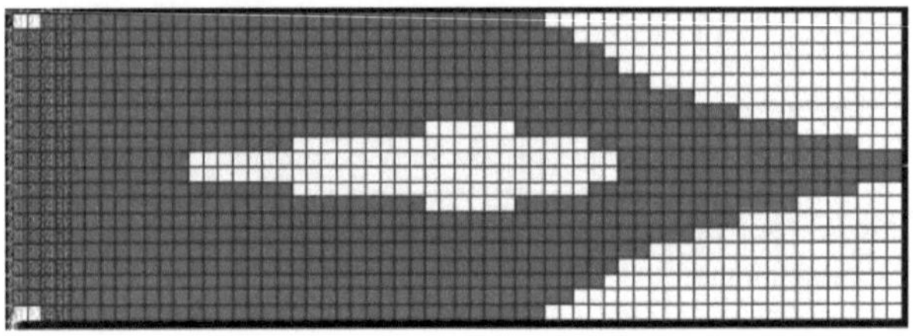

Figure 2.58: Optimal topology structure.

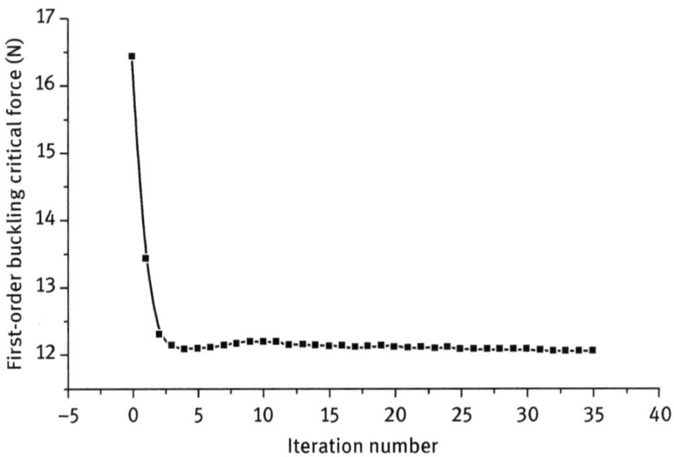

Figure 2.59: Iterative history of the first-order buckling critical load.

is $\bar{\lambda}_1 = 12\,\text{N}$ after 36 iterations of convergence, such as the topology shown in Figure 2.58. The optimized structure quality is 0.055 kg (Figure 2.60). The quality of topology structure quality is 68% of the base structure. The critical buckling force is 12.05 N (Figure 2.59), and the error rate is 0.4%. The critical buckling force is controlled within the effective range, and the error rate is usually defined as 2%.

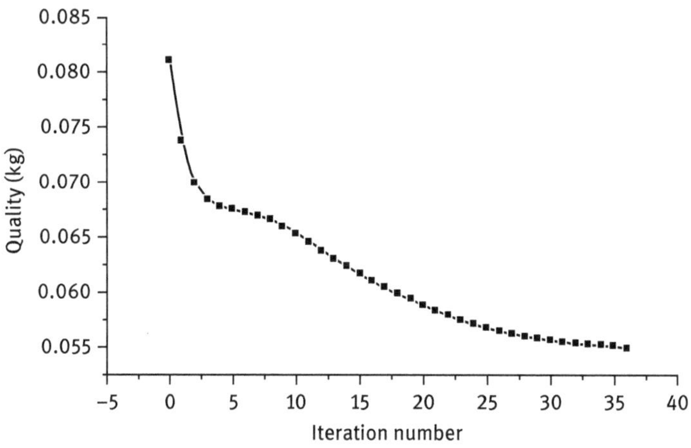

Figure 2.60: Iterative history of the first-order mode quality.

Example of Multiworking Conditions

Example 1

As illustrated in Figure 2.61, the base structure is a plane elastic body with size 600 × 260 × 2.0 mm. The elastic modulus of structure material is E = 68,890 MPa, Poisson's ratio is $\gamma = 0.3$, and the material density is $\rho = 2.7 \times 10^{-6}$ kg/mm^3. The right-hand and left-hand underpart corner points of the structure are clasped. A quadrilateral element with four nodes is used and the number of total elements is 60 × 26. Condition 1: a concentrated force P1 = 15,000 N acting on the upper boundary of the midpoint. Condition 2: two concentrated force P2 = 7,500 acting on the upper boundary point, where the distance is 220 mm to the midpoint. The first-order buckling force of based structure is 112 N, the upper critical buckling is $\bar{\lambda}_1 = 60$ N, the second-order buckling critical force is 792 N and the upper critical force of buckling is $\bar{\lambda}_2 = 400$ N. The base structure's quality is 0.84 kg. The convergence precision is 0.001. The case study is the comparison of the continuum topology optimization for the single and multimodal buckling constraints.

Compared with the single working condition and multicondition of the topology graph (Figure 2.62-2.66), it can be concluded that in this example, the topology of multiworking conditions is not a simple combination of single working conditions, but it has a considerable difference with the topological graph of single working condition.

With the increase in the buckling critical force (Figure 2.67), the topological structure also changes, and the quality of the topological structure is increased obviously (Figure 2.68).

It can be seen from the topological graph that the topology structure of multimodal is identical with the high-order buckling constraint topology, and the low-order constraint does not play a role (Table 2.13).

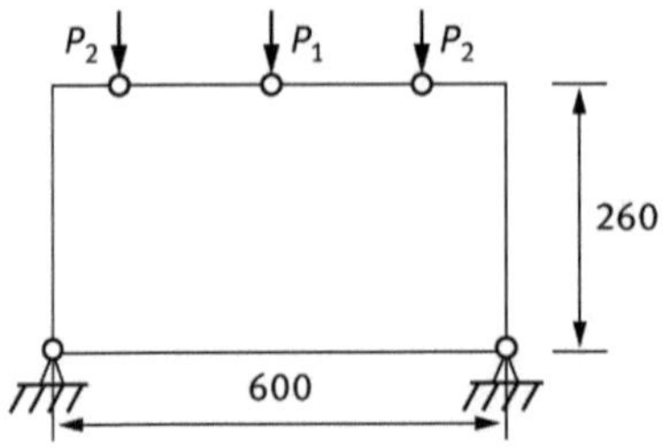

Figure 2.61: Base structure.

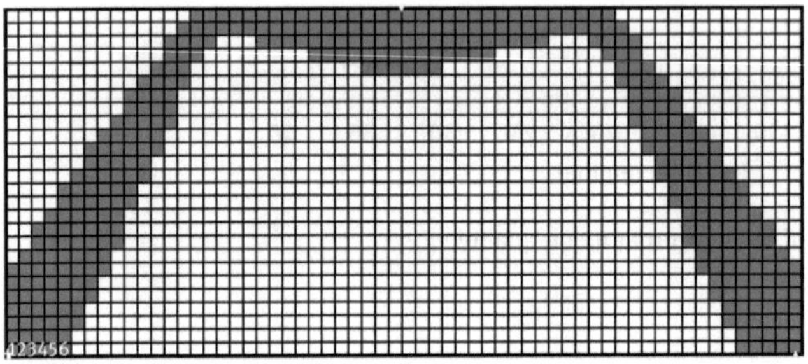

Figure 2.62: Optimal topological configuration of load case 1.

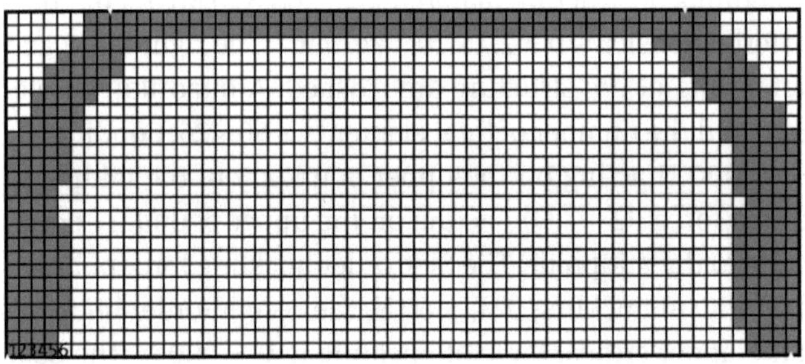

Figure 2.63: Optimal topological configuration of load case 2.

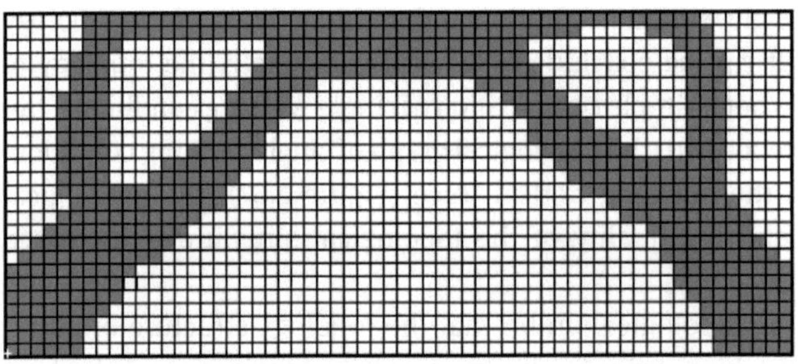

Figure 2.64: Topological configuration with multiple load cases $\bar{\lambda}_1 = 60$ N.

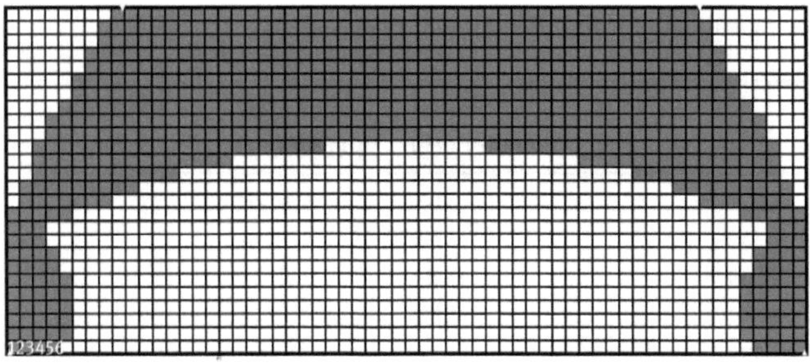

Figure 2.65: Second-order mode topological configuration with multiple load cases λ_2 = 400 N.

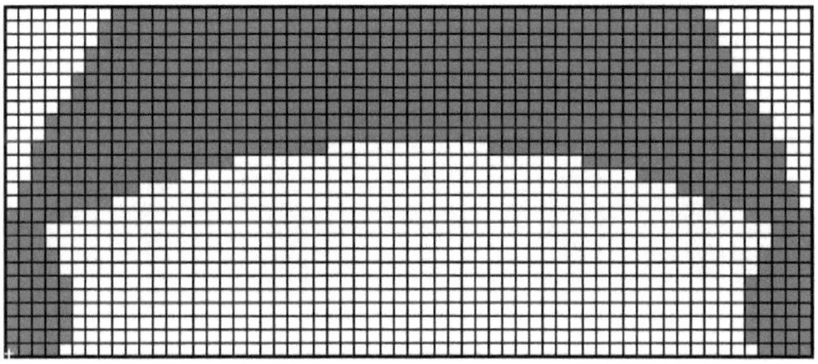

Figure 2.66: First and second-order mode topological configuration with multiple load cases $\bar{\lambda}_1$, 60 N, $\bar{\lambda}_2$ = 400 N.

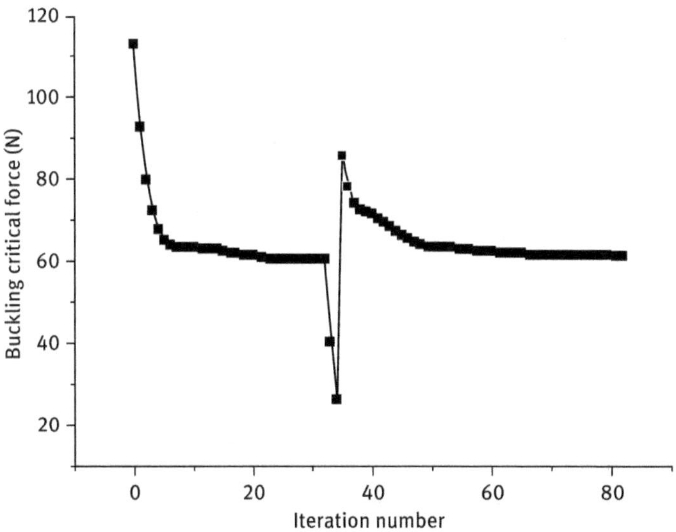

Figure 2.67: First-order mode iterative history of multiple load case critical load.

Table 2.13: Comparison of topological result with different load cases.

Constrain type	Initial quality (kg)	Optimized quality (kg)	Iteration number
Single working condition 1	0.84	0.20	39
Single working condition 2	0.84	0.18	40
First-order buckling + multicondition	0.84	0.23	82
Second-order buckling + multicondition	0.84	0.387	31
The first-second order buckling + multicondition	0.84	0.387	31

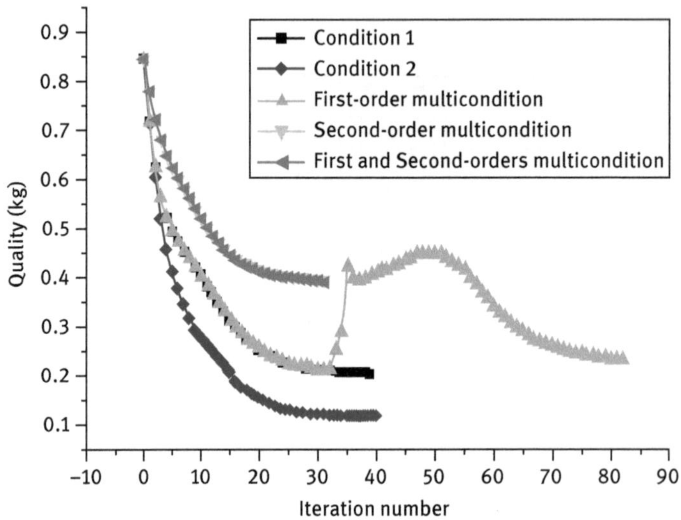

Figure 2.68: Different iterative history of load cases quality.

Example 2

As illustrated in Figure 2.69, the base structure is a plane elastic body with a size 120 × 60 × 6.0 mm^3. The elastic modulus of structure material is E = 210 GPa, Poisson's ratio is γ = 0.25 and the material density is ρ = 7.8 × 10^{-6} kg/mm^3. A quadrilateral element with four nodes is used and the total number of elements is 60 × 30. There are two concentrated force P = 2,000 acting on the upper and lower boundary corners; the left boundary has fixed constraints.

Condition 1: P acting on the upper corner point. Condition 2: P acting on the lower corner point. The base structure's quality is 0.337 kg. The first-order buckling critical force of the base structure is 67,066 N, and the single condition upper limit of the buckling critical force is 30,000 N.

As you can see from the graph, the topology structure of multiworking conditions is not in the superposition as that of the single working condition structure (Figure 2.70-2.74).

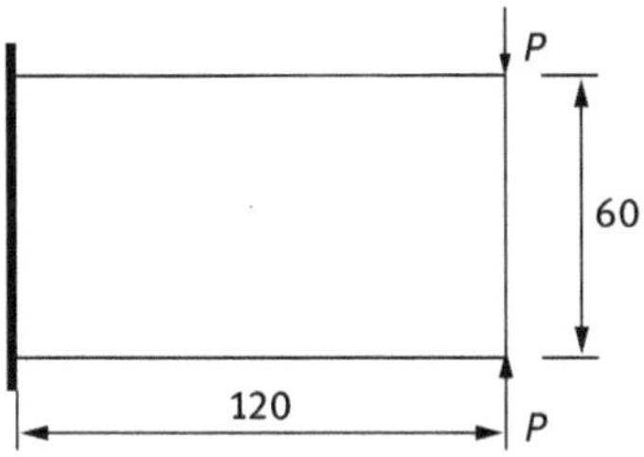

Figure 2.69: Base structure.

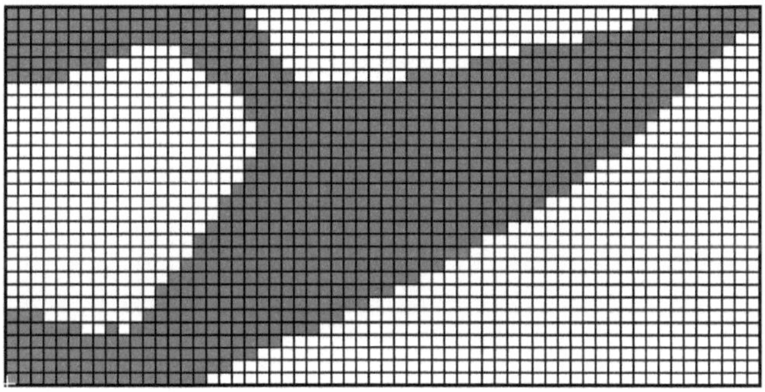

Figure 2.70: Optimal topological configuration of load case 1.

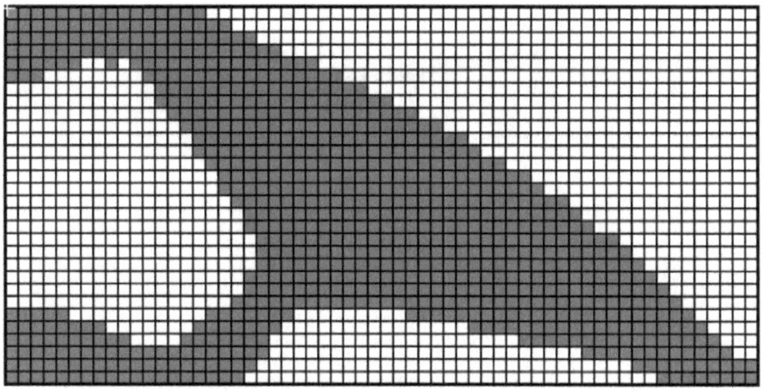

Figure 2.71: Optimal topological configuration of load case 2.

2.9.2 Example of Cylindrical Shell

1. Example of Long Cylindrical Shell

As illustrated in Figure 2.75, the base structure is a half cylindrical shell, with a radius of 50 mm and a length of 300 mm. The elastic modulus of structure material is $E = 210$ GPa, Poisson's ratio is $\gamma = 0.30$ and the material density is $\rho = 7.8 \times 10^{-6}$ kg/mm^3. A concentrated force $P = 9{,}000$ N along the up-to-down ($-y$) direction is located in upper

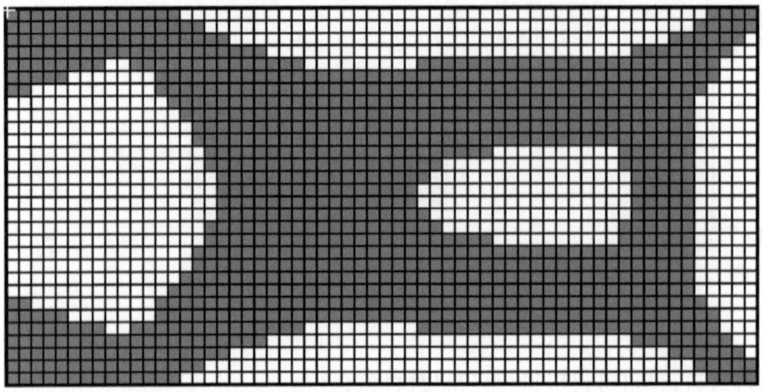

Figure 2.72: Optimal topological configuration of multiple load case.

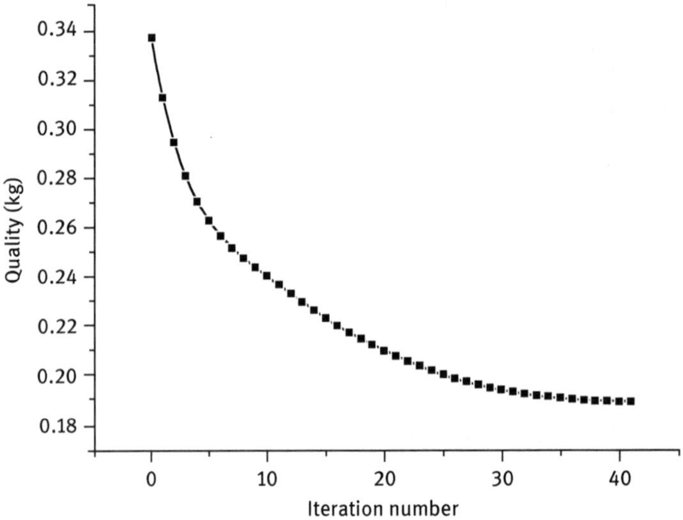

Figure 2.73: Iterative history of multiple load cases quality.

boundary midpoints of the structure. A fixed support is arranged at the bottom left corner and right corner and at the midpoint. A shell element with four-node quadrilateral shell element is used, and the total number of elements is 25 × 72. The base structure's quality is 1.4664 kg.

The base structural first-order buckling factor $\zeta_1 = 29.597$, the first-order buckling critical force is 266,373 N. In this case, the critical buckling upper limit is $\bar{\lambda}_1 = 150{,}000$ N after 29 iterations of convergence; the topology is as shown in Figure 2.76-2.77. The optimized structure quality is 1.014 kg (Figure 2.80) and the buckling critical force is 148,023 N (Figure 2.79). Figure 2.78 shows the deformation of the first-order buckling mode. The numerical example shows that the program can be well applied to the topology optimization of long cylindrical shells.

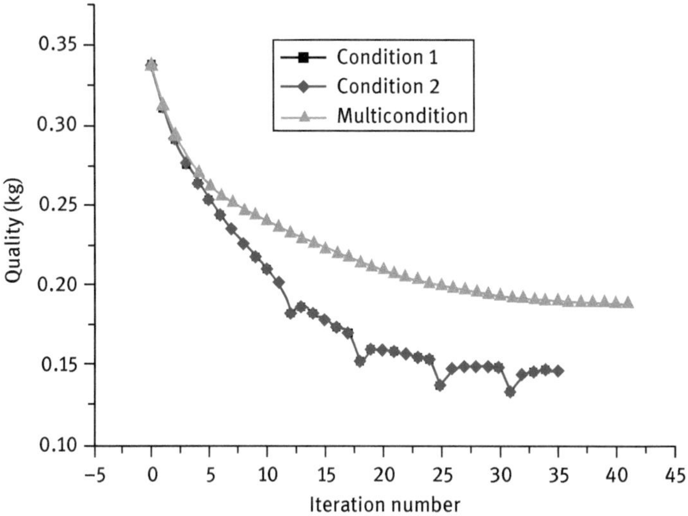

Figure 2.74: Iterative history of single and multiple load cases quality.

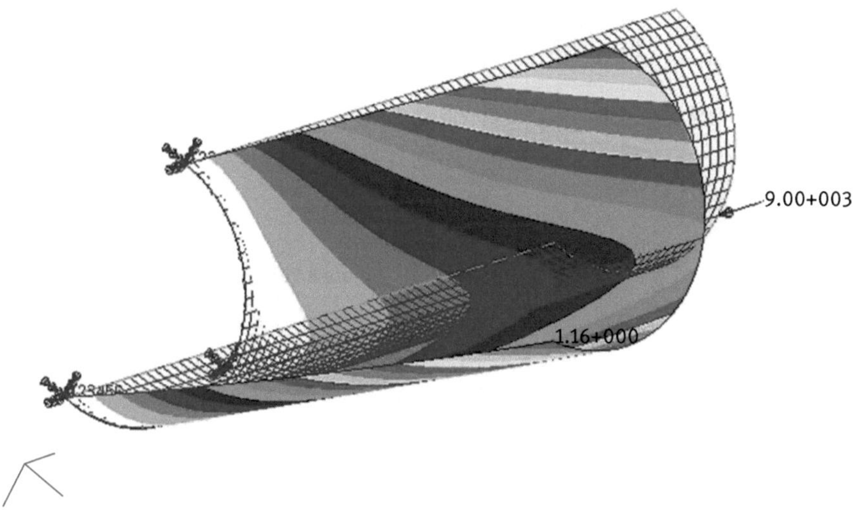

Figure 2.75: Finite element model.

2. Example of short cylindrical shell

Example 1

As illustrated in Figure 2.81, the base structure is a half cylindrical shell, with a radius of 50 mm and a length of 55 mm. The elastic modulus of structure material is E = 210 GPa, Poisson's ratio is γ = 0.30 and the material density is $\rho = 7.8 \times 10^{-6}$ kg mm^3. A concentrated force P = 6 N along the up-to-down (-y) direction is located in upper

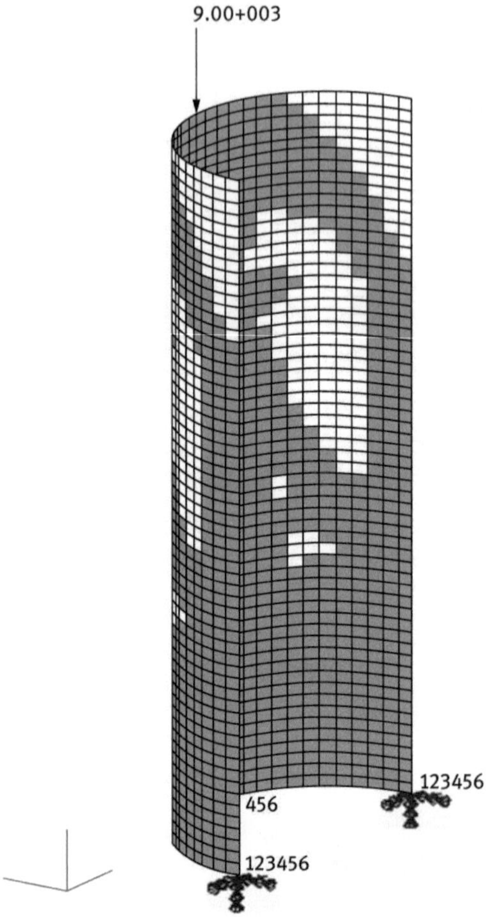

Figure 2.76: Optimal topology structure.

boundary midpoints of the structure. A fixed support is arranged at the bottom left corner and right corner point. A shell element with a four-node quadrilateral shell element is used and the total number of elements is 25×11. The base structure's quality is 0.2691 kg.

The base structural first-order buckling factor is $\zeta_1 = 1.5469$, and the first-order buckling critical force is 9.2814 N. The second-order buckling factor is $\zeta_2 = 1307.2$, and the second-order buckling critical force is 7843.2 N. In this case, the first critical buckling upper limit is $\bar{\lambda}_1 = 5$ N after 32 iterations of convergence. The optimized structure quality is 0.1248 kg, and the critical buckling force of the first-order of the topological structure is 4.98 N, such as the topology shown in Figure 2.83. The second critical buckling upper limit is $\bar{\lambda}_2 = 4{,}700$ N after 15 iterations of convergence. The optimized structure quality is 0.2020 kg, and the critical force of the second-order buckling of the topological structure is 3,897 N, such as the topology shown in Figure 2.87. Mode shape and iterative history show as Figure 2.82 and Figure 2.84-2.86.

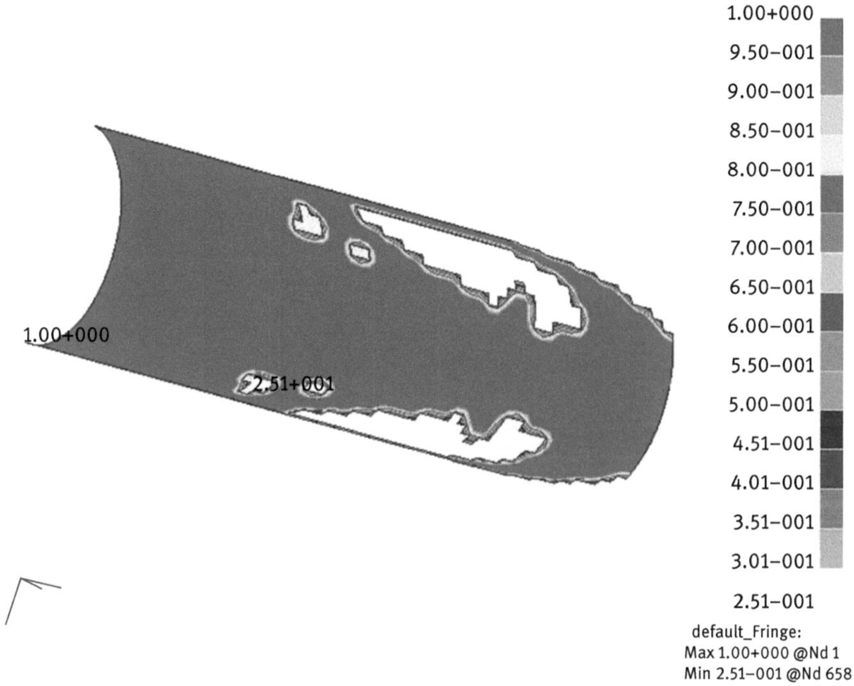

Figure 2.77: Topological variables of topological configuration.

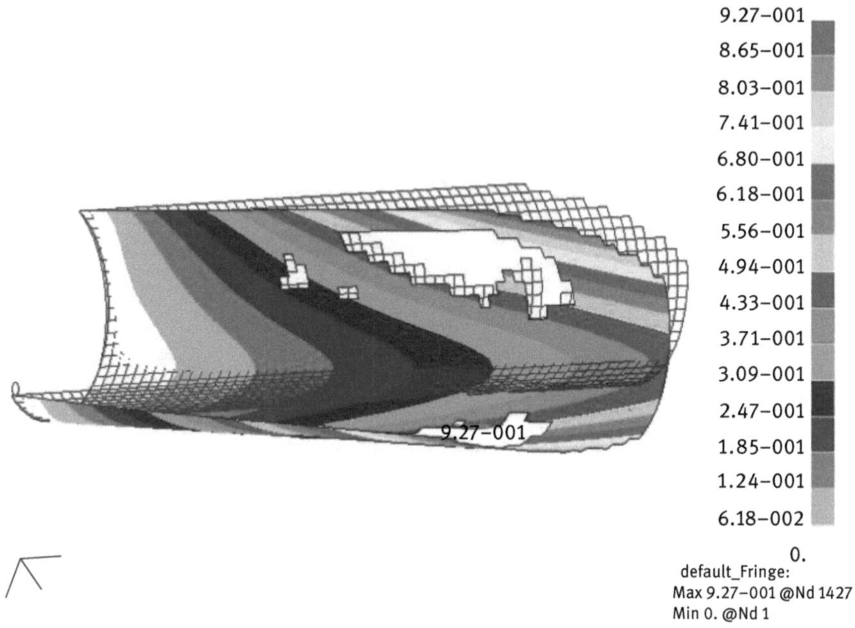

Figure 2.78: The first-order mode shape of topological configuration.

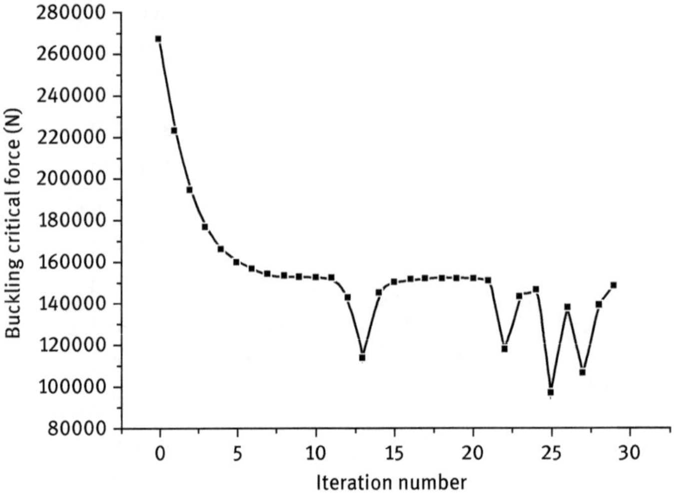

Figure 2.79: Iterative history of the first-order buckling critical load.

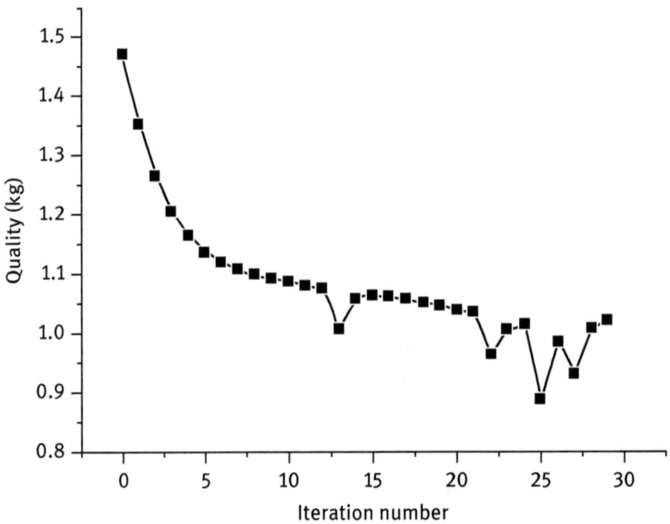

Figure 2.80: Iterative history of the first-order mode quality.

Example 2

In this example, the basic data of the model is the same as example 1, and the difference is that in the middle of the bottom of the cylindrical shell to add a fixed support, the external force is 7,000 N (Figure 2.84-2.91).

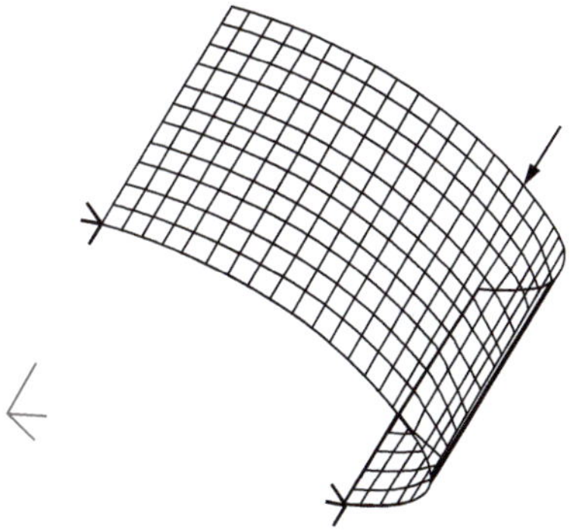

Figure 2.81: Finite element model.

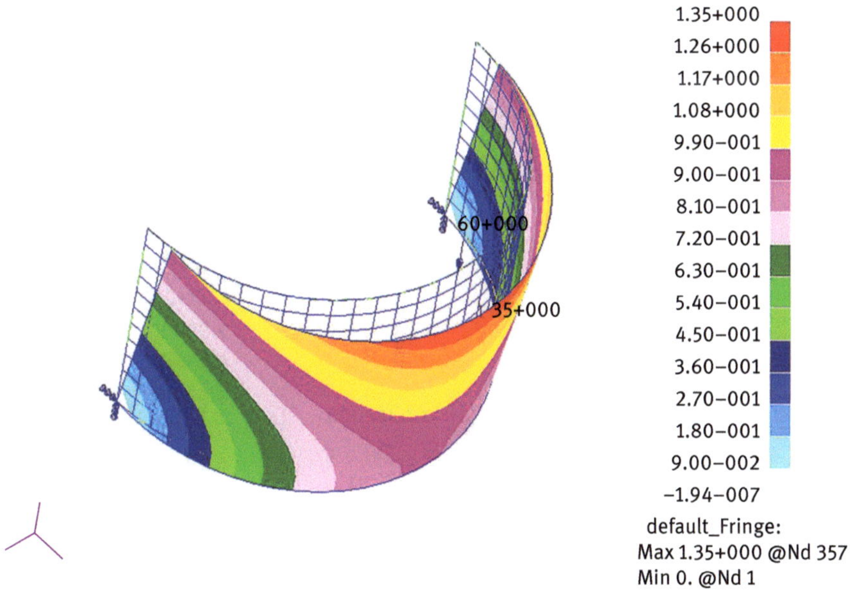

Figure 2.82: The first-order mode shape of topological configuration.

The base structural first-order buckling factor is $\zeta_1 = 42.377$, and the first-order buckling critical force is 296,639 N. In this case, the first critical buckling upper limit is $\bar{\lambda}_1 = 160{,}000$ N after 19 iterations of convergence. The optimized structure quality is 0.1634 kg (Figure 2.94), and the critical buckling force of the first order of the topological structure is 153,720 N (Figure 2.93), such as the topology shown in Figure 2.92. The optimal topology of cylindrical shells and the deformation of the buckling modes are shown. The ICM method can be applied to the topology optimization of cylindrical shells.

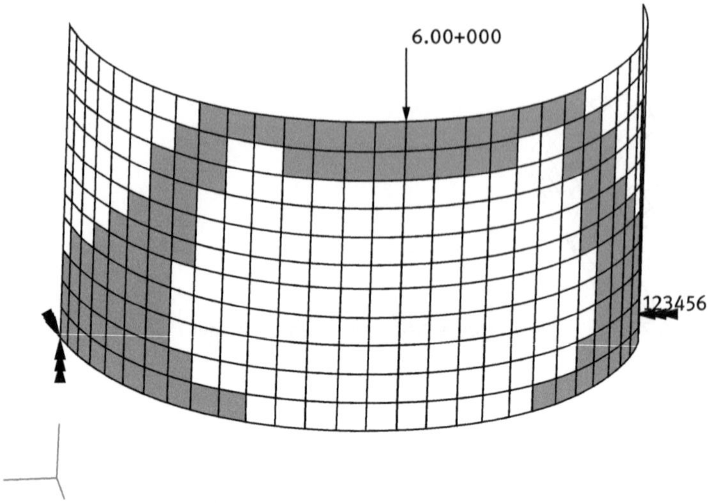

Figure 2.83: The first-order mode of the optimal topology structure.

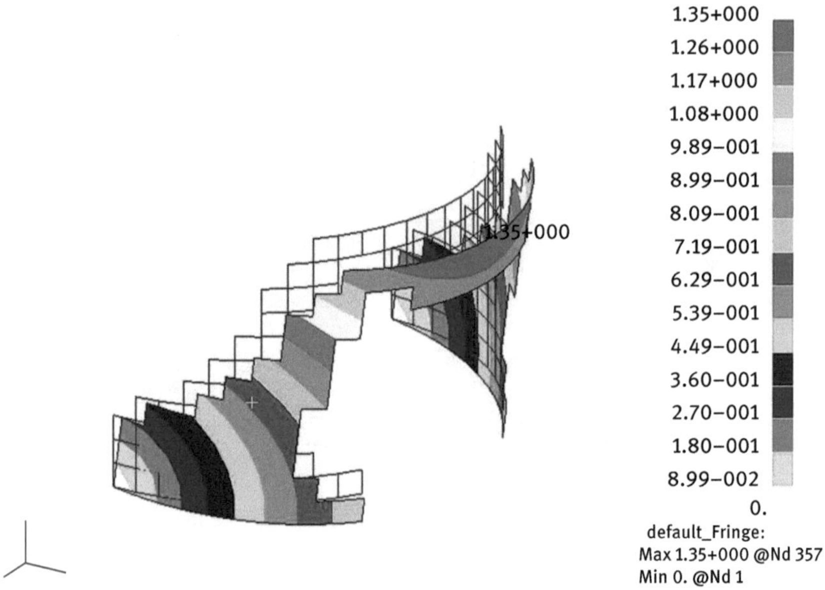

Figure 2.84: The first-order mode shape of topological configuration (Delete elements).

2.9.3 Numerical Example of Three-dimensional Solid Structure

A solid structure is the length, width and height of the same order of magnitude of the structure. For example, when the size difference of the thickness of the

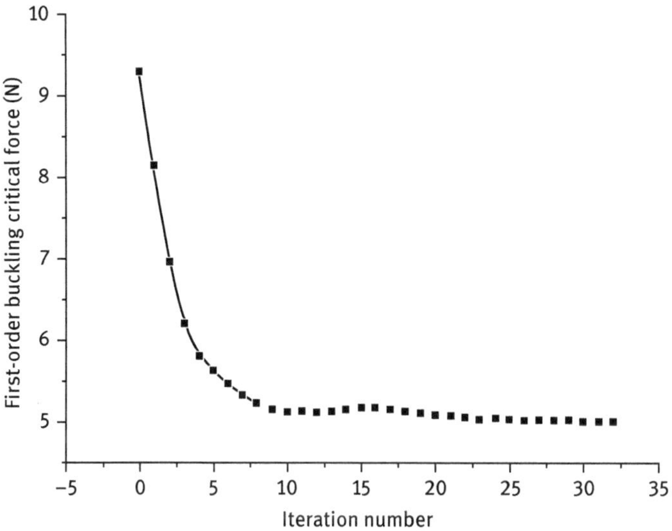

Figure 2.85: Iterative history of the first-order buckling critical load.

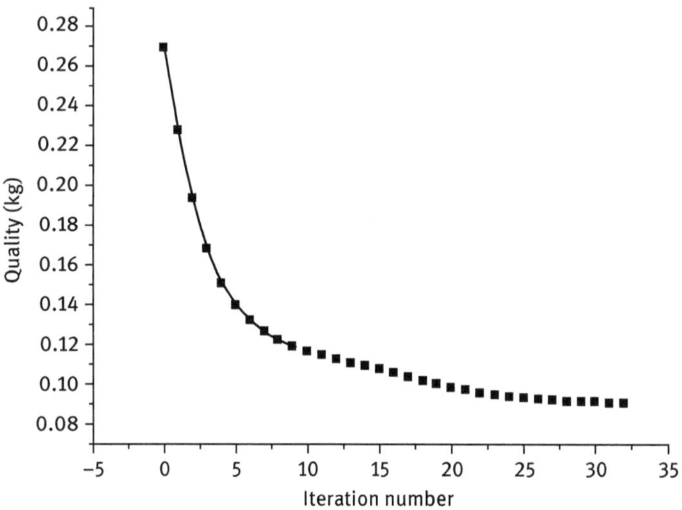

Figure 2.86: Iterative history of the first-order mode quality.

plate and the length and width of the plate is not large, the structure is transformed into a three-dimensional (3D) structure.

1. Concentrated force calculation example

As illustrated in Figure 2.95, the base structure is a 3D elastic body with a size 120 mm × 45 mm × 8 mm. The elastic modulus of structure material is $E = 70 \times 10^3$ MPa, Poisson's

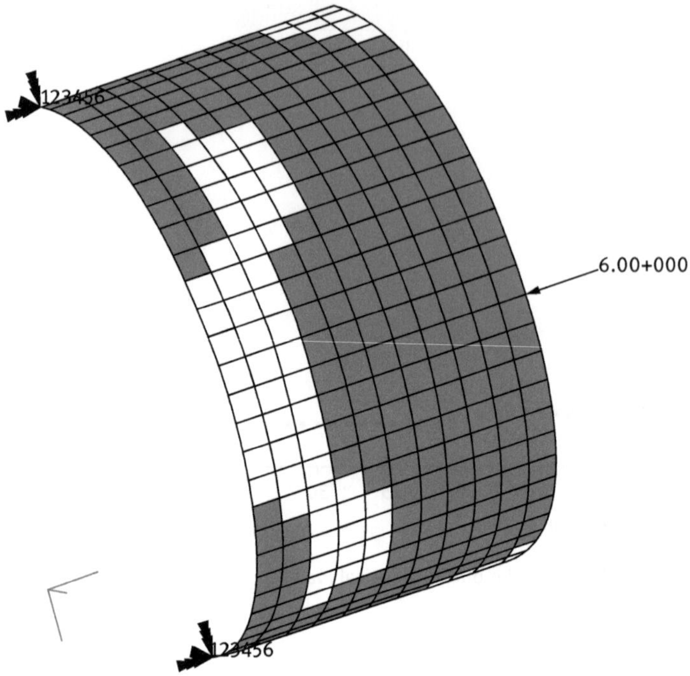

Figure 2.87: The second-order mode of the optimal topology structure.

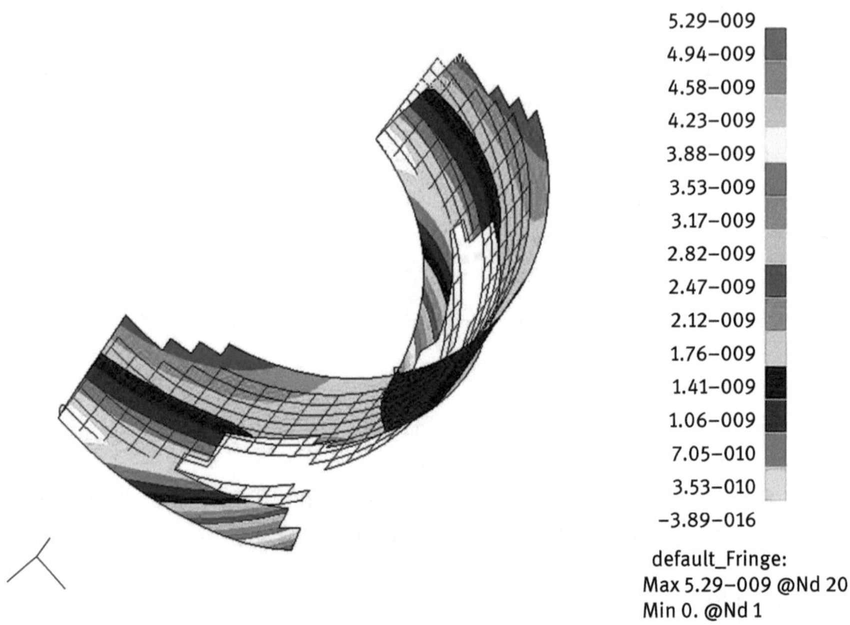

Figure 2.88: The second-order mode shape of topological configuration.

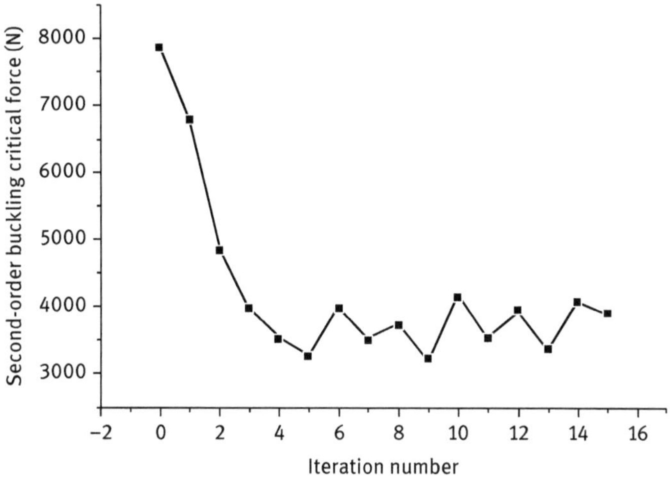

Figure 2.89: Iterative history of the second-order buckling critical load.

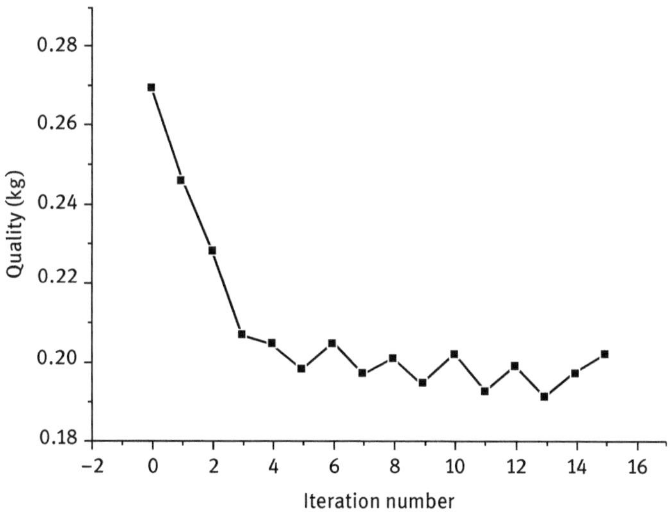

Figure 2.90: Iterative history of the second-order mode quality.

ratio is $\gamma = 0.30$, and the material density is $\rho = 2.7 \times 10^{-6}$ kg/mm^3. A concentrated force P = 12,500 N along the up-to-down (-y) direction is located in the upper boundary midpoints of the structure. The force is distributed on the five nodes in the middle of the upper boundaries. A fixed support is arranged at the lower left corner and lower right corner points. A solid element with eight nodes is used and the total number of elements is 40 × 15 × 4. The base structure's quality is 0.11664 kg.

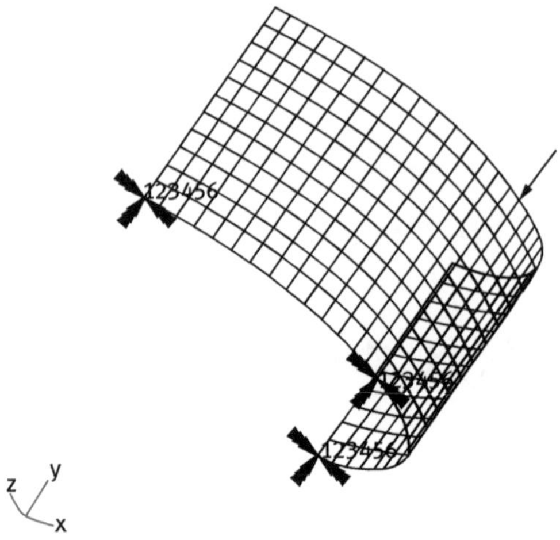

Figure 2.91: Finite element model.

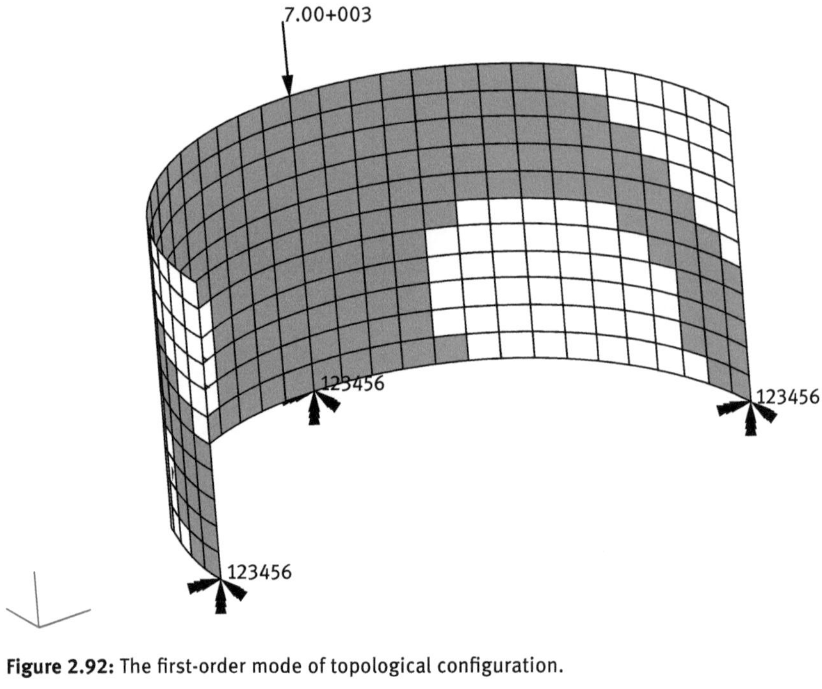

Figure 2.92: The first-order mode of topological configuration.

The base structural first-order buckling factor is ζ_1 = 2.7551, and the first-order buckling critical force is 34438.75 N. In this case, the first critical buckling upper limit is $\bar{\lambda}_1$ = 28,000 N after 32 iterations of convergence. The optimized structure

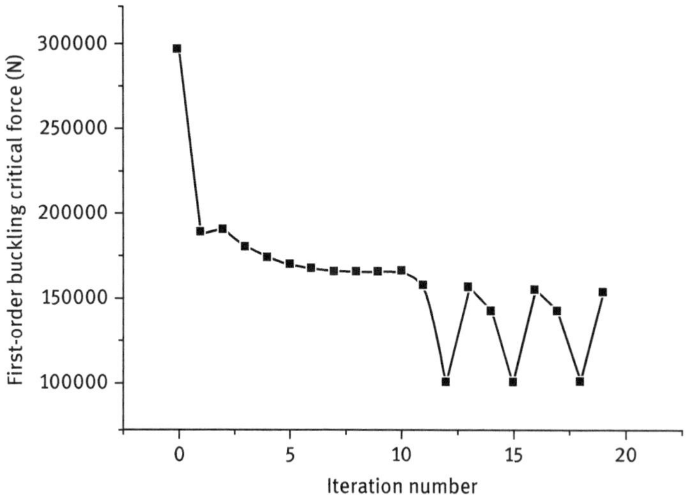

Figure 2.93: Iterative history of the first-order buckling critical load.

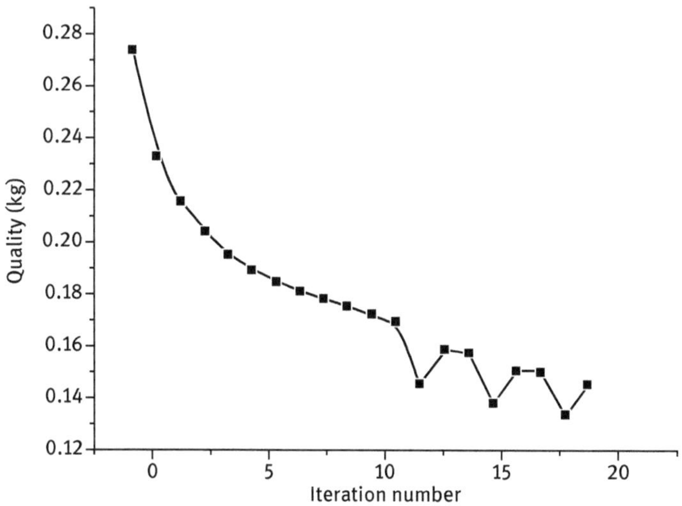

Figure 2.94: Iterative history of the first-order mode quality.

quality is 0.05167 kg. We need to obtain a topological structure as shown in Figure 2.96-2.97. Figure 2.98 shows the first-order mode shape of the topological structure, Figure 2.99 shows the first-order buckling critical force iteration history and Figure 2.100 shows the quality of iteration history.

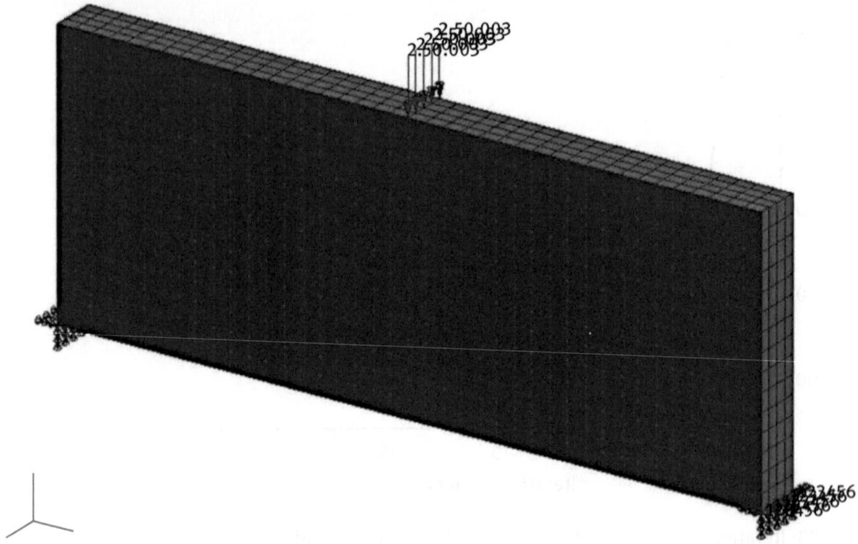

Figure 2.95: Finite element model.

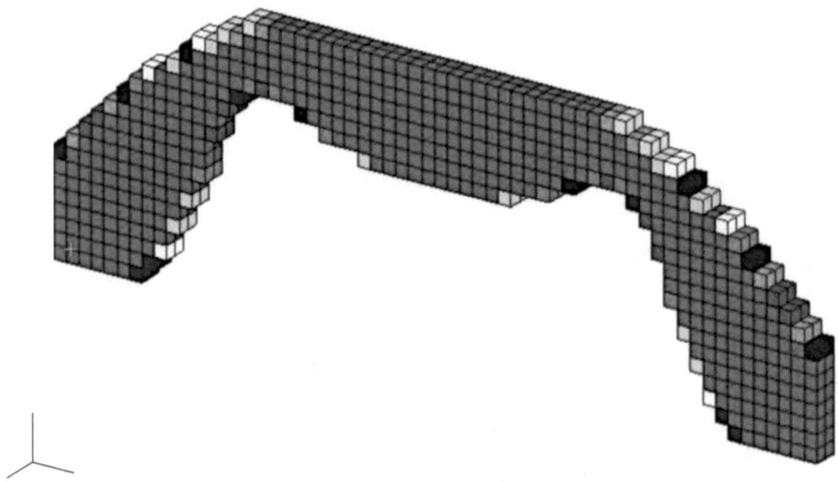

Figure 2.96: The first-order buckling mode of topological configuration.

2. Distributed force calculation example

As illustrated in Figure 2.101, the base structure is a 3D elastic body with a size of 300 mm × 100 mm × 10 mm. The elastic modulus of structure material is $E = 70 \times 10^3$ MPa, Poisson's ratio is $\gamma = 0.30$ and the material density is $\rho = 2.7 \times 10^{-6}$ kg/mm^3. A distributed force P = 12,810 N along the up-to-down (-y) direction is

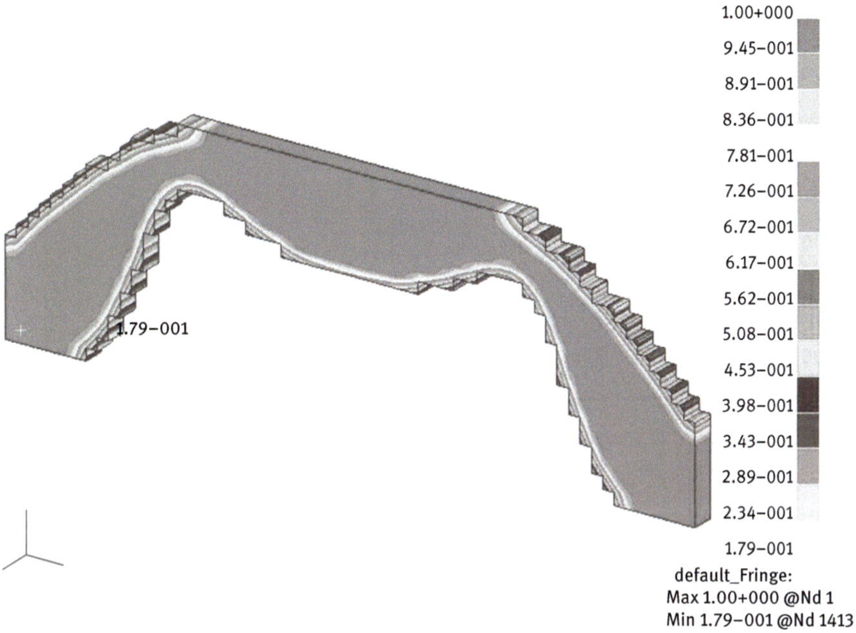

Figure 2.97: Topological variables of topological configuration.

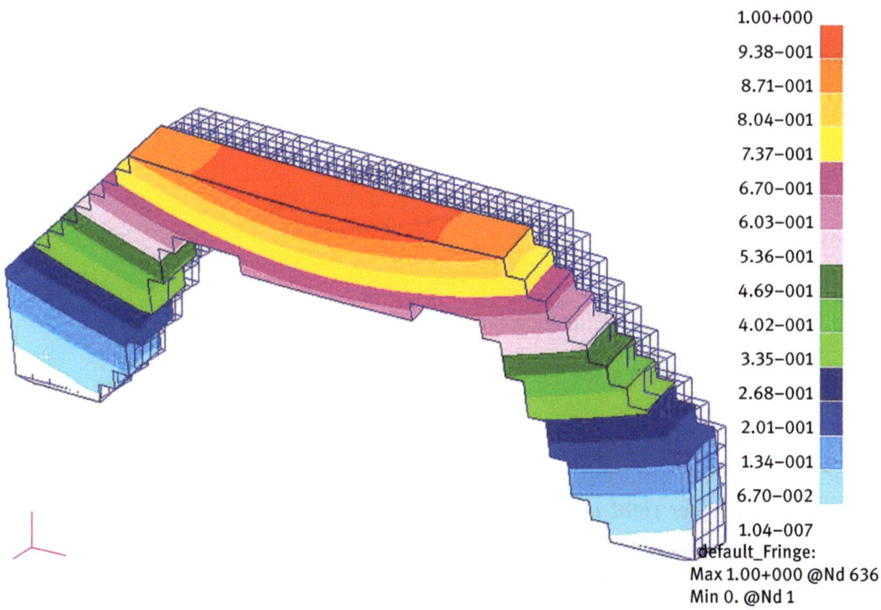

Figure 2.98: The first-order mode shape of topological configuration.

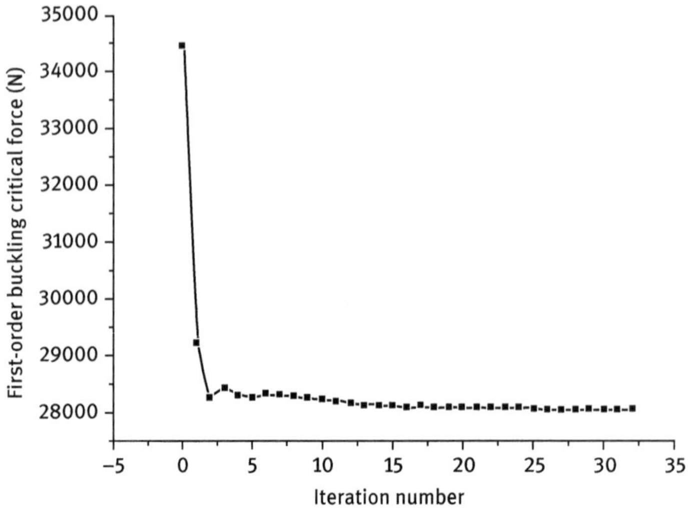

Figure 2.99: Iterative history of the first-order buckling critical load.

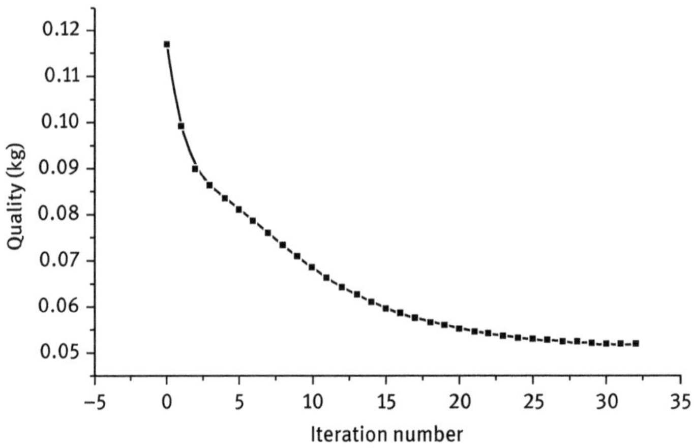

Figure 2.100: Iterative history of the first-order mode quality.

located in the upper surface of the structure. A fixed support is arranged at the lower left corner and lower right corner points. A solid element with eight nodes is used and the total number of elements is 60 × 20 × 2. The base structure's quality is 0.81 kg. Because the upper boundary of the structure is applied force, the units of the upper two layers of the model are reserved, and the remaining area is the design area.

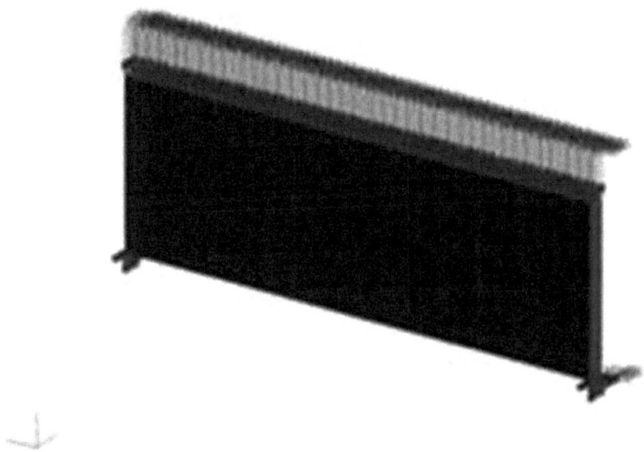

Figure 2.101: Finite element model.

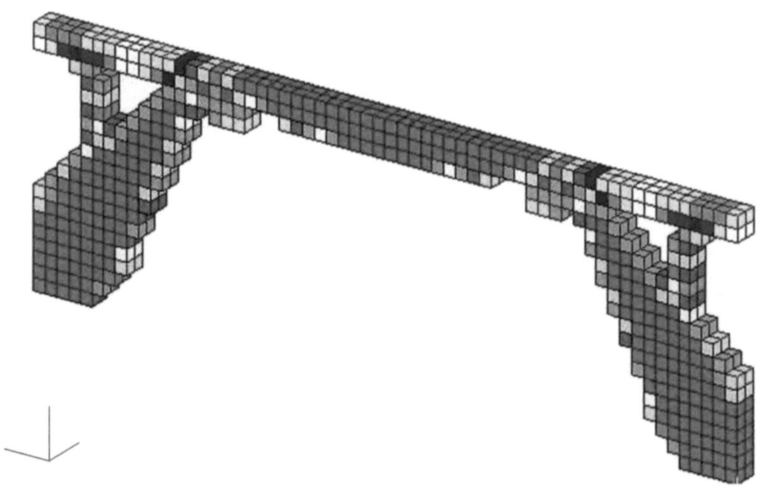

Figure 2.102: The first-order mode of optimal topological configuration.

The base structural first-order buckling factor is $\zeta_1 = 2.6136$, and the first-order buckling critical force is 33,480 N. In this case, the first critical buckling upper limit is $\bar{\lambda}_1 = 28{,}000$ N after 45 iterations of convergence. The optimizedstructure quality is 0.24 kg. We need to obtain the topological structure as shown in Figure 2.102-2.103. Figure 2.104 shows the first-order mode shape of the topological structure, Figure 2.105 shows the first-order buckling critical force iteration history, and Figure 2.106 shows the quality of iteration history.

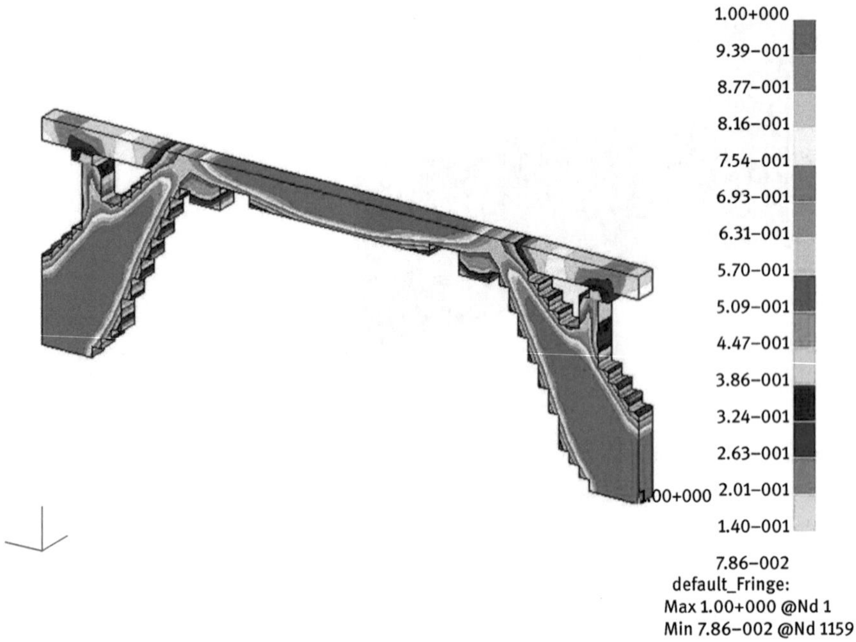

Figure 2.103: Topological variables of topological configuration.

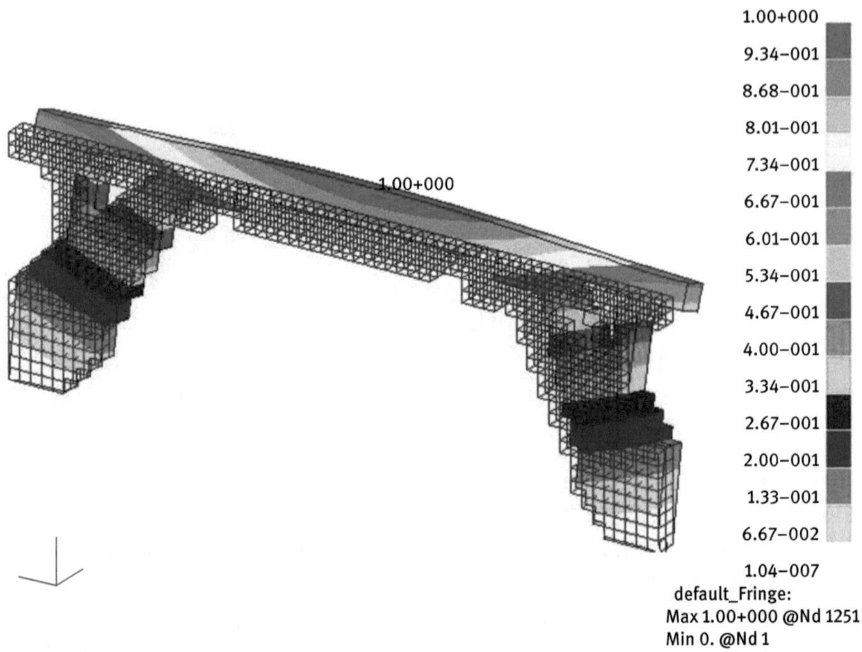

Figure 2.104: The first-order mode shape of topological configuration.

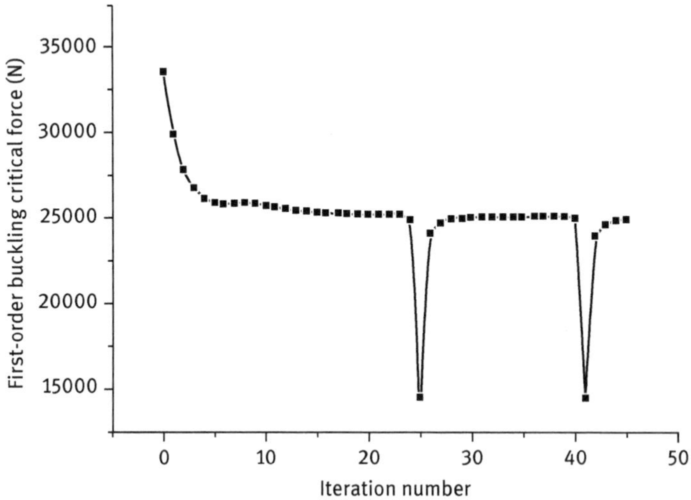

Figure 2.105: Iterative history of the first-order buckling critical load.

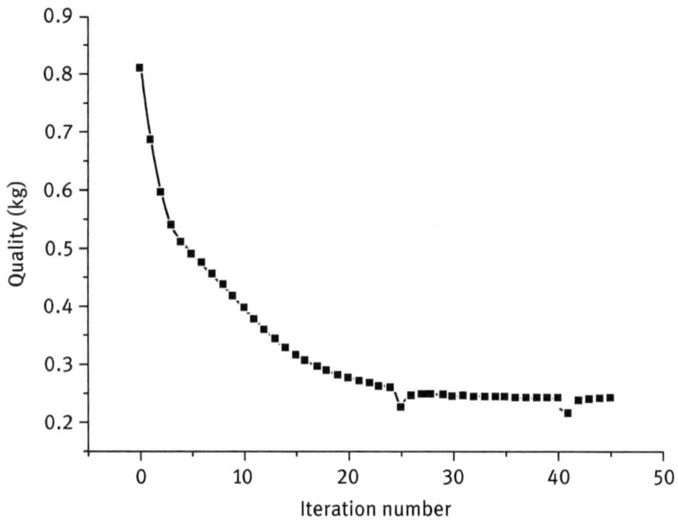

Figure 2.106: Iterative history of the first-order mode quality.

3. Cube example

As illustrated in Figure 2.107, the base structure is a 3D elastic body with a size $100 \times 100 \times 100$ mm^3. The elastic modulus of structure material is $E = 210 \times 10^3$ MPa, Poisson's ratio is $\gamma = 0.30$ and the material density is $\rho = 7.8 \times 10^{-6}$ kg/mm^3. A distribution force $P = 30{,}000$ N along the up-to-down (-y) direction is located at the middle five points in upper surface of the structure. A fixed support

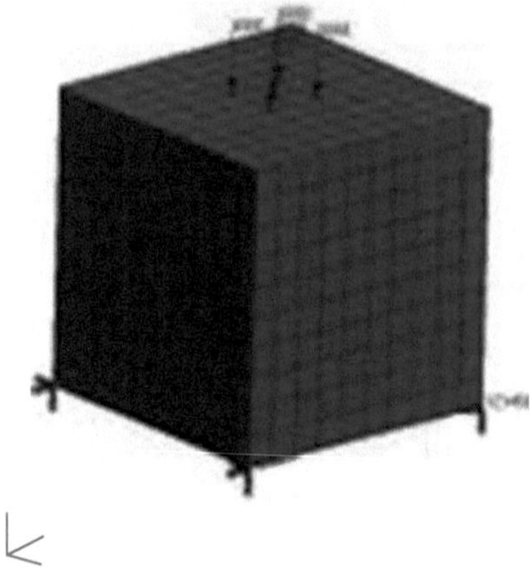

Figure 2.107: Finite element model.

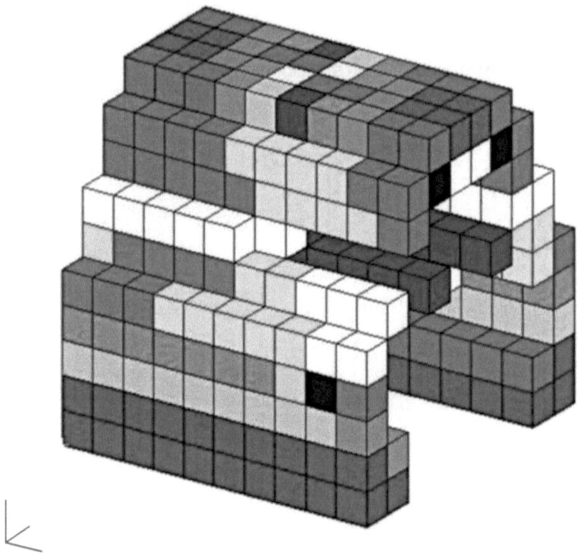

Figure 2.108: The first-order mode of optimal topological configuration.

is arranged at the four corners of the cube. A solid element with eight nodes is used, and the number of total elements is 10 × 10 × 10. The base structure's quality is 7.8 kg.

The base structural first-order buckling factor is ζ_1 = 544.94, and the first-order buckling critical force is 1.63482 × 10^7 N. In this case, the first critical buckling upper limit is $\bar{\lambda}_1$ = 9,500,000 N after 63 iterations of convergence. The optimized structure quality is 3.12 kg. We need to obtain the topological structure as shown in Figure 2.108-2.112. In this figure, the first-order modal shapes and quality iteration history are given.

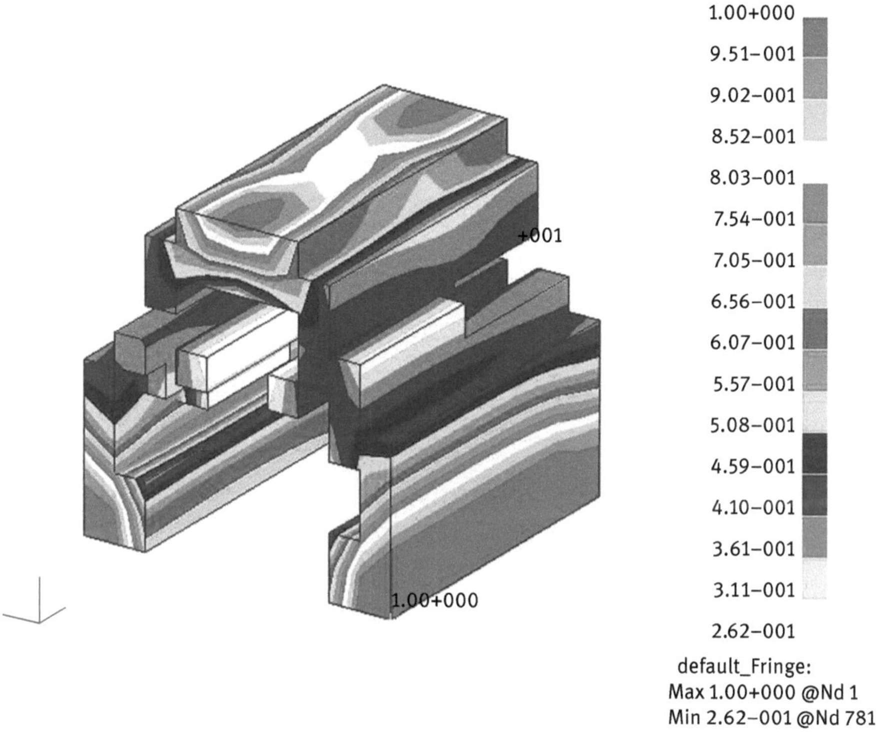

Figure 2.109: Topological variables of topological configuration.

4. Square column calculation example
As illustrated in Figure 2.113, the base structure is a 3D elastic body with a size 100 × 100 × 600 mm^3. The elastic modulus of structure material is $E = 210 \times 10^3$ MPa, Poisson's ratio is $\gamma = 0.30$ and the material density is $\rho = 7.8 \times 10^{-6}$ kg/mm^3. A distribution force $P = 50{,}000$ N along the up-to-down (-y) direction is located at the middle five points in the upper surface of the structure. A fixed support is arranged at the four corners of the cube. A solid element with eight nodes is used and the total number of elements is 10 × 10 × 60. The base structure's quality is 46.8 kg.

The base structural first-order buckling factor is $\zeta_1 = 69.932$, and the first-order buckling critical force is $\lambda_1 = 3{,}496{,}600$ N. The second-order buckling factor is $\zeta_2 = 69.932$, and the second-order buckling critical force is $\lambda_2 = 3{,}496{,}600$ N. The third-order buckling factor is $\zeta_3 = 901.09$, and the third-order buckling critical force is $\lambda_3 = 45{,}054{,}500$ N. In this case, the first critical buckling upper limit is $\bar{\lambda}_1 = 21{,}00{,}000$ N after 55 iterations of convergence. The optimized structure quality is 26.988 kg, and the critical buckling force of the first order of the topological structure is 20,82,550 N. The second critical buckling upper limit is $\bar{\lambda}_2 = 2{,}100{,}000$ N after 29 iterations

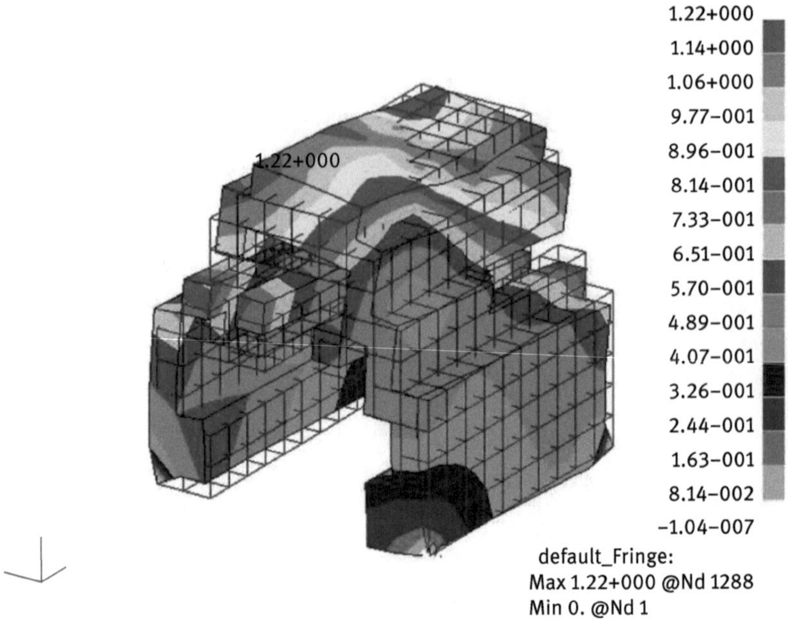

Figure 2.110: The first-order mode shape of topological configuration.

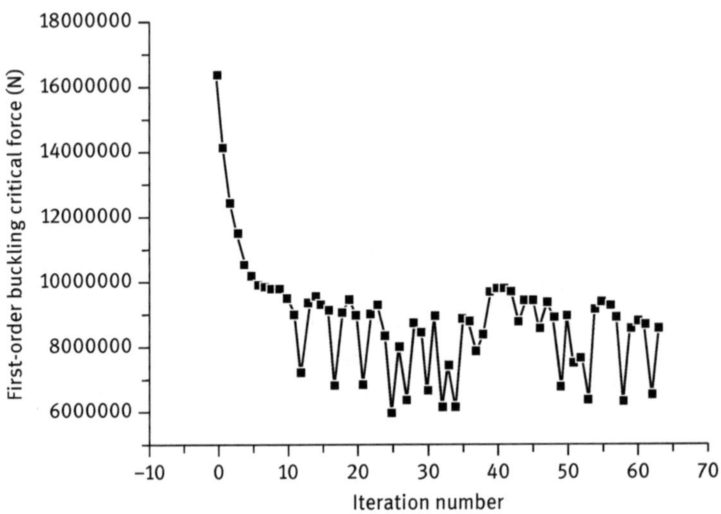

Figure 2.111: Iterative history of the first-order buckling critical load.

of convergence. The optimized structure quality is 18.642 kg, and the critical force of the second-order buckling of the topological structure is 21,07,500 N. The third critical buckling upper limit is $\bar{\lambda}_3 = 27{,}000{,}000$ N after 19 iterations of convergence.

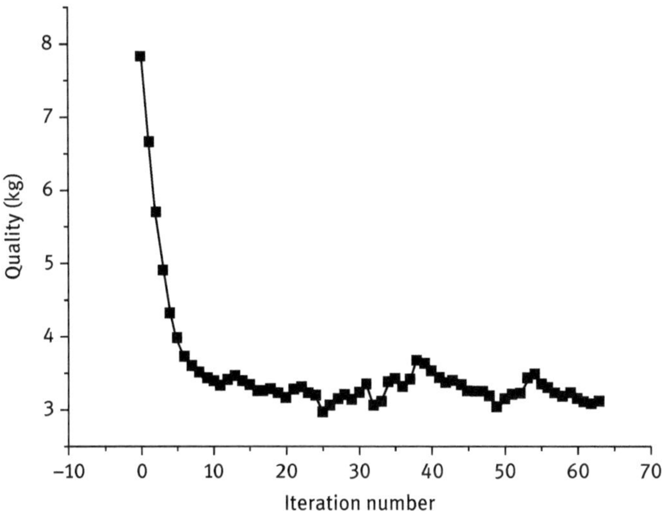

Figure 2.112: Iterative history of the first-order mode quality.

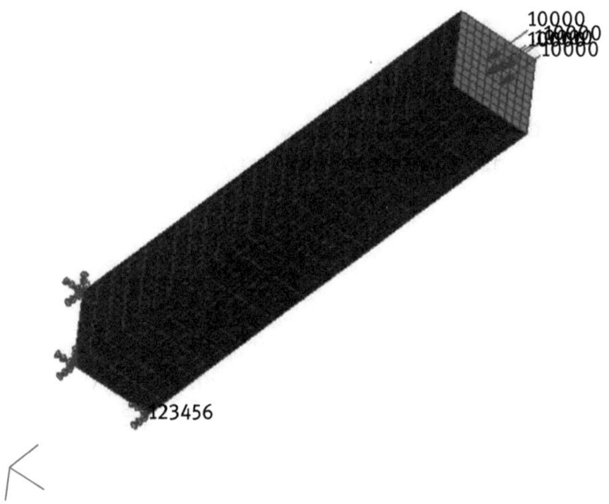

Figure 2.113: Finite element model.

The optimized structure quality is 33.54 kg, and the critical force of the third-order buckling of the topological structure is 26,616,500 N. In figure 2.114-2.120, the optimal topology and the deformation of the buckling modes are given.

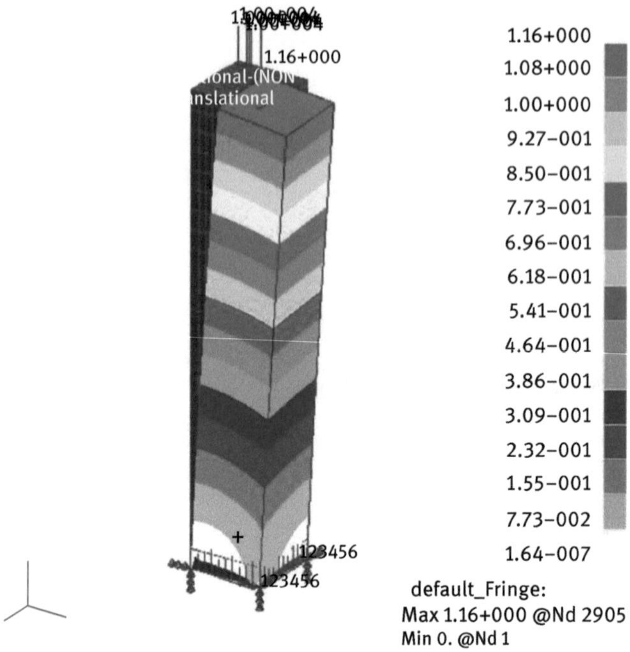

Figure 2.114: Analysis result of the first-order mode.

It can be seen from the figure 2.114-2.120 that the buckling deformation is in the direction of the rectangular columns, and to resist the deformation of the structure, the materials in this direction were held, and the materials in the perpendicular direction were deleted. The hollow part of the topological structure was obtained.

5. Cylindrical Example
(1) Concentrated force model
As illustrated in Figure 2.121, the base structure is a 3D elastic body with a radius of 20 mm, and has a oblong column of height 600 mm. The elastic modulus of structure material is $E = 210 \times 10^3$ MPa, Poisson's ratio is $\gamma = 0.30$ and the material density is $\rho = 7.8 \times 10^{-6}$ kg/mm^3. A concentrated force $P = 20{,}000$ N along the up-to-down (-y) direction is located in upper surface center points of the structure. The four ends of the two diameter of the lower boundary are fixed and supported.

A solid element with eight nodes is used and the total number of elements is 2,560. The base structure's quality is 5.733 kg.

The base structural first-order buckling factor is $\zeta_1 = 5.97$, and the first-order buckling critical force is $\lambda_1 = 1{,}34{,}390$ N. In this case, the first critical buckling upper limit is $\bar{\lambda}_1 = 70{,}000$ N after 10 iterations of convergence. The

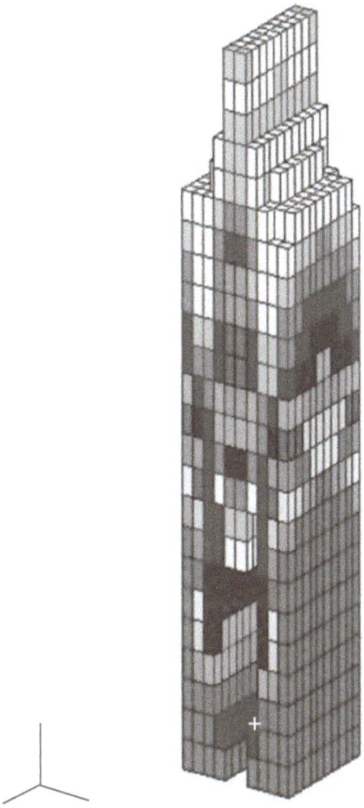

Figure 2.115: The first-order mode of optimal topological configuration.

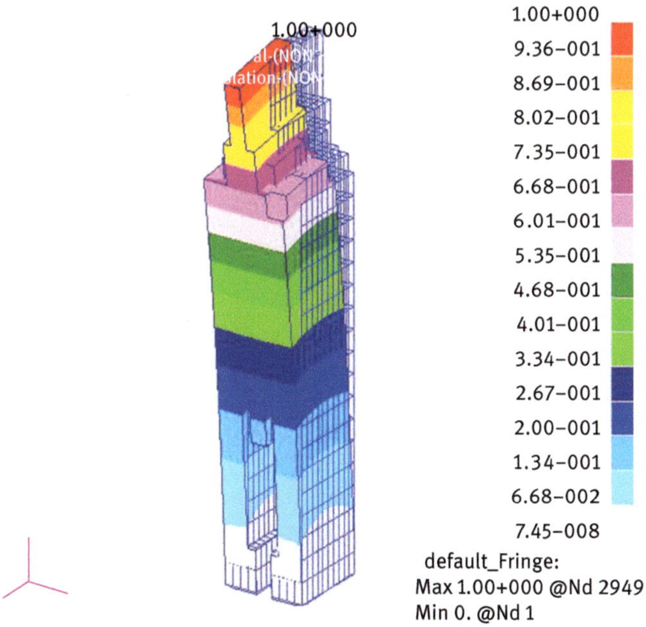

Figure 2.116: The first-order mode shape of topological configuration.

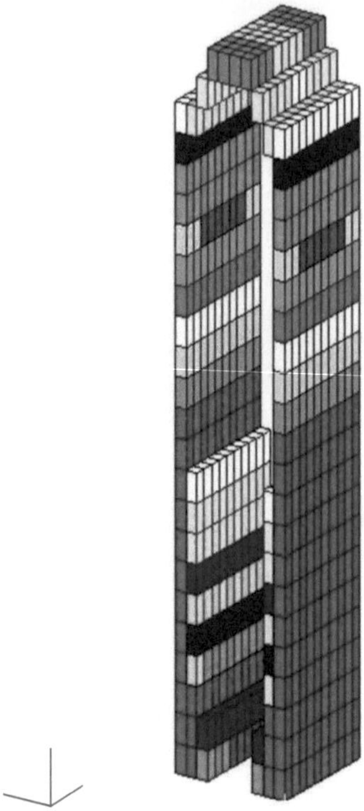

Figure 2.117: The second-order mode of optimal topological configuration.

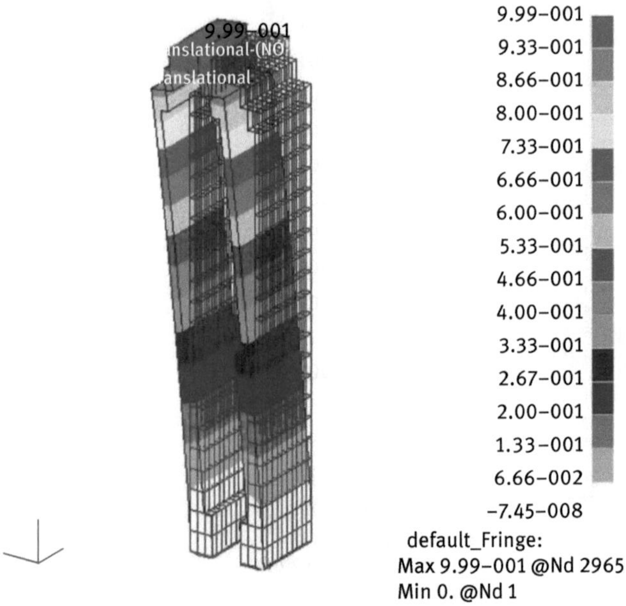

Figure 2.118: The second-order mode shape of topological configuration.

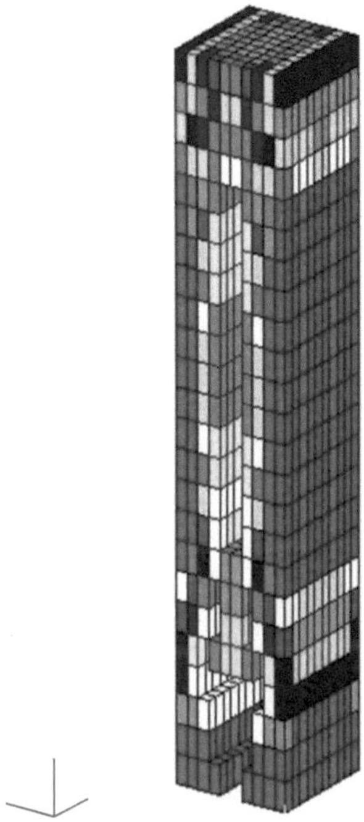

Figure 2.119: The third-order mode of optimal topological configuration.

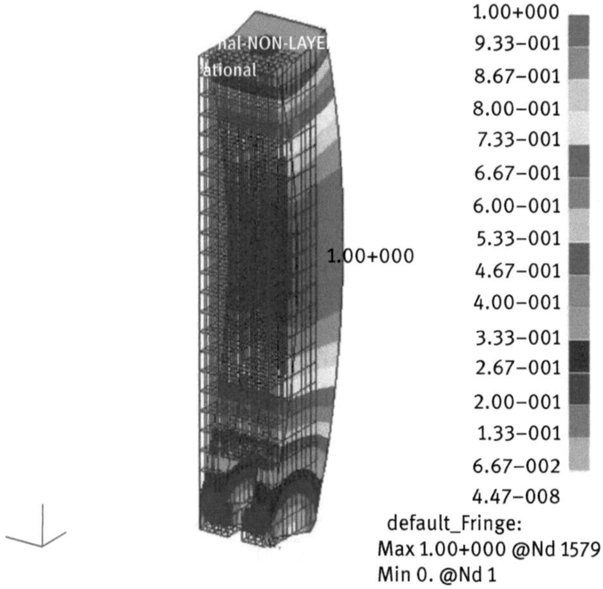

Figure 2.120: The third-order mode shape of topological configuration.

Figure 2.121: Finite element model.

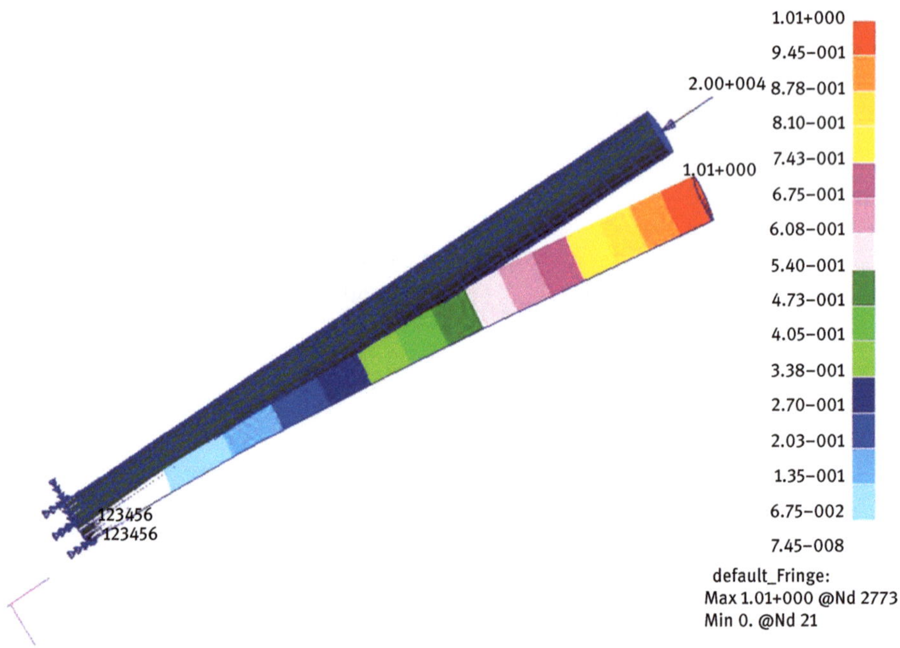

Figure 2.122: Analysis result of the first-order mode.

optimized structure quality is 4.368 kg. The first critical buckling force of the topological structure is 69,866 N. In figure 2.122-2.126, the optimal topology of the first-order buckling constraints and the deformation of the buckling modes are presented.

Figure 2.123: The first-order mode of optimal topological configuration.

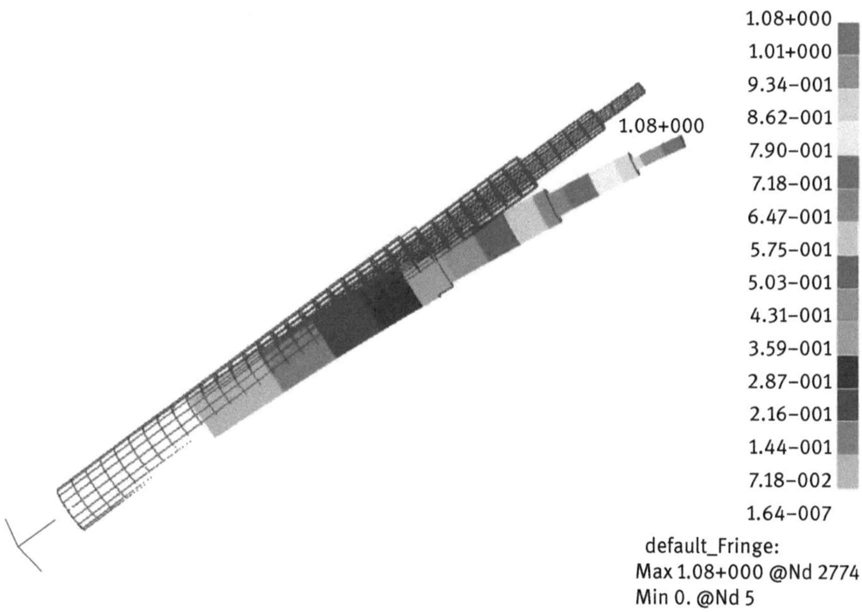

Figure 2.124: The first-order mode shape of topological configuration.

It can be seen from the figure 2.122-2.124 that circular section columns are different from the rectangular ones, and buckling deformations are likely to occur in the circular section in all the radius directions, and so the topology in the circumference of the radial is consistent.

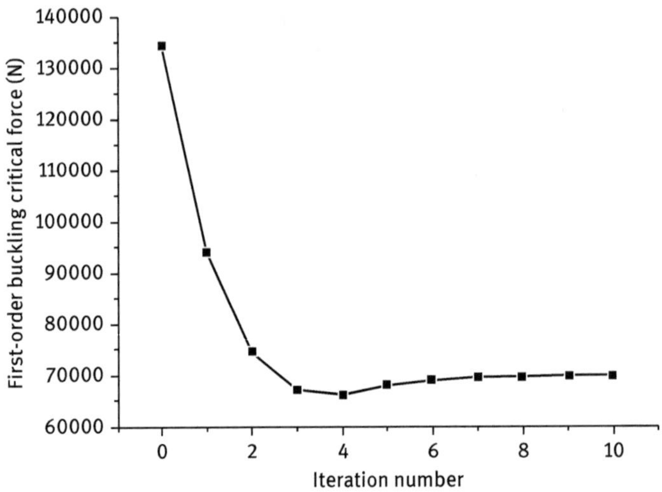

Figure 2.125: Iterative history of the first-order buckling critical load.

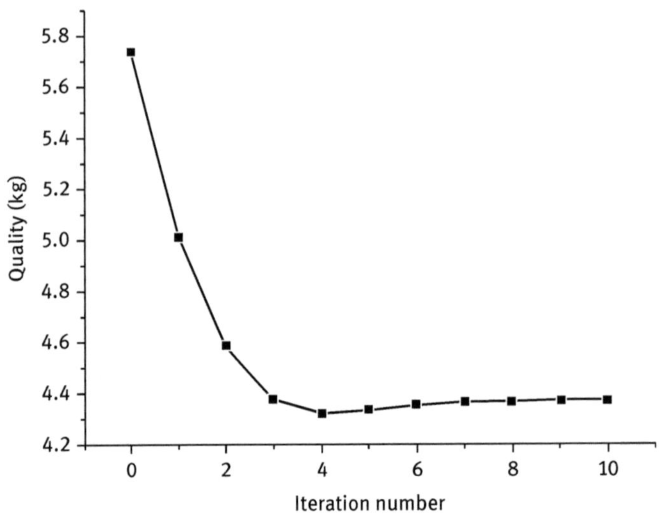

Figure 2.126: Iterative history of the first-order mode quality.

The top of the structure is relatively small, and so it will have a larger buckling deformation, but with support of the bottom, it will produce a smaller deformation; therefore, the structure is basically remain unchanged.

(2) Distributed force model
As shown in Figure 2.127, the model is same as discussed in the above example. The distribution force is P = 20,010 N on the upper circular section. Because the

Figure 2.127: Finite element model.

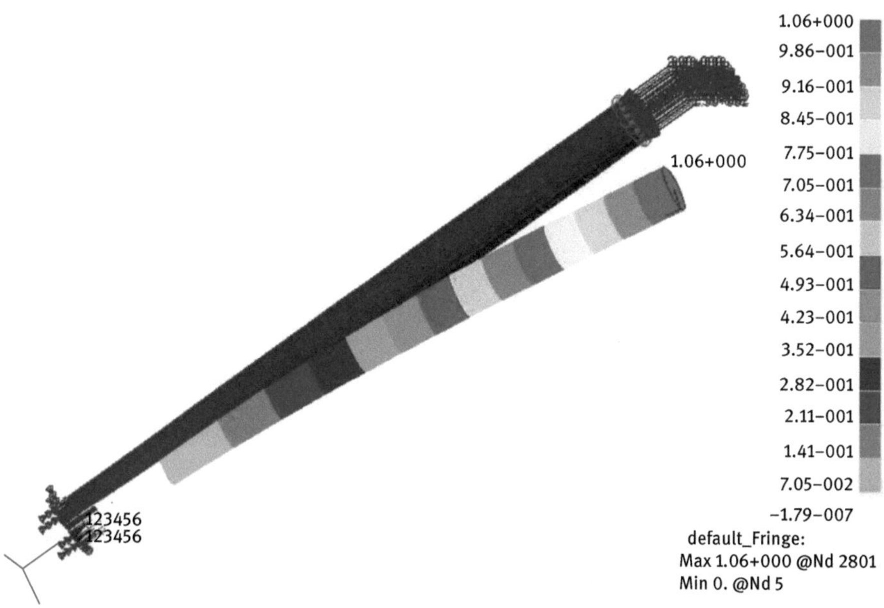

Figure 2.128: Analysis result of the first-order mode.

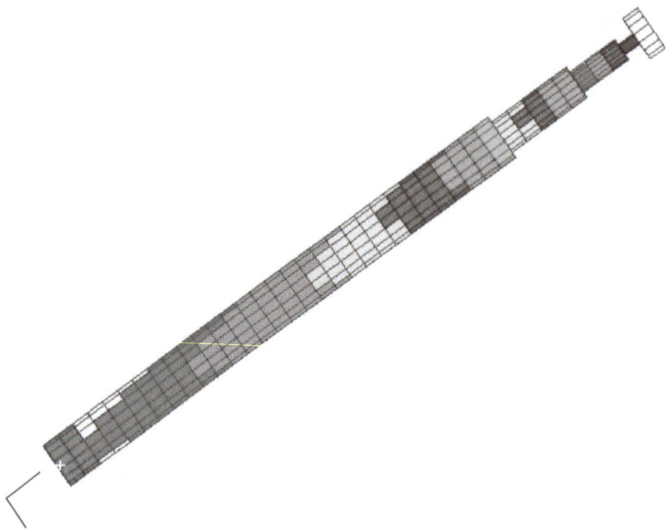

Figure 2.129: The first-order mode of optimal topological configuration.

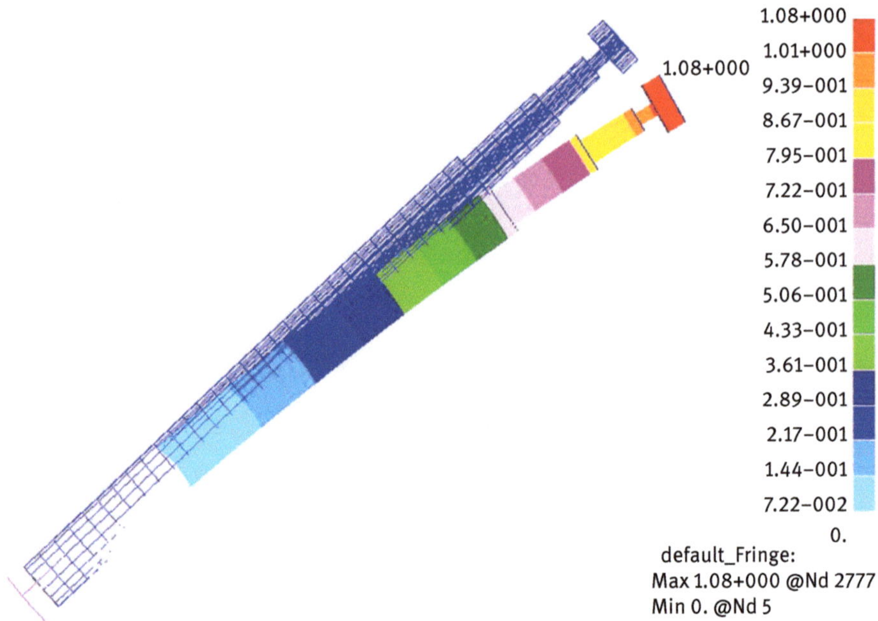

Figure 2.130: The first-order mode shape of topological configuration.

upper circular section is applied to the distribution force, the elements are the retaining unit, and the remaining is the design area.

The base structural first-order buckling factor is $\zeta_1 = 5.97$, and the first-order buckling critical force is $\lambda_1 = 134{,}390$ N. In this case, the first critical buckling upper

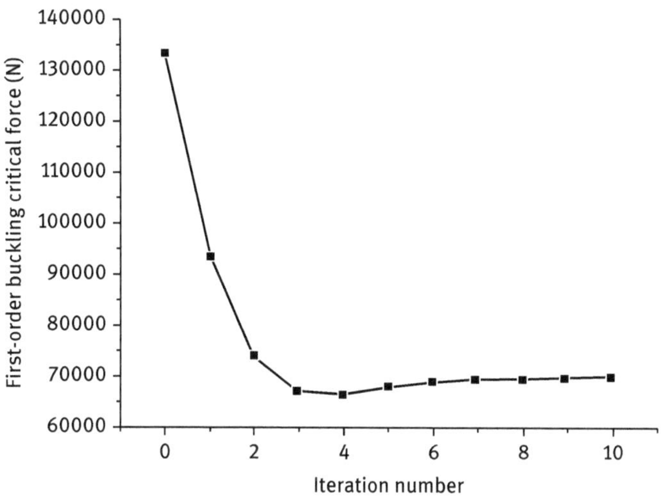

Figure 2.131: Iterative history of the first-order buckling critical load.

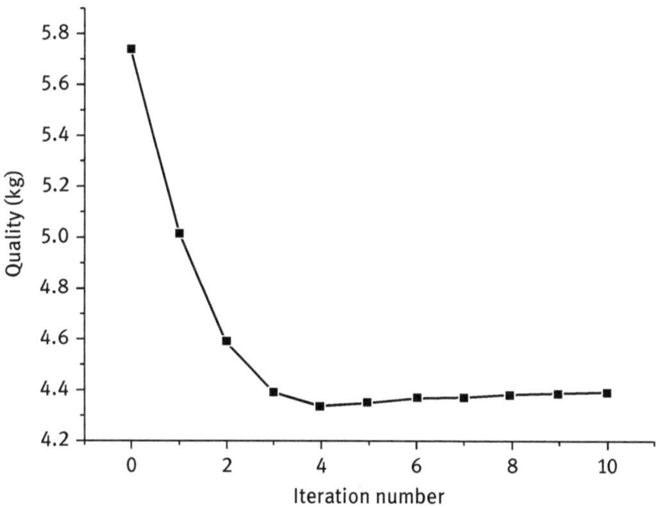

Figure 2.132: Iterative history of the first-order mode quality.

limit is $\bar{\lambda}_1$ = 70,000 N after 10 iterations of convergence. The optimized structure quality is 4.3836 kg. The first critical buckling force of the topological structure is 69,887 N. In this figure 2.128-2.132, the optimal topology of the first-order buckling constraints and the deformation of the buckling modes are presented.

As can be seen from the graph, the constraint conditions are all the same, the topology structure is different and the quality of the optimized topology is not the same.

2.10 Summary

In this chapter, according to the ICM method, the topology optimization model for the continuum structure was constructed. The model had the minimized quality as the objective function was subjected to the buckling constraints. On the basis of idea of homogenization, the strain energy expression for the filter function of the element stiffness and the element geometrical stiffness were derived. A numerical example was given to calculate the element quality, element stiffness and element geometric stiffness filter function expression. The concept of “quality ratio” and “buckling constraint ratio” was defined. The one-order, two-order and three-order function curves of the “quality ratio” and “quality-buckling constraints ratio” of topology optimization structure were established. The least square method was used to fit the curve of “quality-buckling constraints ratio,” which provides a reference for determining the value of the constraint in the topology optimization of buckling constraints.

In this chapter, a number of examples are discussed that imply the optimal topology under multiconditions is not the superposition of the single working condition topological structure, which is caused by the interaction between the two conditions. At the same time, with an increase in the upper limit of buckling critical force, the optimal topology will be changed, the change trend will be strengthened and the quality of the whole topology optimization structure will be increased.

To solve the optimization model, the objective function and the buckling critical force constraints are approximated and solved by the dual programming method. The reasonable topology shape and clear topology structure were obtained. Finally, the method is successfully applied to the topology optimization problem of 2D flat plate, cylindrical shell and 3D continuum structure. The results show that the method is effective and feasible.

3 Topological Optimization for Continuous Structure with Bulking and Displacement Constraint

3.1 Introduction

In the topology optimization of continuum structures, the constraints can be considered in many forms such as dimension constraints, displacement constraints, volume constraints, frequency constraints, stability constraints, etc. In practical engineering, to make the structure work normally and safely and get a more reasonable structural design, designers often have to consider a variety of constraints in the structural design. The main focus of this chapter is the topology optimization problem of displacement constraint and buckling constraint. As the displacement is the structural rigidity characterization parameters, the structure is required to meet the requirements of the stiffness, that is, the displacement (deformation) of one or more points on the structure or component does not exceed the allowable value. The stability of a structure (buckling) can be defined as the critical force of the structure that does not exceed the allowable value.

At present, the topology optimization method to solve the displacement constraint of the continuum structure mainly considers the method of homogenization optimization proposed by Bendsoe and Kikuchi, variable density method, variable thickness method, evolutional structural optimization method, bidirectional evolutionary structural optimization method, metamorphic development method and other methods. The compliance is take as the objective function, the structure of the volume is take as the constraint of the continuum structure topology optimization model in these methods. At the same time, displacement constraints and buckling constraints are not usually taken into account.

There are two kinds of constraints state quantities in the structural optimum: global quantity and local quantity. The global state quantity mainly includes displacement, frequency and global stability. The local state quantity mainly includes stress and local stability. In this chapter, we consider displacement constraints and global stability constraints for topology optimization model.

In this chapter, the topology optimization of continuum structures with displacement and buckling constraints was established based on the independent continuous mapping (ICM) method. The displacement constraint is explicitly typed by the Moore theorem method.

Rayleigh quotient, Taylor expansion and the filter function are applied to approximate explicit the buckling constraint which can avoid the sensitivity of the calculation. Finally, the optimization model is transformed into dual programming, the number of design variables is reduced, the solving scale of the model is reduced

https://doi.org/10.1515/9783110462272-003

and the computational efficiency is improved by using sequential quadratic programming.

3.2 Model Establishment and Solution

3.2.1 Continuum Topology Optimization Model

As in Chapter 2, the buckling problem eigenvalue equation is given as

$$(\boldsymbol{K}+\lambda_j\boldsymbol{G})\boldsymbol{u}_j=\mathbf{0} \quad (j=1,\ \cdots,J) \tag{3.1}$$

The symbols $\boldsymbol{K}$, $\boldsymbol{G}$, λ_j and $\boldsymbol{u}_j$ denote the stiffness matrix, geometric stiffness matrix, jth eigenvalue and corresponding eigenvector, respectively.

The relationship between critical load and outer load can be rewritten as

$$\lambda_j=\zeta_j P \tag{3.2}$$

where ζ_j is buckling factor and P is outer load. The buckling critical force constraint of the structure can be expressed as

$$\lambda_j(f_k(t_i),f_g(t_i))\le\bar{\lambda}_j \tag{3.3}$$

In this study, a form of power function is used in the filter function. The filter functions of quality, stiffness matrix and geometric stiffness matrix for one element are denoted as follows:

$$f_w(t_i)=t_i^a,f_k(t_i)=t_i^b,f_g(t_i)=t_i^c \tag{3.4}$$

where a and b are constants. The numerical example demonstrates that the optimal results with most of the elements having value 0 or 1 can be obtained using the filter functions, where $a=1$, $b=c=3.3$.

Stiffness is one of the important aspects of the structure's security. The structure's displacement, which is the key point's displacement, must be limited. Thus, the displacement constraints can be represented as

$$u_r\le\bar{u}_r \tag{3.5}$$

where u_r is the rth displacement and $\bar{u}_r$ is upper limit of rth displacement.

Finally, the topology optimization model for the continuum structure is constructed, where the minimized quality as the objective function has been subjected to the buckling and displacement constraints.

$$\begin{cases} \text{find} \quad \boldsymbol{t} \in E^N \\ \min \quad W = \sum_{i=1}^{N} f_w(t_i) w_i^0 \\ \text{s.t.} \quad \lambda_{jl}(f_k(t_i), f_g(t_i)) \le \bar{\lambda}_{jl} \\ \qquad \sum_{i=1}^{N} u_{irjl}(f_k(t_i)) \le \bar{u}_{rjl} \\ \qquad 0 \le t_i \le 1 \qquad (i=1,\ldots,N; r=1,\ldots,R; j=1,\ldots,J; l=1,\ldots,L) \end{cases} \tag{3.6}$$

where $\boldsymbol{t}$ is the topology design variables, W is the total quality of the structure, w_i and N are the quality of ith element and the total number of element, $\bar{\lambda}_{jl}$ is the upper limit of the critical load, J is the total number of the buckling mode, u_{irjl} is the ith element's r under the l condition and j mode displacement and $\bar{u}_{rjl}$ is upper limit of r under the l condition and j mode displacement.

3.2.2 Explicit Approximate Function of the Constraint

(1) Explicit approximate function of the buckling constraint

The buckling constraint's approximate explicit function is same as in Chapter 2. Uniform expression formula of buckling constraints can be obtained:

$$\sum_{i=1}^{N} d_{ijl} x_i \le e_{jl} \tag{3.7}$$

If we define $A_{ijl}^{(k)} = (U_{ijl}^{(k)} + \lambda_{jl} V_{ijl}^{(k)}) / V_{\Sigma jl}$, then left-hand side of inequality (3.7) is equal to

$$\sum_{i=1}^{N} d_{ijl} x = \sum_{i=1}^{N} x_i (U_{ijl}^{(k)} + \lambda_{jl} V_{ijl}^{(k)}) / V_{\Sigma jl} x_i^{(k)} = \sum_{i=1}^{N} A_{ijl}^{(k)} x_i / x_i^{(k)} = \sum_{i=1}^{N} (t_i^{(k)})^b A_{ijl}^{(k)} x_i \tag{3.8}$$

If we define $d_{ijl} = (t_i^{(k)})^b A_{ijl}^{(k)}$, then the left-hand side of equation (3.3.7) can be rewritten as

$$e_{jl} = \bar{\lambda}_{jl} - \lambda_{jl}(\boldsymbol{x}^{(k)}) + \sum_{i=1}^{N} (U_{ijl}^{(k)} + \lambda_{jl} V_{ijl}^{(k)}) / V_{\Sigma jl} = \bar{\lambda}_{jl} - \lambda_{jl}(\boldsymbol{x}^{(k)}) + \sum_{i=1}^{N} A_{ijl}^{(k)} \tag{3.9}$$

where $U_{ijl} = \boldsymbol{u}_{jl}^{\mathrm{T}} \boldsymbol{k}_i \boldsymbol{u}_{jl}/2$ is the mode strain energy for the ith element under the lth condition in jth mode. $V_{ijl} = \boldsymbol{u}_{jl}^{\mathrm{T}} \boldsymbol{g}_i \boldsymbol{u}_{jl}/2$ is the mode geometric strain energy for the ith element under the lth condition in jth mode. $V_{\Sigma jl} = \boldsymbol{u}_{jl}^{\mathrm{T}} \boldsymbol{g}_i \boldsymbol{u}_{jl}/2$ is the total mode geometric strain energy for the structure under the lth condition in jth mode.

(1) Explicit approximate function of the displacement constraint
According to the Moore theorem, the generalized displacement of one node can be expressed as

$$u_r = \sum_{i=1}^{N} \int (\boldsymbol{\sigma}_i^{\mathrm{V}})^{\mathrm{T}} (\varepsilon_i^{\mathrm{R}}) dv \tag{3.10}$$

where $\boldsymbol{\sigma}_i^{\mathrm{V}}$ is the stress component of ith element under the unit virtual load. $\boldsymbol{\varepsilon}_i^{\mathrm{R}}$ is the strain component of ith element under the unit real load.

According to the principle of virtual work, we can obtain

$$\int (\boldsymbol{\sigma}_i^{\mathrm{V}})^{\mathrm{T}} (\boldsymbol{\varepsilon}_i^{\mathrm{R}}) dv = (\boldsymbol{F}_i^{\mathrm{R}})^{\mathrm{T}} \boldsymbol{u}_i^{\mathrm{V}} \tag{3.11}$$

where $\boldsymbol{F}_i^{\mathrm{R}}$ is the nodal force vector of ith element under the real load case. $\boldsymbol{u}_i^{\mathrm{V}}$ is the nodal displacement vector of ith element under the virtual load case. Thus, eq. (3.10) can be written as follows:

$$u_r = \sum_{i=1}^{N} \int (\boldsymbol{\sigma}_i^{\mathrm{V}})^{\mathrm{T}} (\boldsymbol{\varepsilon}_i^{\mathrm{R}}) dv = \sum_{i=1}^{N} (\boldsymbol{F}_i^{\mathrm{R}})^{\mathrm{T}} \boldsymbol{u}_i^{\mathrm{V}} \tag{3.12}$$

According to the total stiffness equation, formula (3.12) can be written as

$$u_r = \sum_{i=1}^{N} \left(\boldsymbol{F}_i^{\mathrm{R}}\right)^{\mathrm{T}} \boldsymbol{u}_i{}^{\mathrm{V}} = \sum_{i=1}^{N} (\boldsymbol{u}_i)^{\mathrm{T}} \boldsymbol{k}_i \boldsymbol{u}_i^{\mathrm{V}} \tag{3.13}$$

According to the theory of ICM method, the filter function is introduced into the displacement constraint, and formula (3.13) transform is:

$$u_r = \sum_{i=1}^{N} \left(\boldsymbol{F}_i^{\mathrm{R}}\right)^{\mathrm{T}} \boldsymbol{u}_i{}^{\mathrm{V}} = \sum_{i=1}^{N} \frac{\left(t_i^{(k)}\right)^b}{t_i^b} \left(\boldsymbol{F}_i^{\mathrm{R}}\right)^{\mathrm{T}} \boldsymbol{u}_i^{\mathrm{V}} \tag{3.14}$$

Also based on $x_i = \dfrac{1}{t_i^b}$

$$u_r = \sum_{i=1}^{N} \left(\boldsymbol{F}_i^{\mathrm{R}}\right)^{\mathrm{T}} \boldsymbol{u}_i{}^{\mathrm{V}} = \sum_{i=1}^{N} \frac{x_i}{x_i^{(k)}} \left(\boldsymbol{F}_i^{\mathrm{R}}\right)^{\mathrm{T}} \boldsymbol{u}_i^{\mathrm{V}} \tag{3.15}$$

Let $(\boldsymbol{F}_i^{\mathrm{R}})^{\mathrm{T}} \boldsymbol{u}_i^{\mathrm{V}} / x_i^{(k)} = c_{ir}$, and the displacement constraint can be expressed as an explicit approximate function:

$$u_r = \sum_{i=1}^{N} \frac{x_i}{x_i^{(k)}} \left(\boldsymbol{F}_i^{\mathrm{R}}\right)^{\mathrm{T}} \boldsymbol{u}_i{}^{\mathrm{V}} = \sum_{i=1}^{N} c_{ir} x_i \le \bar{u}_r \tag{3.16}$$

3.2.3 Explicit Approximate Function of the Objective Function

The object function can be expressed as

$$w = \sum_{i=1}^{N} f_w(t_i) w_i^0 \tag{3.17}$$

where the filter function is $f_w(t_i) = t_i^a$. On the basis of the relationship between topological variables and design variables, we can obtain $t_i = 1/x_i^{1/b}$ and the filter function can be rewritten as

$$f_w(t_i) = 1/x_i^{a/b} \tag{3.18}$$

If we define $a/b = \beta$, then eq. (3.18) can be written as $f_w(t_i) = 1/x_i^{\beta}$. When we substitute it into eq. (3.17), the object function can be rewritten as

$$w = \sum_{i=1}^{N} w_i^0 / x_i^{\beta} \tag{3.19}$$

The object function is translated into explicit expression by using second-order Taylor's expansion:

$$W = \sum_{i=1}^{N} \{[w_i^0 \beta(\beta+1)/2x_i^{(k)\beta+2}]x_i^2 + [-w_i^0 \beta(\beta+2)/x_i^{(k)\beta+1}]x_i\} \tag{3.20}$$

where $x_i^{(k)}$ is the initial design variable of ith element at the kth iteration.

We define $A_i = w_i^0 \beta(\beta+1)/2x_i^{(k)\beta+2}$, $B_i = -w_i^0 \beta(\beta+2)/x_i^{(k)\beta+1}$

The objective function can be written as $W = \sum_{i=1}^{N}(A_i x_i^2 + B_i x_i)$. The whole topology optimization model is changed into

$$\begin{cases} \text{find} & \boldsymbol{x} \in E^N \\ \min & W = \sum\limits_{i=1}^{N} (A_i x_i^2 + B_i x_i) \\ \text{s.t.} & \sum\limits_{i=1}^{N} d_{ijl} x_i \le e_{jl} \quad (j = 1, \ldots, J; l = 1, \ldots, L) \\ & \sum\limits_{i=1}^{N} c_{irjl} x_i \le \bar{u}_{rjl} \quad (r = 1, \ldots, R) \\ & 1 \le x_i \le \bar{x}_i \quad (i = 1, \ldots, N) \end{cases} \tag{3.21}$$

3.2.4 Solution of the Topological Optimization Model

On the basis of the relationship between the topology and the design variables, the topology optimization model is transformed into a model with the design variables. To reduce the number of the variables and the model's scale, the optimization model is translated into a dual programming:

$$\begin{cases} \text{find} & \boldsymbol{\chi} \\ \max & \boldsymbol{\Phi}(\boldsymbol{\chi}) \\ \text{s.t.} & \boldsymbol{\chi} \geq 0 \end{cases} \tag{3.22}$$

In equation $\boldsymbol{\chi} = \begin{bmatrix} \boldsymbol{\mu} \\ \boldsymbol{\eta} \end{bmatrix}$, $\Phi(\boldsymbol{\mu}, \boldsymbol{\eta}) = \min_{1 \leq x_i \bar{x}_i} L(\boldsymbol{x}, \boldsymbol{\mu}, \boldsymbol{\eta})$,

$$L(\boldsymbol{x}, \boldsymbol{\mu}, \boldsymbol{\eta}) = \sum_{i=1}^{N} (A_i x_i^2 + B_i x_i) + \sum_{j=1}^{J} \sum_{l=1}^{L} \mu_{jl} (\sum_{i=1}^{N} d_{ijl} x_i - e_{jl}) + \sum_{j=1}^{J} \sum_{l=1}^{L} \eta_{jl} (\sum_{i=1}^{N} c_{irjl} x_i - \bar{u}_{rjl}) \tag{3.23}$$

According to the Kutta condition, the derivative of augmented Lagrangian function with respect to the design variable can be written as follows:

$$\partial L(\boldsymbol{x}, \boldsymbol{\mu}, \boldsymbol{\eta}) / \partial x_i = 2A_i x_i^* + B_i + \sum_{j=1}^{J} \sum_{l=1}^{L} \boldsymbol{\mu}_{jl} d_{ijl} + \sum_{j=1}^{J} \sum_{l=1}^{L} \boldsymbol{\eta}_{jl} c_{irjl} \begin{cases} \leq 0 & (x_i^* = \bar{x}_i) \\ = 0 & (1 < x_i^* < \bar{x}_i) \\ \geq 0 & (x_i^* = 1) \end{cases} \tag{3.24}$$

Making $I_a = \{i | 1 < x_i^* < \bar{x}_i\}$ the active variable set, the following equation is established:

$$2A_i x_i^* + B_i + \sum_{j=1}^{J} \sum_{l=1}^{L} \boldsymbol{\mu}_{jl} d_{ijl} + \sum_{j=1}^{J} \sum_{l=1}^{L} \boldsymbol{\eta}_{jl} c_{irjl} \equiv 0$$

For μ_{ml}, μ_{jm}, η_{nl}, η_{jn}, respectively, we get following the derivation:

$$\begin{aligned} \partial x_i^* / \partial \mu_{ml} &= -d_{iml}/2A_i, \partial x_i^* / \partial \eta_{nl} = -c_{inl}/2A_i, \\ \partial x_i^* / \partial \mu_{jm} &= -d_{ijm}/2A_i, \partial x_i^* / \partial \eta_{jn} = -c_{ijn}/2A_i \end{aligned} \tag{3.25}$$

Because it is difficult to get an explicit expression for $\Phi(\chi)$, it needs be transformed into a two-order Taylor expansion approximation. According to the dual theory, the objective function of formula (3.23) for the design of a variable's one- and two-order derivative is

$$\partial\boldsymbol{\Phi}/\partial\mu_{jl}=\sum_{i=1}^{N}d_{ijl}x_i^*-e_{jl}\,,\ \ \partial\boldsymbol{\Phi}/\partial\eta_{jl}=\sum_{i=1}^{N}c_{irjl}x_i^*-\bar{u}_{rjl} \tag{3.26}$$

$$\begin{aligned}\partial^2\boldsymbol{\Phi}/(\partial\mu_{jl}\partial\mu_{ml})&=\sum_{i=1}^{N}d_{ijl}\partial x_i^*/\partial\mu_{ml}\,,\ \ \partial^2\boldsymbol{\Phi}/(\partial\eta_{jl}\partial\eta_{nl})=\sum_{i=1}^{N}c_{irjl}\partial x_i^*/\partial\eta_{nl}\,,\\ \partial^2\boldsymbol{\Phi}/(\partial\mu_{jl}\partial\mu_{jm})&=\sum_{i=1}^{N}d_{ijl}\partial x_i^*/\partial\mu_{jm}\,,\ \ \partial^2\boldsymbol{\Phi}/(\partial\eta_{jl}\partial\eta_{jn})=\sum_{i=1}^{N}c_{irjl}\partial x_i^*/\partial\eta_{jn}\end{aligned} \tag{3.27}$$

By substituting Formula (3.25) formula (3.27), we get

$$\begin{aligned}\partial^2\boldsymbol{\Phi}/(\partial\mu_{jl}\partial\mu_{ml})&=-\sum_{i=1}^{N}d_{ijl}\,d_{iml}/2A_i,\ \partial^2\boldsymbol{\Phi}/(\partial\eta_{jl}\partial\eta_{nl})=-\sum_{i=1}^{N}c_{irjl}\,c_{irnl}/2A_i\\ \partial^2\boldsymbol{\Phi}/(\partial\mu_{jl}\partial\mu_{ml})&=-\sum_{i=1}^{N}d_{ijl}\,d_{iml}/2A_i,\ \partial^2\boldsymbol{\Phi}/(\partial\eta_{jl}\partial\eta_{nl})=-\sum_{i=1}^{N}c_{irjl}\,c_{irnl}/2A_i\end{aligned} \tag{3.28}$$

$$\partial^2\boldsymbol{\Phi}/(\partial\mu_{jl}\partial\eta_{jl})=\partial^2\boldsymbol{\Phi}/(\partial\eta_{jl}\partial\mu_{jl})=-\sum_{i=1}^{N}d_{ijl}c_{irjl}/2A_i \tag{3.29}$$

$\boldsymbol{\Phi}(\boldsymbol{\chi})$ will be a two-order approximation, and omitting the constant term, we can get the second-order programming model:

$$\begin{cases}\text{find} & \boldsymbol{\chi}\\ \min & -\boldsymbol{\Phi}(\boldsymbol{\chi})=\dfrac{1}{2}\boldsymbol{\chi}^{\mathrm{T}}\boldsymbol{D}\boldsymbol{\chi}+(\boldsymbol{H}^{\mathrm{T}}-\boldsymbol{\chi}^{0^{\mathrm{T}}}\boldsymbol{D})\boldsymbol{\chi}\\ \text{s.t.} & \boldsymbol{\chi}\ge 0\end{cases} \tag{3.30}$$

where $\boldsymbol{D}=-\nabla^2\boldsymbol{\Phi}$ and $\boldsymbol{H}=-\nabla\boldsymbol{\Phi}$. By solving the second-order programming model, and we can get $\boldsymbol{\chi}$. The active variable set was updated. Therefore, this process needs to be repeated until

$$\|\boldsymbol{x}^{(k+1)}-\boldsymbol{x}^{(k)}\|/\|\boldsymbol{x}^{(k)}\|\le\varepsilon$$

The optimum values of design variables $\boldsymbol{x}^*$ are solved, and the optimum values of topological variables $\boldsymbol{t}^*$ are accordingly obtained. With the new topological variables $\boldsymbol{t}^{(k+1)}=\boldsymbol{t}^*$, the structure is modified. Then the structural analysis and topological optimization are renewable and should satisfy the following convergence condition:

$$\Delta W=|(W^{(k+1)}-W^{(k)})/W^{(k+1)}|\le\varepsilon \tag{3.31}$$

where $W^{(k)}$ and $W^{(k+1)}$ are the structural quality of previous iteration and current iteration, respectively, and ε is the convergence precision. In this study, $\varepsilon=0.001$ is employed.

3.3 Optimization Strategy

3.3.1 Inversion Criterion of Element Topological Variable

In the iterative process, the topological variables are continuous variables, which guarantee the continuity of the topological structure. In the optimization process, the topological variables are gradually close to 0 or 1. At the last step of the iteration, the topology optimization structure must satisfy the convergence precision, and the topological structure is obtained. In the optimal topology, there are still a few nonwhite nonred regions, namely, the topological variable values between 0 and 1 of the elements. To get a clearer topology, we have given a deletion threshold T. When the topology variable is lower than the threshold value, the topology variable is 0. When the topology is higher than the threshold value, the topology variable is 1. The threshold T generally is taken in the 0.3–0.7 range, so that the continuous topological variables were inversed to 0–1 type of discrete topology, and one can get a clear topology.

3.3.2 Grid Dependence and the Solution of the Phenomenon of the Chess Board

To solve the problem of the grid phenomenon and the grid dependence, the method of image filtering is used, which was introduced by Xirong [62]. It does not filter the sensitivity of the objective function, but filters the displacement coefficient of the element.

Defining the geometric centroid distance function of element i and j is $d(i,j)$:

$$d(i,j) = max(0, r_{\min} - \mathrm{dis}(i,j)) \tag{3.32}$$

Here $\mathrm{dis}(i,j)$ is the distance between the center of element i and j. The distance function $d(i,j)$ has symmetry on elements i and j.

In the displacement constraint, to filter the unit contribution coefficient (subscript j is the modal number)

$$A_{ij} = \sum_{k=1}^{N} \boldsymbol{\Phi}_{ik} A_{kj}$$

In the quality coefficient matrix $\boldsymbol{\Phi}$ elements,

$$\boldsymbol{\Phi}_{ij} = \frac{d(i,j)}{\sum_{j=1}^{N} d(i,j)} \tag{3.33}$$

The quality coefficient matrix $\boldsymbol{\Phi}$ has the following attributes: $\sum_{j=1}^{N} \boldsymbol{\Phi}_{ij} = 1$. The meaning is: for the i element, the sum of all element weight coefficients is 1

3.4 Constrain Selection

3.4.1 Constraint Primary select

In sections discussing optimization, shape optimization and multiple constraints, to improve efficiency and to reduce the number of structure, the effective constraint primaries select method is considered, so as to reduce the number of constraints and the cost of computation [59]. In topology optimization with buckling and displacement constraint, this strategy can also be into account to remove unnecessary constraints, to reduce the amount of calculation model and to improve the calculation speed, so that a satisfactory result to topology optimization can be obtained quickly.

3.4.2 Selection Criterion of Displacement Upper Limit

The upper limit of displacement can also affect the optimal topology shape and quality of the optimal topology. Accurate selection of the upper limit of the displacement and the upper limit of the critical buckling force should be quick and effective so as to obtain a reasonable topology. Similarly, the size of the upper bound can be estimated by the quality of the topology. The function correspondence will give the designer more convenience to apply the topology optimization at the product conceptual design stage. At the same time, in the joint action of buckling constraints and displacement constraints, if the critical buckling force is too large, the displacement constraints cannot be useful, but at the same time, if the displacement upper limit is too large, the buckling constraints will not play a role. How to select the size of these two kinds of constraints? What is the relationship between the two kinds of constraints?

According to a specific numerical example, the corresponding relationship between the quality ratio and the displacement ratio of the topological structure is obtained. To ensure the comparability, the numerical example calculated is same as in Chapter 2.

Numerical example: As illustrated in Figure 3.1, the base structure is a plane elastic body with a size $520 \times 260 \times 2.0$ mm^3. The elastic modulus of structure material is $E = 68.89 \times 10^3$ MPa, Poisson's ratio is $\gamma = 0.3$, and the material density is $\rho = 2.7 \times 10^{-6}$ kg/mm^3. The right-hand and left-hand underpart corner points of the structure are clasped. A concentrated force $P = 15{,}000$ N along the up-to-down (-y)

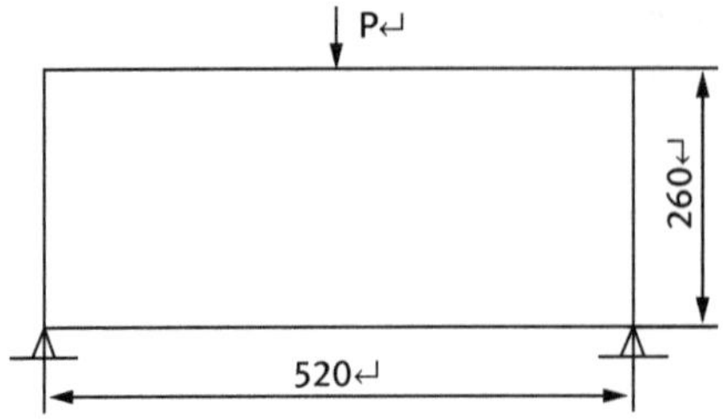

Figure 3.1: Base structure.

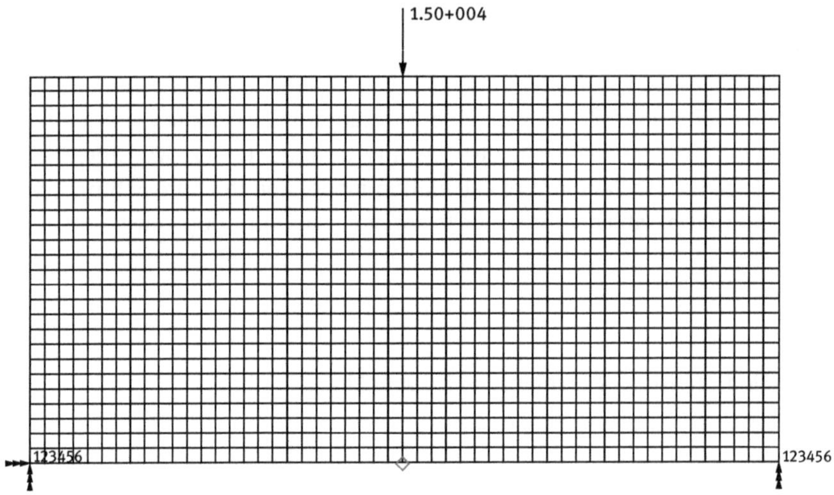

Figure 3.2: Finite element model.

direction is located at the midpoint of the upper boundary of the structure. A quadrilateral element with four nodes is used, and the total number of elements is 52 × 26. The base structure's quality is 0.73 kg (Figure 3.2).

The displacement of the applied force node is 0.746 mm, the buckling factor is 0.00794 and the buckling critical force is 119 N.

To eliminate the influence of the corresponding relationship between the displacement constraint and the quality of the topological structure, the concept of "displacement constraint ratio" is introduced here. With an increase in the displacement of the key points of the structure, the quality of the topological structure will be reduced. In general, the maximum elastic deformation of the plate structure cannot exceed three times the thickness of the plate. Hence, the displacement constraint ratio is defined as the ratio of displacement to three times the thickness of the base structure.

The displacement constraint and topological quality of each numerical experiment and the topology optimization structure are listed in Table 3.1.

Table 3.2 lists the "quality ratio" and "displacement constraint ratio" of the 17 times topology optimization.

Table 3.1: Displacement constraint and quality of optimal structure.

Test number	Displacement constraint	Topological quality	Repeated analysis times	Topological graph
1	0.746	0.73	1	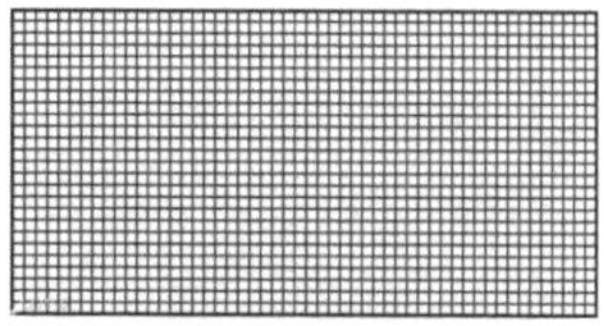
2	0.821	0.39	31	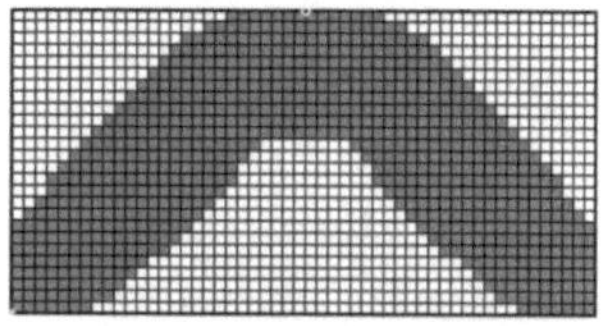
3	0.896	0.31763	31	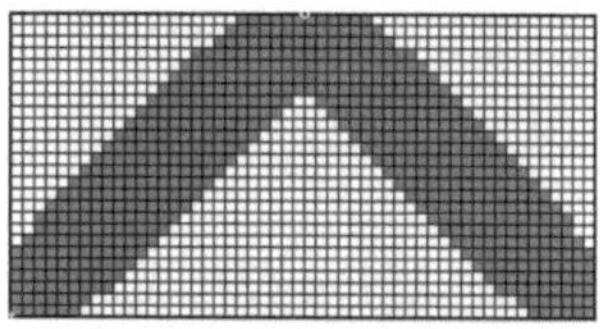
4	0.970	0.2754	30	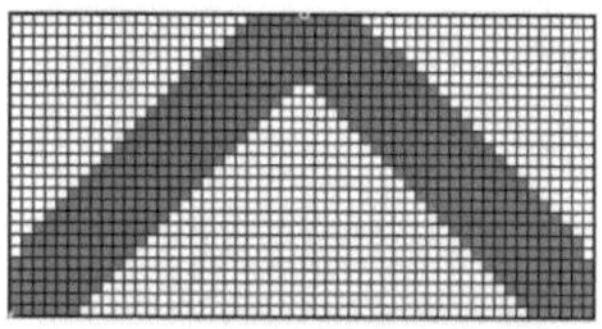
5	1.194	0.20796	32	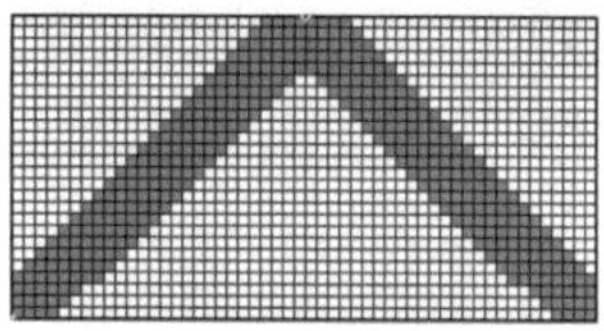
6	1.418	0.17283	33	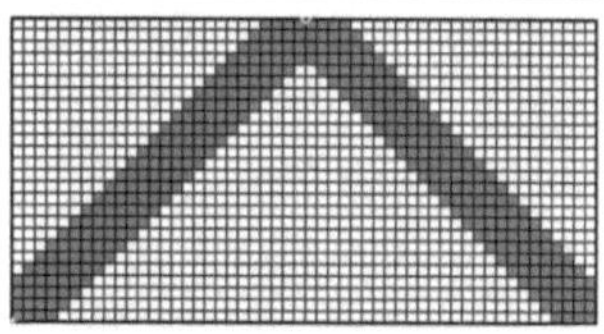
7	1.642	0.15036	34	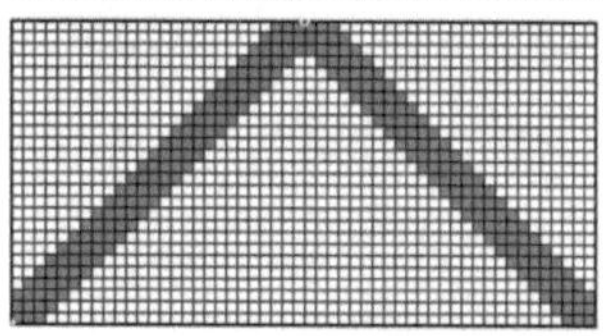

(continued)

Table 3.1: (continued)

Test number	Displacement constraint	Topological quality	Repeated analysis times	Topological graph
8	1.866	0.13489	36	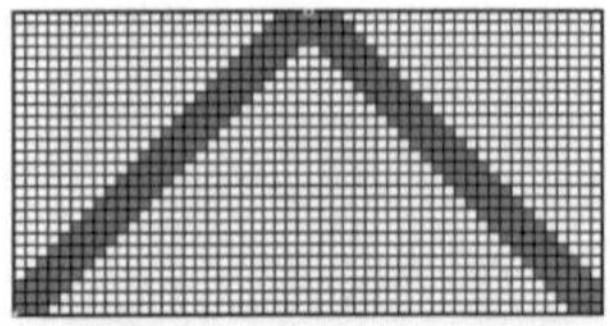
9	2.090	0.12454	36	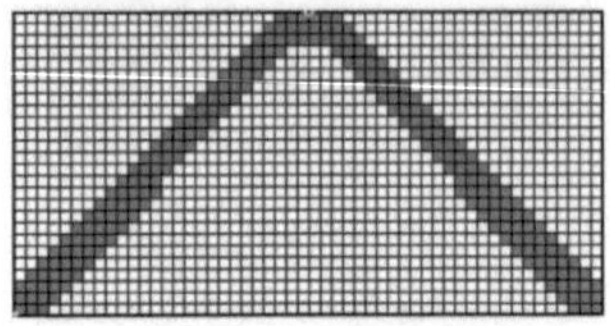
10	2.314	0.11628	36	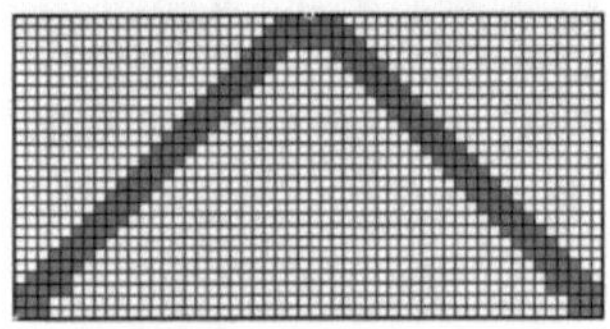
11	2.538	0.11319	37	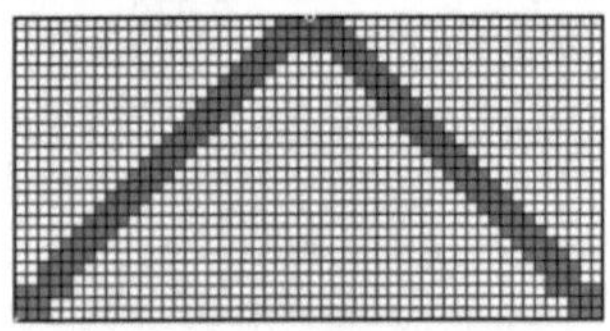
12	2.762	0.11058	37	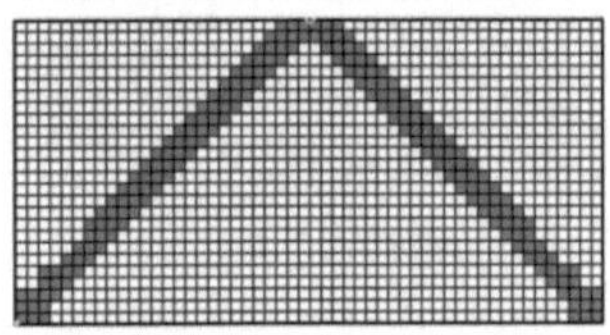
13	2.986	0.1089	38	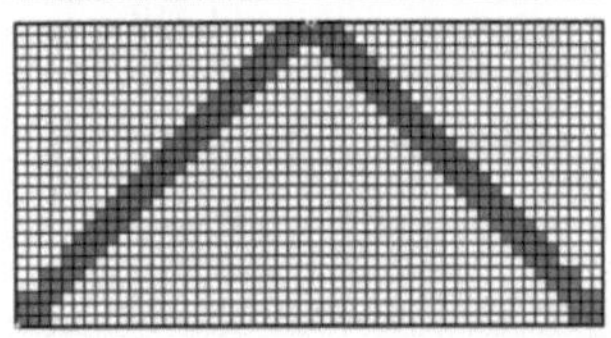
14	3.210	0.10725	37	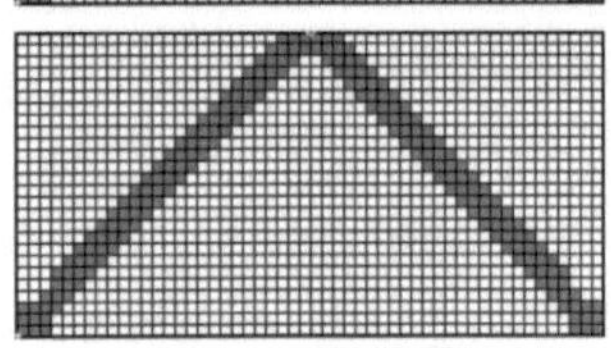

Table 3.1: (continued)

Test number	Displacement constraint	Topological quality	Repeated analysis times	Topological graph
15	3.434	0.10545	37	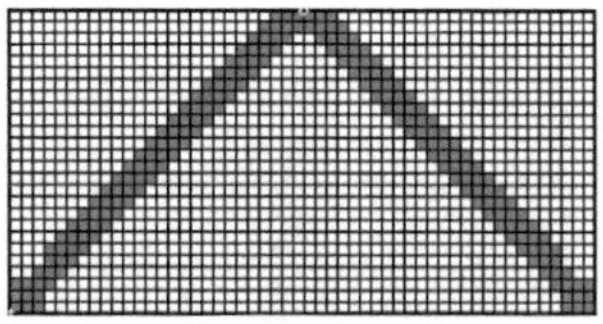
16	4.000	0.10171	38	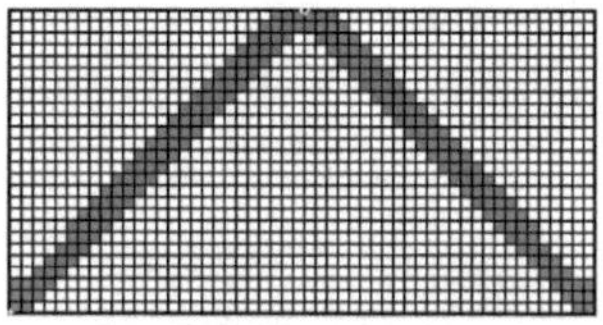
17	5.000	0.0964	38	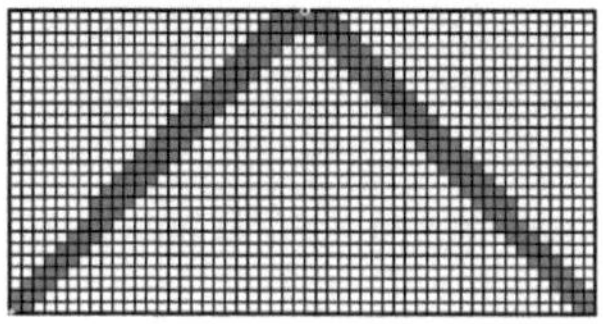

Table 3.2: The ratio of displacement constraint and ratio of quality.

Test number	Displacement constraint ratio	Quality ratio of topological structure
1	0.1243	1
2	0.1368	0.5340
3	0.1493	0.4351
4	0.1617	0.3772
5	0.199	0.2849
6	0.2363	0.2367
7	0.2736	0.2060
8	0.311	0.1848
9	0.3483	0.1706
10	0.3857	0.1593
11	0.423	0.1550
12	0.4603	0.1515
13	0.4977	0.1492
14	0.535	0.1469
15	0.5723	0.1444
16	0.6667	0.1393
17	0.8333	0.1320

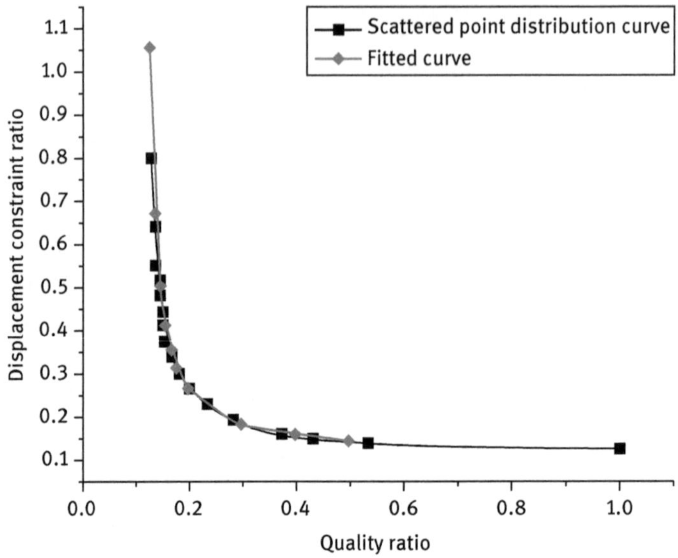

Figure 3.3: A curve of quality ratio and displacement constraint ratio.

The "quality ratio" is on the horizontal coordinate X axis, and the "displacement constraint ratio" is on the vertical coordinate Y axis as shown in Figure 3.3. It can be seen from the graph that the curve is nonlinear. To get the function relationship between the quality ratio and displacement constraint ratio, the least square method is used to fit the curve. A hyperbolic function is selected according to the distribution of the discrete points:

$$\frac{1}{y} = a + \frac{b}{x} \tag{3.34}$$

Using the data in Table 3.3 to write equations:

$$\begin{bmatrix} \sum_{i=1}^{17} 1 & \sum_{i=1}^{17} X_i \\ \sum_{i=1}^{17} X_i & \sum_{i=1}^{17} X_i^2 \end{bmatrix} \begin{bmatrix} a \\ b \end{bmatrix} = \begin{bmatrix} \sum_{i=1}^{17} Y_i \\ \sum_{i=1}^{17} X_i Y_i \end{bmatrix} \tag{3.35}$$

$$\begin{cases} 17a + 86.2035b = 62.69071 \\ 86.2035a + 506.9389b = 245.031 \end{cases} \tag{3.36}$$

We get $a = 8.98$ and $b = -1.0436$. Substituting these into the hyperbolic equation,

$$y = \frac{x}{8.98x - 1.0436} \tag{3.37}$$

Table 3.3: The simulation result of displacement constraint curve.

Serial number	x	y	$X = 1/x$	$Y = 1/y$	X^2	XY
1	1	0.1243	1	8.04505	1	8.04505
2	0.5340	0.1368	1.8728	7.30994	3.5070	13.69006
3	0.4351	0.1493	2.2983	6.69792	5.2822	15.39383
4	0.3772	0.1617	2.6511	6.18429	7.0283	16.39517
5	0.2849	0.199	3.5100	5.02513	12.3200	17.63821
6	0.2367	0.2363	4.2248	4.23191	17.8489	17.87897
7	0.2060	0.2736	4.8544	3.65497	23.5652	17.74269
8	0.1848	0.311	5.4113	3.21543	29.2822	17.39966
9	0.1706	0.3483	5.8617	2.87109	34.3595	16.82947
10	0.1593	0.3857	6.2775	2.59269	39.4070	16.27561
11	0.1550	0.423	6.4516	2.36407	41.6231	15.25203
12	0.1515	0.4603	6.6007	2.1725	43.5692	14.34002
13	0.1492	0.4977	6.7024	2.00924	44.9222	13.46673
14	0.1469	0.535	6.8074	1.86916	46.3407	12.72412
15	0.1444	0.5723	6.9252	1.74734	47.9584	12.10068
16	0.1393	0.6667	7.1786	1.49993	51.5323	10.7674
17	0.1320	0.8333	7.5758	1.20005	57.3927	9.09134
$\sum$			86.2035	62.69071	506.9389	245.031

To facilitate comparison, the "quality ratio–buckling constraint ratio" curve and "quality ratio–displacement constraint ratio" curve were drawn in the same coordinate system, as shown in Figure 3.4.

From the graph, we can see that the constraint ratio is in the range of 0.2–0.5, the change of quality ratio is between 0.12 and 0.22, with little change in quality and the quality changes are in the range of 0.5–1; the structure topology will have a greater change. Therefore, to obtain a better topology, the upper limit of constraint is appropriately selected from 20% to 50% of the base structure.

Similarly, if the expected quality of the optimized structure is given in advance, the approximate value of the corresponding constraints can be obtained according to the quality ratio and the constraint ratio.

3.5 Optimization Algorithms

The optimization algorithm of the displacement and buckling constraints is based on the second development platform of MSC.Patran/Nastran software, the use of the development tool PCL language to design the optimization algorithm program and user input interface. The specific steps of optimization are as follows:

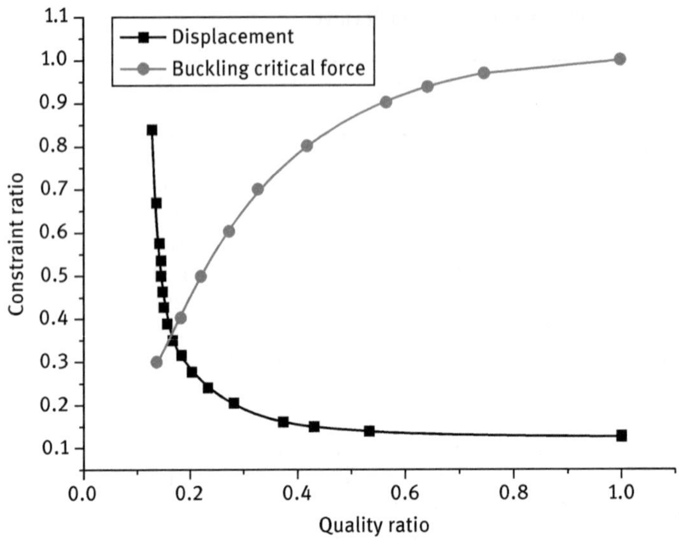

Figure 3.4: A curve of quality ratio and buckling displacement constraint ratio.

(1) Finite element model: The geometric model of a continuum structure is established on the MSC.Patran platform, which is divided into finite element mesh, which defines the properties of the material, boundary conditions and the properties of the element. Because of the structural topology optimization by buckling constraints, the properties of planar elements are set to shell elements.

(2) Basic optimization parameter input: This includes the convergence accuracy, the radius of the filter, the external force, etc.

(3) Displacement constrained optimization parameter input: This includes the constraint name, the displacement of the control node, displacement constraints, etc.

(4) Buckling constraint optimal parameter input: This includes the sum of the total number of buckling modes, the name of the modal constraint, the numerical value of the buckling constraint, the number of corresponding buckling modes and the condition name.

(5) Establish virtual load condition.

(6) Submit MSC. Nastran solution: This includes the linear static finite element analysis of the structure and the finite element analysis of the buckling mode. The linear static finite element analysis solver is SOL101, and the buckling mode finite element analysis solver is SOL105.

(7) Calculation and extraction of the corresponding finite element analysis results.

(8) The effective displacement constraint and the buckling constraint are selected.

(9) Generate input files required for the dual quadratic programming.

(10) Call dual quadratic programming to solve optimization model.

(11) According to the convergence criterion formula, we determine whether to satisfy the convergence condition. If it do not meet, it will go back to (6), if it satisfy, then stop the iteration. Record the data, determine the optimal topology, inversion topology variables, get the optimal topology graph and end the optimization calculation.

The block diagram of the optimization program is shown in Figure 3.5.

3.6 Numerical Simulation and Case Study

3.6.1 Numerical Example of Two-dimensional Plane Continuum

Example 1

As illustrated in Figure 3.6, the base structure is a plane elastic body with a size $100 \times 240\times 2.0$ mm^3. The elastic modulus of structure material is E = 68.89×10^3 MPa, Poisson's ratio is $\gamma = 0.3$ and the material density is $\rho = 2.7 \times 10^{-6}$ kg/mm^3. The left-hand points of the structure are clasped. A concentrated force P = 15,600 N along the up-to-down (-y) direction is located in the right edge midpoint of the structure. To avoid the effect of stress concentration, the force is distributed on the three nodes in the middle of the right side boundary. A quadrilateral element with four nodes is used, and the total number of elements is 48×20. The base structure's quality is 0.1296 kg.

The vertical downward displacement of the center point is 0.335 mm. The buckling factor is 0.4347. The initial critical load is 6,782 N. The upper limit of the displacement of the left boundary midpoint is 0.72 mm. In this example, the upper limit maximal value of the critical load is 2,618 N. Convergence precision is 0.001 after 29 iterations.

The reliability of the function expression of the "quality displacement ratio" is verified after fitting.

The quality of the base structure is 0.1296 kg, when only the displacement constraints are considered; after 33 iterations of convergence, the topology structure's quality is 0.0415 kg. It is 32% of the base structure's quality. The maximum allowable vertical displacement at the center point of the base structure is 0.335 mm, and after optimization, the control point displacement is 0.711 mm, and the displacement ratio is 0.12 (Figure 3.7-3.13). Substituting the quality ratio into the quality–displacement constraint ratio formula, we can get the constraint ratio as 0.17 (Table 3.4). The results obtained for the function relation calculation and numerical calculation have some errors, which are mainly due to the following reasons: the calculation error, fitting error quality inversion topology and curve caused.

The topological structure of Yunkang and Xirong's [61] displacement constraints is relatively small, the distance between the two branch is large and the stability of this structure is poor. The buckling constraint topological structure becomes shorter, and the structure branch adds one, the stability of the structure is better. When the

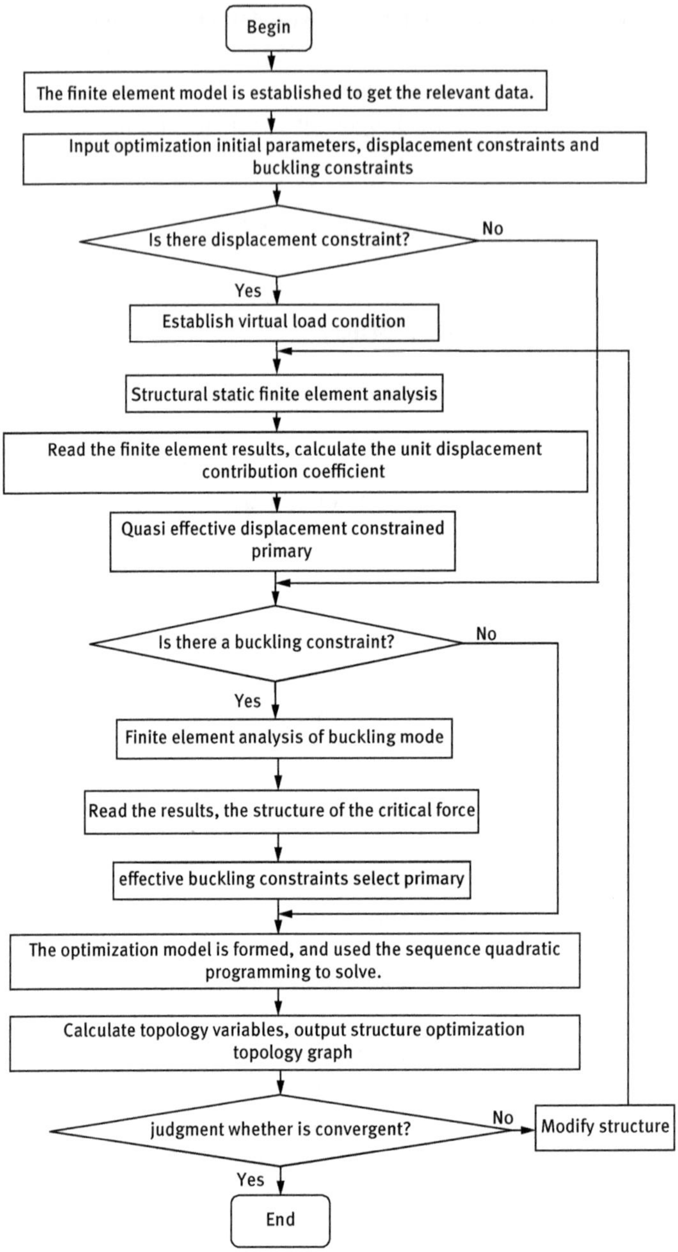

Figure 3.5: Flow chart for the algorithm of topology optimization of continuum structure subjected to buckling constraints and displacement constraints.

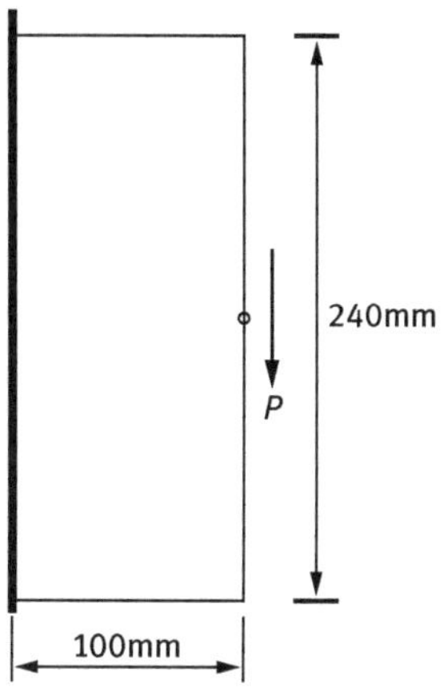

Figure 3.6: Base structure.

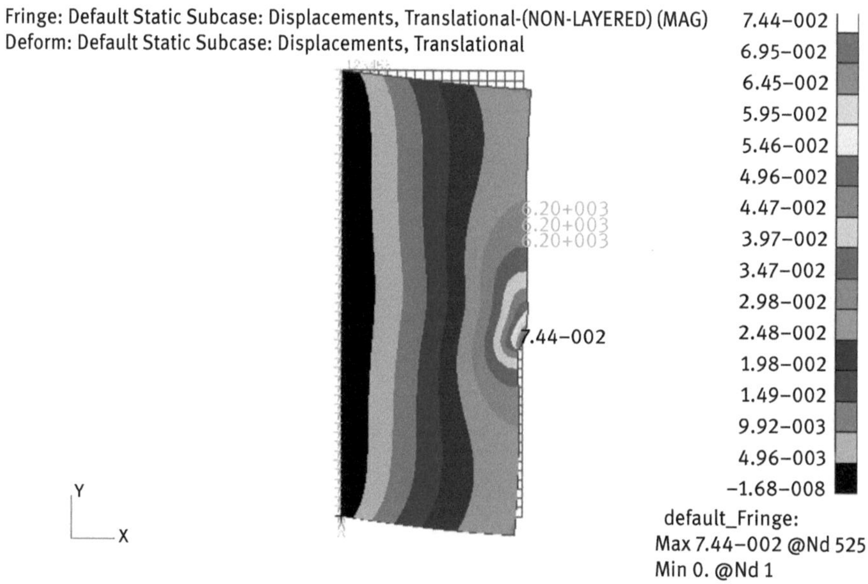

Figure 3.7: Displacement fringe result of the base structure.

displacement constraints and buckling constraints are considered, the topology changes with respect to the single constraint, but the simple sum of the two topologies remains the same. The support of the structure is changed in two branches, the distance between the branches is reduced and the width of the branch is strengthened. On the basis of the displacement, the stability of the structure is strengthened after increasing the buckling constraint.

Figure 3.8: Optimal topological configuration with displacement constraint in reference [61].

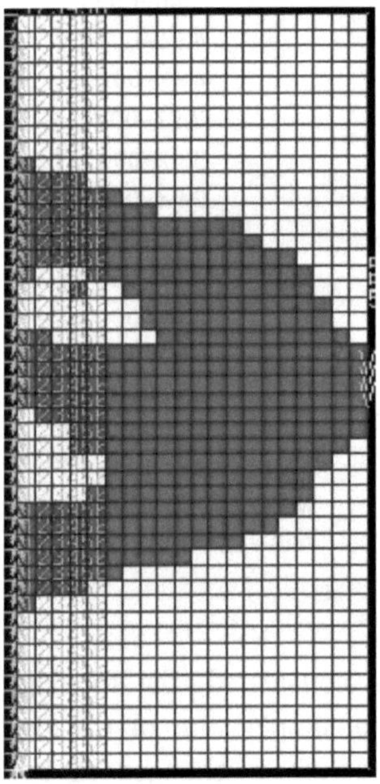

Figure 3.9: Optimal topological configuration with buckling constraint.

Example 2

As illustrated in Figure 3.14, the base structure is a plane elastic body with a size $520 \times 260 \times 2.0\ mm^3$. The elastic modulus of structure material is $E = 68.89 \times 10^3$ MPa, Poisson's ratio is $\gamma = 0.3$ and the material density is $\rho = 2.7 \times 10^{-6}\ kg/mm^3$. The left-hand and

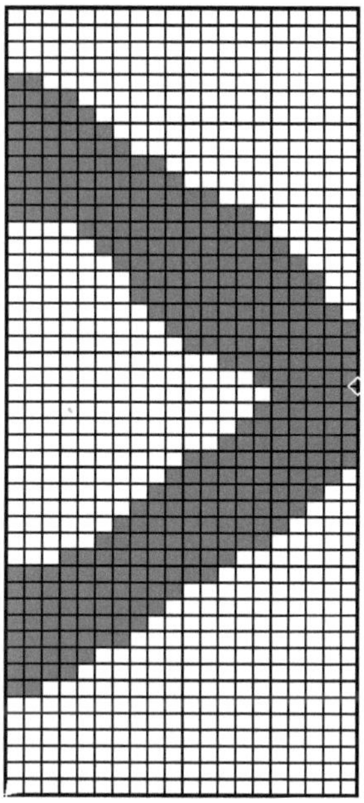

Figure 3.10: Optimal topological configuration with buckling and displacement constraints.

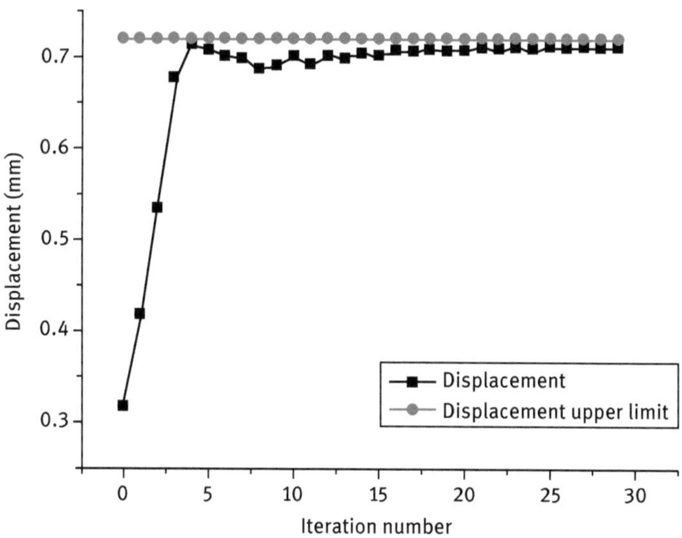

Figure 3.11: Iterative history of displacement constraint.

right-hand points of the structure are clasped. A concentrated force $P = 15{,}000$ N along the up-to-down (-y) direction is located in the upper edge midpoint of the structure. A quadrilateral element with four nodes is used and the total number of elements is 52×26. The base structure's quality is 0.73 kg.

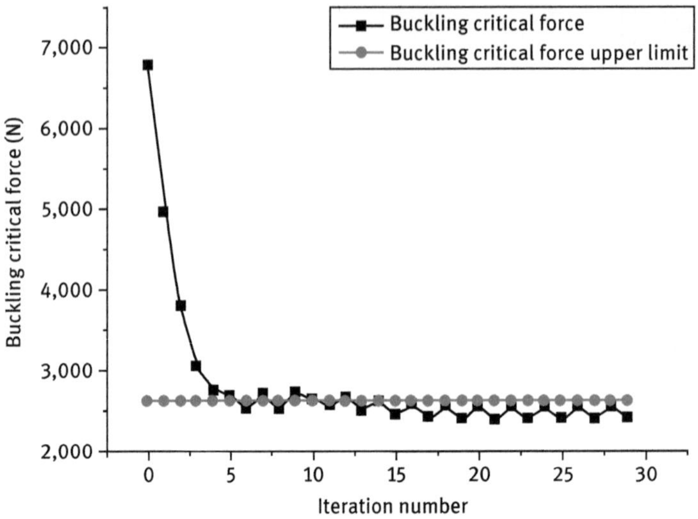

Figure 3.12: Iterative history of buckling critical load.

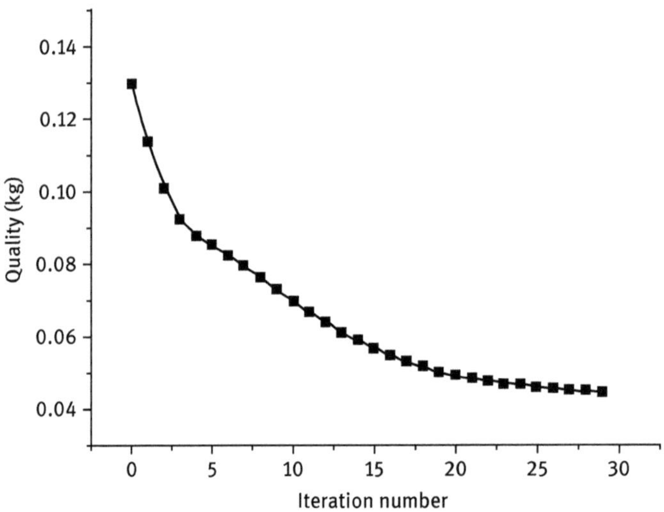

Figure 3.13: Iterative history of quality.

The vertical downward displacement of the center point is 0.746 mm, the buckling factor is 0.0079363, the initial critical load is 119.0445 N and the upper limit of the displacement of the upper boundary midpoint is 1.5 mm. In this example, the upper limit maximal value of the critical load is 70 N. Convergence precision is 0.001 after 36 iterations. The critical force of the topological structure is 70.7 N.

Table 3.4: Comparison of topological results with different constraints.

Constraint type	Initial quality (kg)	Optimized quality (kg)	Iteration number
Displacement	0.1296	0.0415	33
Buckling	0.1296	0.0484	46
Displacement + buckling	0.1296	0.1966	29

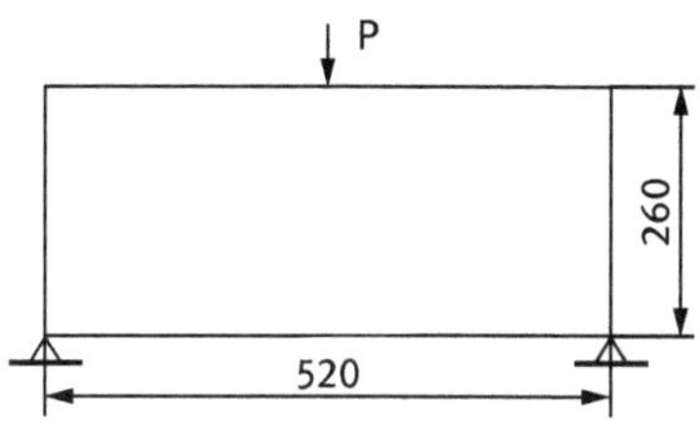

Figure 3.14: Base structure.

Table 3.5: Comparison of topological results with different constraints.

Constraint type	Initial quality (kg)	Optimized quality (kg)	Iteration number
Displacement	0.73	0.160	32
Buckling	0.73	0.195	43
Displacement + buckling	0.73	0.184	36

The reliability of the function expression of the "quality displacement ratio" is verified after fitting.

The quality of the base structure is 0.73 kg, when only the displacement constraints are considered, and after 32 iterations of convergence, the topology structure's quality is 0.16 kg. It is 22% of the base structure's quality. The maximum allowable vertical displacement at the center point of the base structure is 0.746 mm, and after optimization, the displacement control point displacement is 1.554 mm and the displacement ratio is 0.48. Substituting the quality ratio into the quality–displacement constraint ratio formula, we can get the constraint ratio as 0.39 (Table 3.5). The results of function relation calculation and numerical calculation have certain errors, and the results show that the "quality displacement constraint ratio" function expression has a certain reliability for different structure models.

As can be seen from the graph Figure 3.15 to 3.21, the topological structure of the displacement constraint is relatively uniform. At the top of the topological structure of the buckling constraints, the two ends of the horizontal section are provided with a finer topology. At the same time, considering the effect of buckling constraints and displacement constraints, the bottom of the topology is strengthened. To prevent the

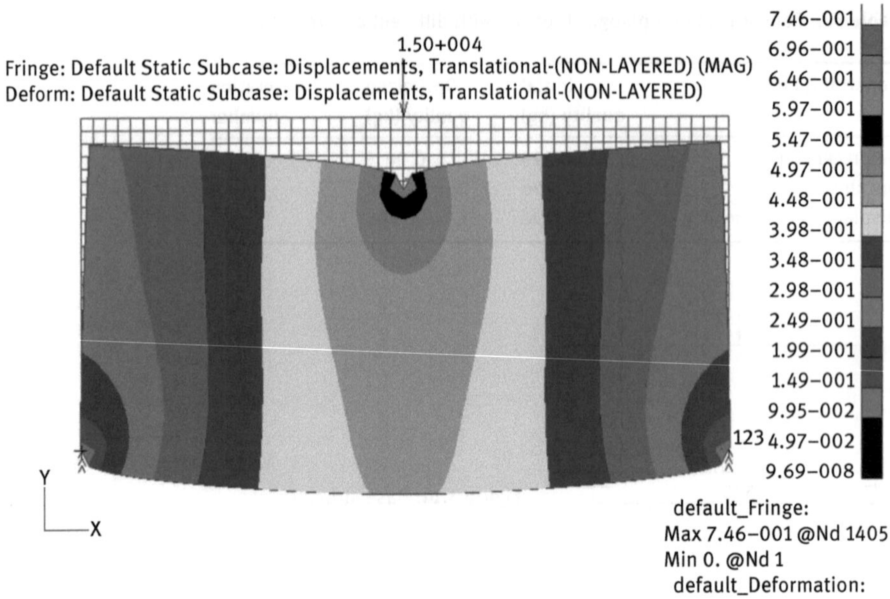

Figure 3.15: Displacement fringe result of the base structure.

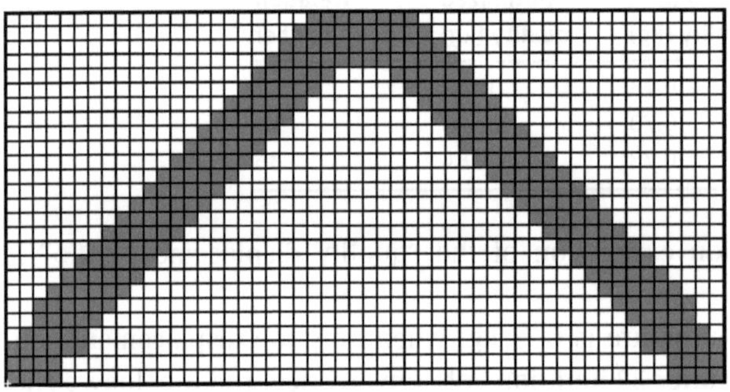

Figure 3.16: Optimal topological configuration with displacement constraint.

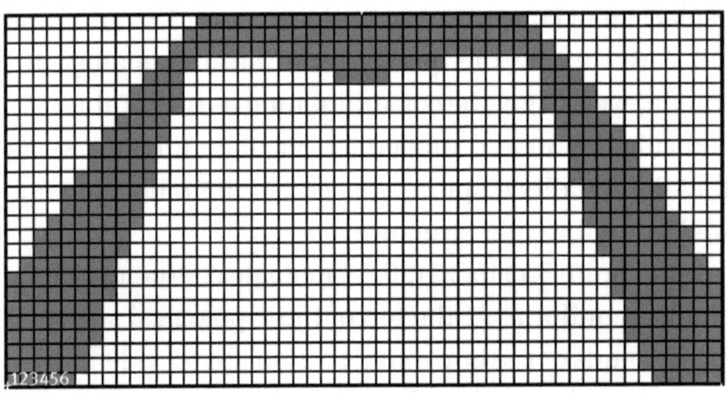

Figure 3.17: Optimal topological configuration with buckling constraint.

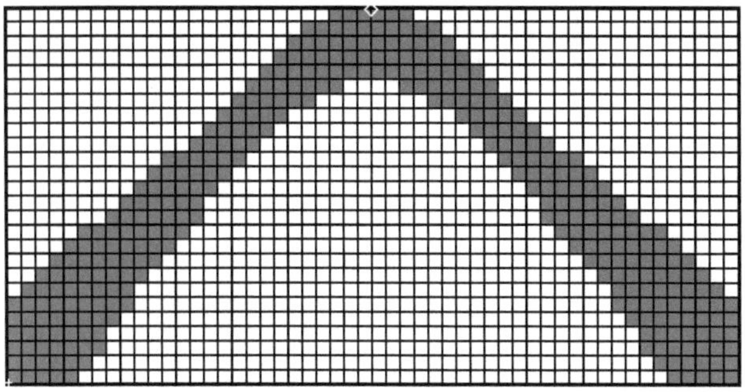

Figure 3.18: Optimal topological configuration with buckling and displacement constraints.

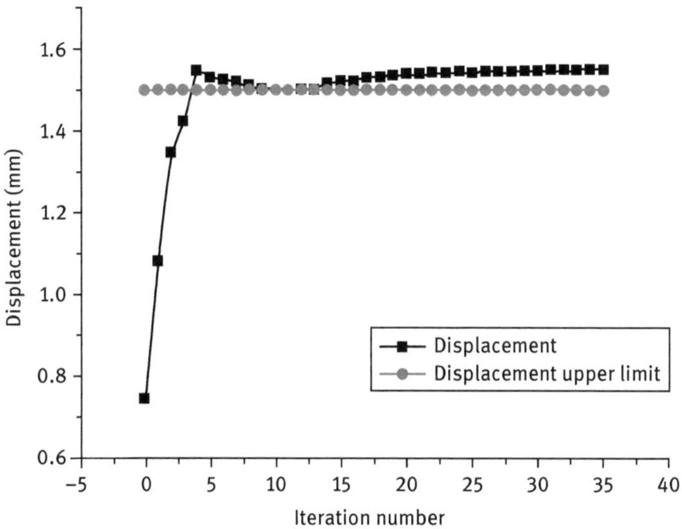

Figure 3.19: Iterative history of displacement constraint.

buckling of the structure, it is necessary to strengthen the structure, which is consistent with the measures to strengthen the stability of the practice engineering.

Example 3

As illustrated in Figure 3.22, the base structure is a plane elastic body with a size $80 \times 50 \times 1.0$ mm^3. The elastic modulus of structure material is E = 100 GPa, Poisson's ratio is γ = 0.3 and the material density is $\rho = 2.7 \times 10^{-6}$ kg/mm^3. The left-hand points of the structure are clasped. A concentrated force P = 9,000 N along the up-to-down (-y) direction is located at the upper boundary right corner of the structure. A quadrilateral element with four nodes is used and the total number of elements is 48×30. The base structure's quality is 0.0108 kg.

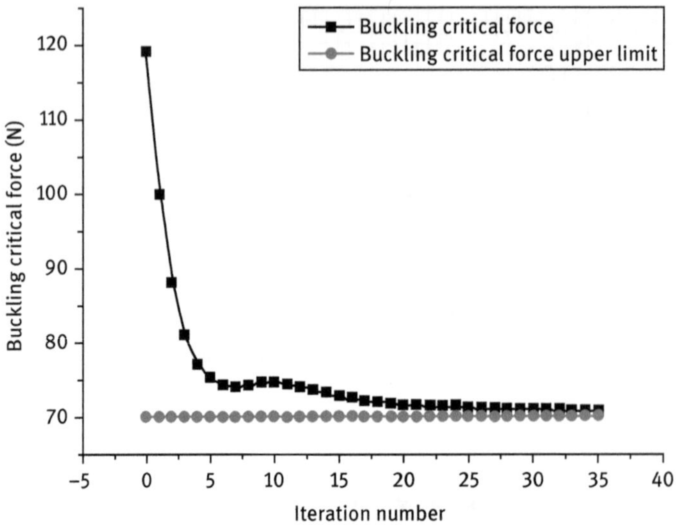

Figure 3.20: Iterative history of buckling critical load.

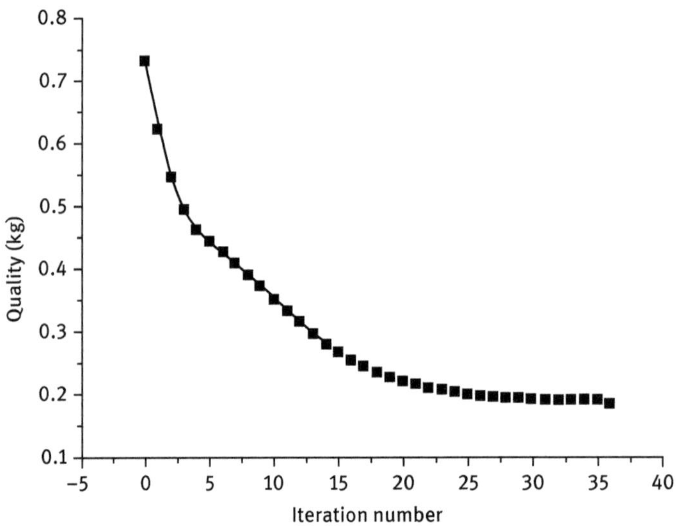

Figure 3.21: Iterative history of quality.

The vertical downward displacement of the center point is 2.81 mm, the buckling factor is 0.030398 and the initial critical load is 273.562 N. The upper limit of the displacement of the upper boundary midpoint is 6 mm. The upper limit maximal value of the critical load is 110 N. Convergence precision is 0.001, after 47 iteration. The critical force of the topological structure is 111.591 N. The quality of the optimized structure is 0.00437 kg (Table 3.6).

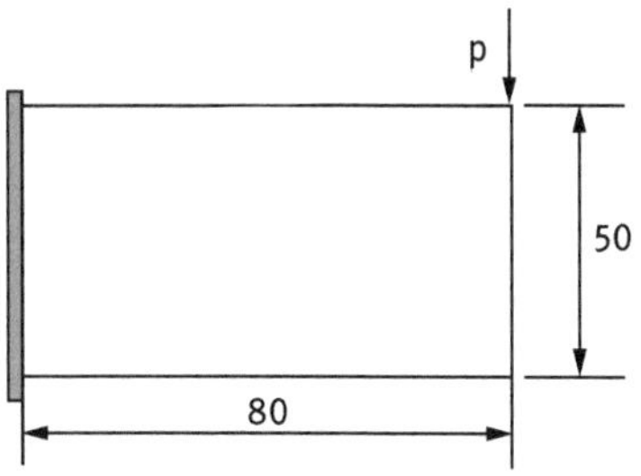

Figure 3.22: Base structure.

Table 3.6: Comparison of topological results with different constraints.

Constraint type	Initial quality (kg)	Optimized quality (kg)	Iteration number
Displacement	0.0108	0.0039	48
Buckling	0.0108	0.00394	38
Displacement + buckling	0.0108	0.00437	48

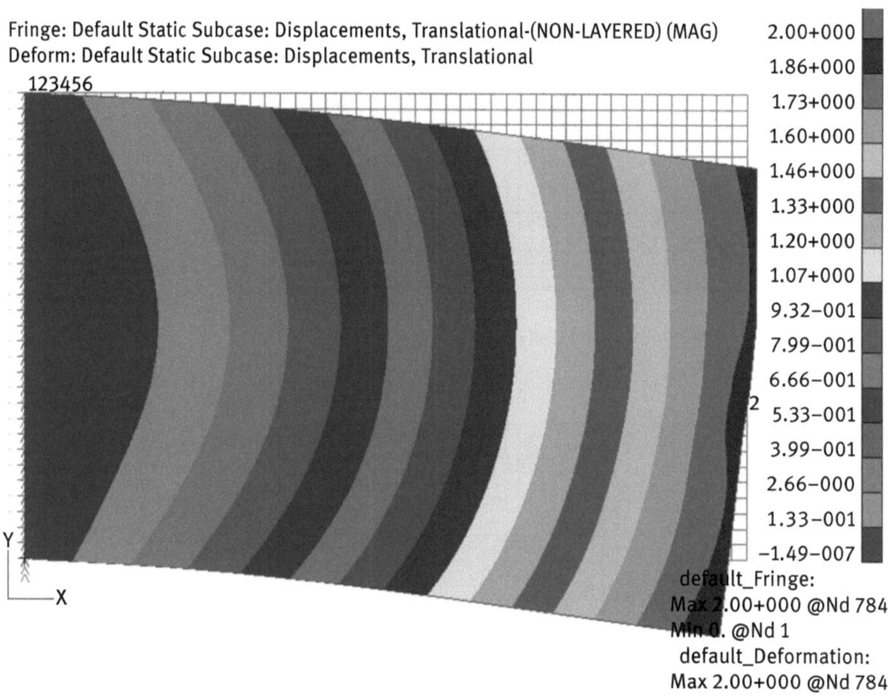

Figure 3.23: Displacement fringe of the base structure.

As can be seen from the graph Figure 3.23 to Figure 3.29, the topological structure of the displacement constraint is relatively thin and uniform. The topological

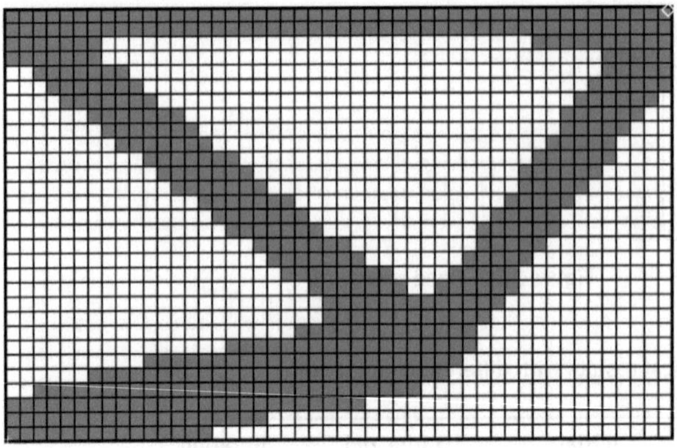

Figure 3.24: Optimal topological configuration with displacement constraint.

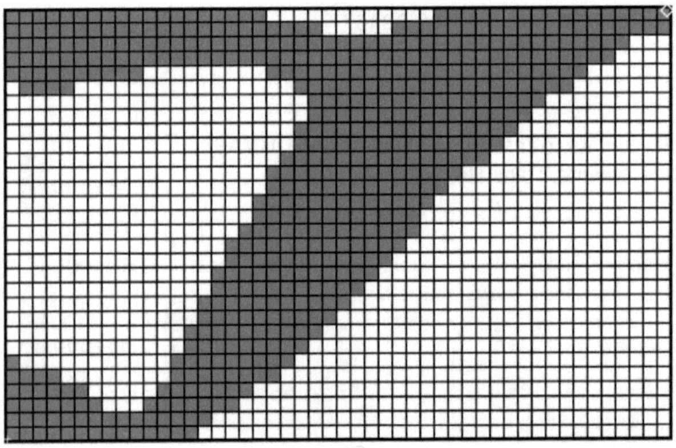

Figure 3.25: Optimal topological configuration with buckling constraint.

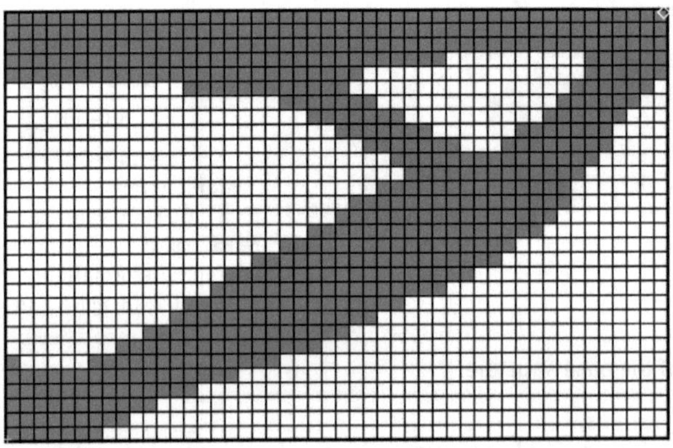

Figure 3.26: Optimal topological configuration with buckling and displacement constraints.

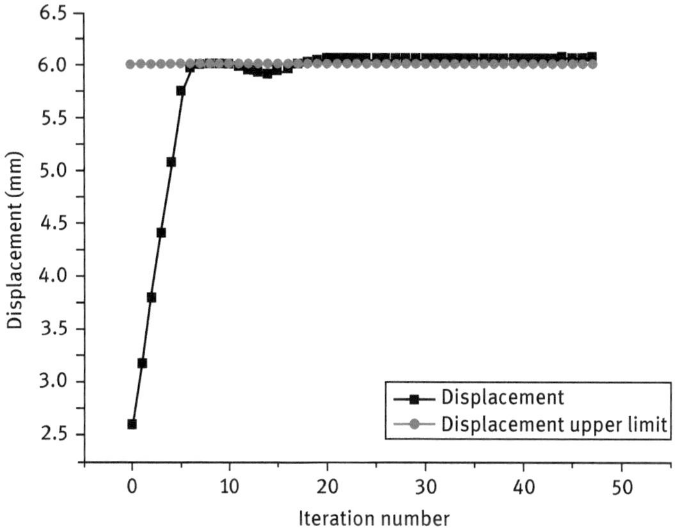

Figure 3.27: Iterative history of displacement constraint.

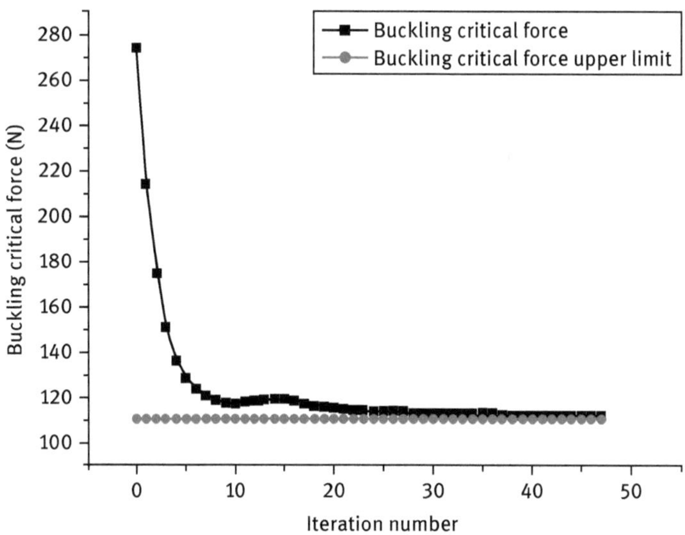

Figure 3.28: Iterative history of buckling critical load.

structure of the buckling restrained is larger than that of the displacement constraint, the thin rod is removed and the other parts are strengthened. At the same time, At the same time, the effect of buckling constraints and displacement constraints were considered, the fine rod was increased in the topological structure, but it is shorter, and it's the position has changed, the other parts have become rough

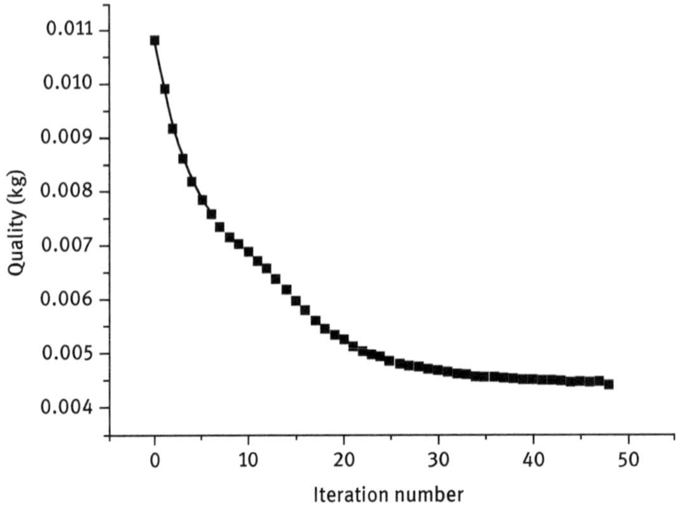

Figure 3.29: Iterative history of quality.

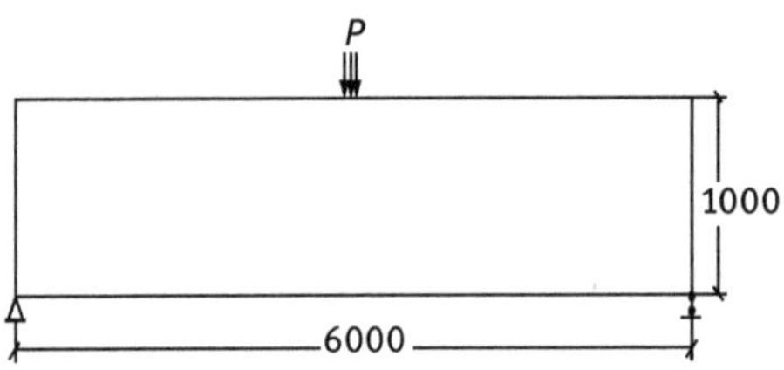

Figure 3.30: Base structure.

strengthened. This is consistent with the measures to improve the stability of the pressure bar in the mechanics of materials.

Example 4

As illustrated in Figure 3.30, the base structure is a plane elastic body with a size 6,000 × 1,000 × 10 mm^3. The elastic modulus of structure material is E = 2.1 × 10^5 MPa, Poisson's ratio is γ = 0.29 and the material density is $\rho = 7.8 \times 10^{-6}$ kg/mm^3. The lower left leg is fixed with a fixed hinge, and the lower right leg is a sliding hinge. A concentrated force P = 30,000 N along the up-to-down (-y) direction is located at the middle of the three points' upper bounds of the structure. A quadrilateral element with four nodes is used and the total number of elements is 109 × 36. The base structure's quality is 468.0 kg (Figure 3.31-3.37).

The vertical downward displacement of the center point is 0.917 mm. The buckling factor is 0.1805928. The initial critical load is 5,417.784 N. The upper limit of the displacement of the upper boundary midpoint is 1.9 mm. The upper limit maximal value of the critical load is 3,000 N. Convergence precision is 0.001 after 48 iteration.

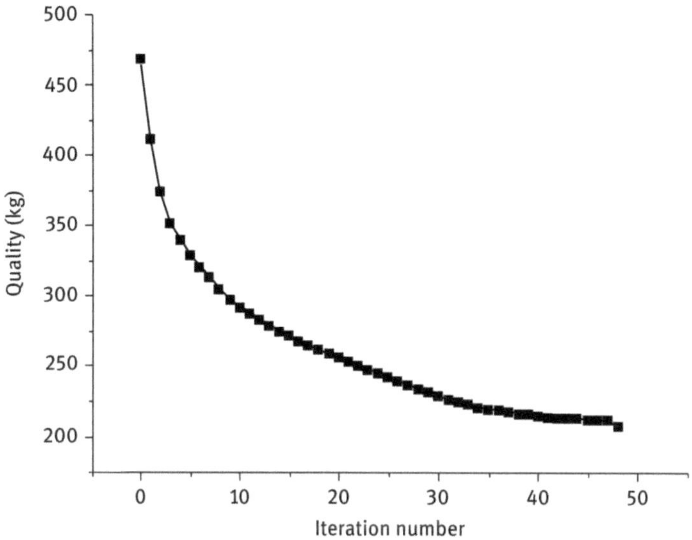

Figure 3.31: Iterative history of quality.

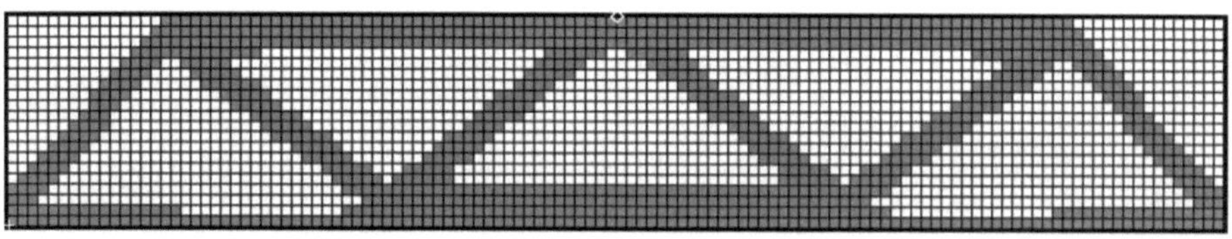

Figure 3.32: Optimal topological configuration with displacement constraint in ref. [62].

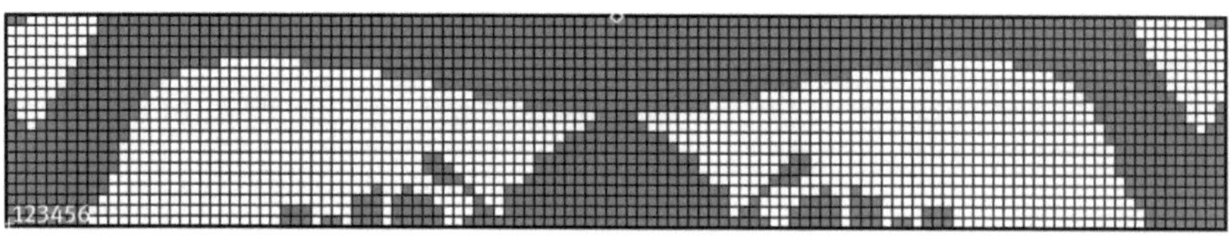

Figure 3.33: Optimal topological configuration with buckling constraint.

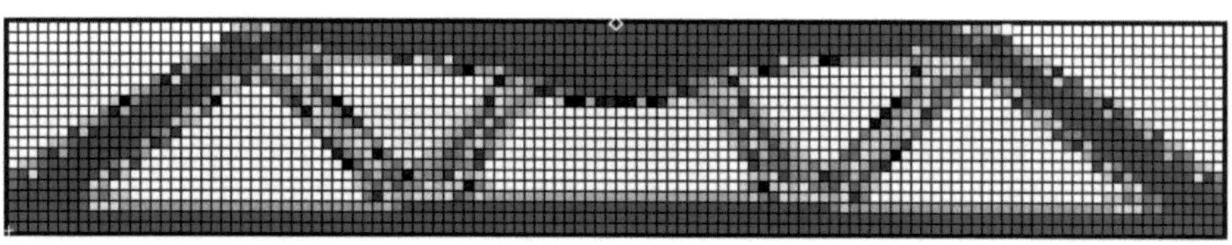

Figure 3.34: Topological configuration with buckling and displacement constraints before inversion.

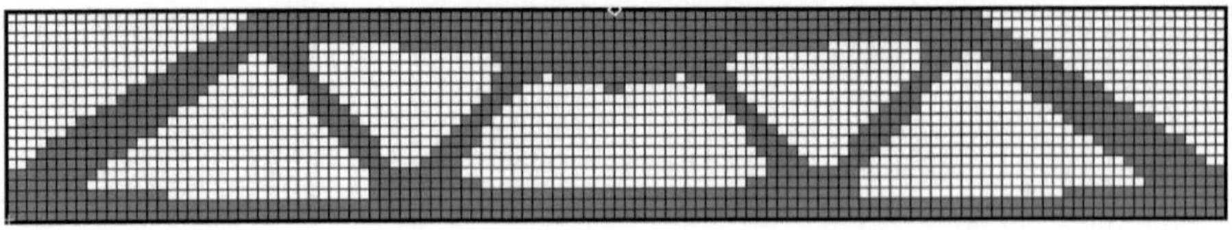

Figure 3.35: Topological configuration with buckling and displacement constraints after inversion.

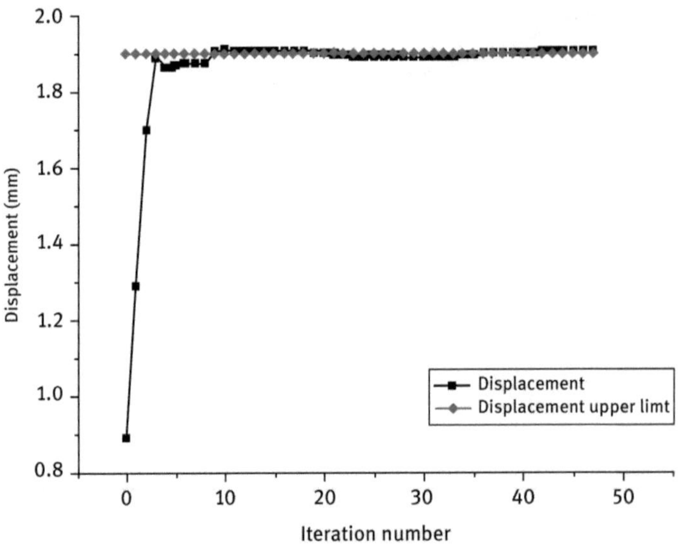

Figure 3.36: Iterative history of displacement constraint.

The critical force of the topological structure is 3,027.6 N. The quality of the optimized structure is 207.0 kg (table 3.7).

After comparing with the topological structure of the displacement constraint in literature [62], we can see that the topological structure is strengthened at the connection point, the width of the upper part of the whole structure is reduced and the stability is enhanced.

Example 5

As illustrated in Figure 3.38, the base structure is a plane elastic body with a size 520 × 260 × 6 mm^3. The elastic modulus of structure material is E = 68.89 GPa, Poisson's ratio is γ = 0.3 and the material density is $\rho = 2.7 \times 10^{-6}$ kg/mm^3. Fixed constraints on the left and right feet of the lower boundary. A concentrated force P = 21,000 N along the up-to-down (-y) direction is located at the midpoint of the lower boundary of the

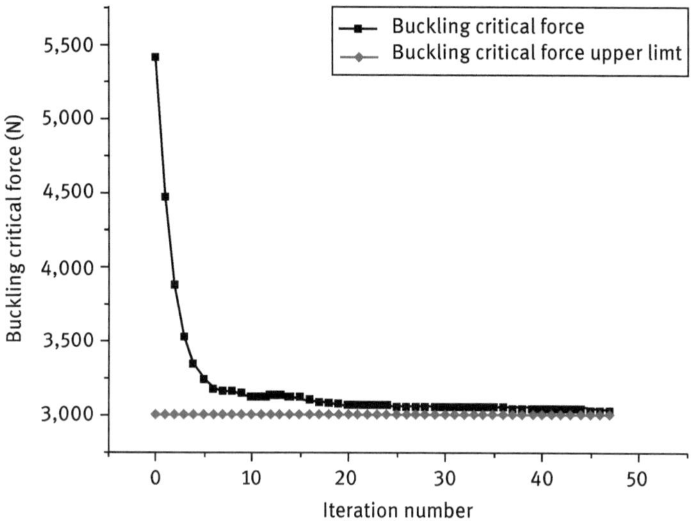

Figure 3.37: Iterative history of buckling critical load.

Table 3.7: Comparison of topological results with different constraints.

Constraint type	Initial quality (kg)	Optimized quality (kg)	Iteration number
Displacement	468.0	234.6	53
Buckling	468.0	190.2	54
Displacement + buckling	468.0	207.0	48

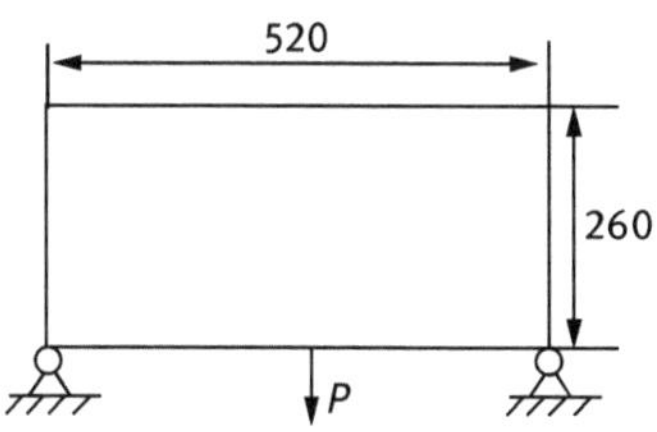

Figure 3.38: Base structure.

structure. A quadrilateral element with four nodes is used and the total number of elements is 52 × 26. The base structure's quality is 2.19 kg and convergence accuracy is 0.001.

The vertical downward displacement of the center point is 0.511 mm. The buckling factor is −0.3601822. The initial critical load is −7,563.78 N. The upper limit of the displacement of the upper boundary midpoint is 1.2 mm. The upper limit maximal value of the critical load is −5,000 N. After 51 iterations, the critical force of the topological structure is –5,018.16 N. The displacement of the topological structure is 1.213 mm. The quality of the optimized structure is 0.89 kg (Figure 3.39-3.45).

The first-order buckling factor of this problem is negative. In the buckling analysis, the negative eigenvalue implies that the force acting on the structure is in opposite direction to the direction of the force. There is a significant difference between the topological structure of the buckling constraint and the shape of the displacement constraint. The joint of the topological structure of buckling and displacement constraints is strengthened, and the stability of the structure is strengthened (Table 3.8).

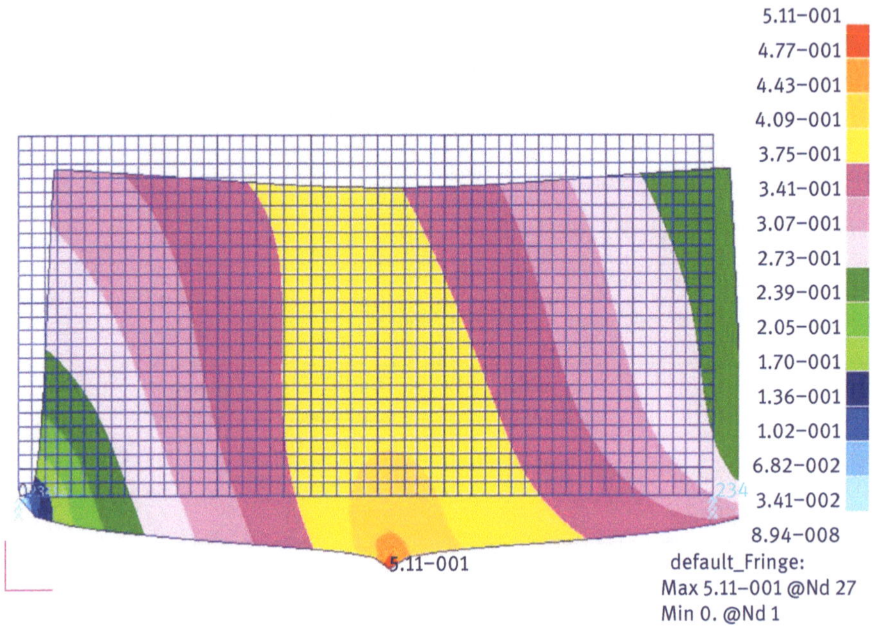

Figure 3.39: Displacement fringe result of the base structure.

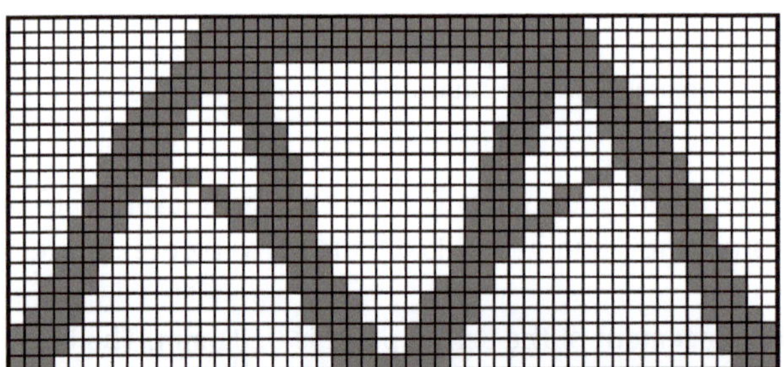

Figure 3.40: Optimal topological configuration with displacement constraint.

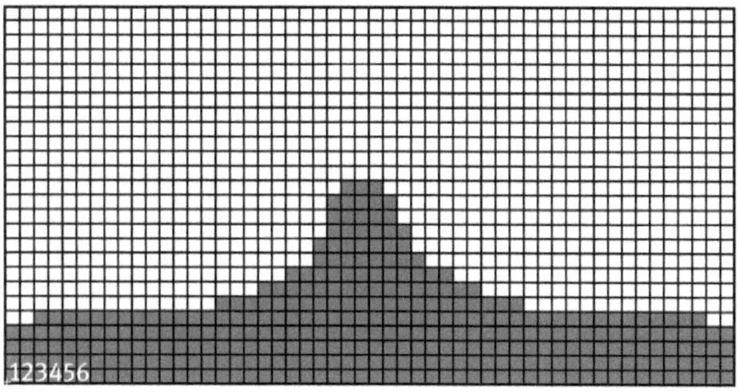

Figure 3.41: Optimal topological configuration with buckling constraint.

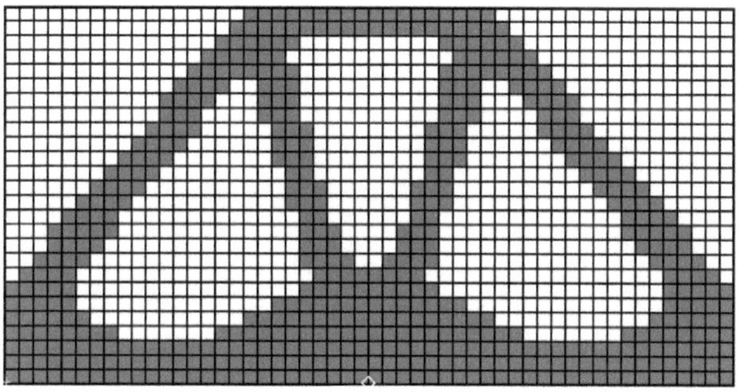

Figure 3.42: Optimal topological configuration with buckling and displacement constraints.

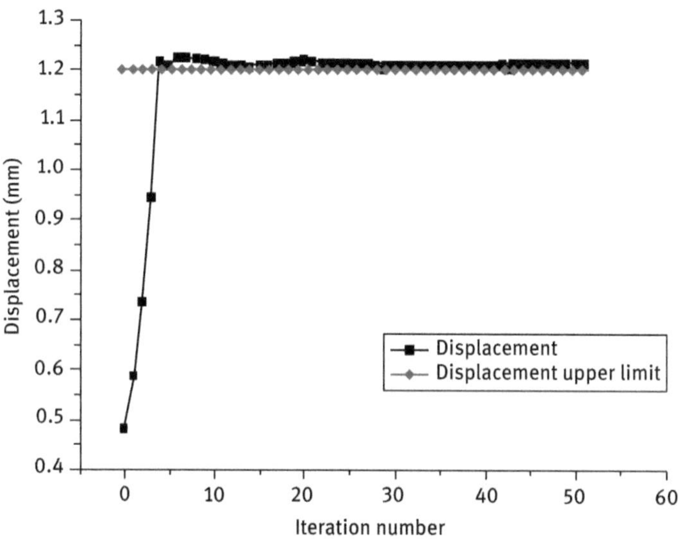

Figure 3.43: Iterative history of displacement constraint.

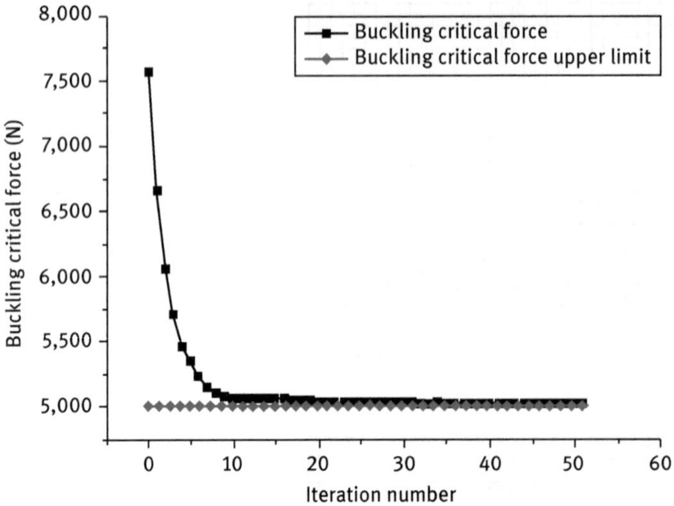

Figure 3.44: Iterative history of buckling critical load.

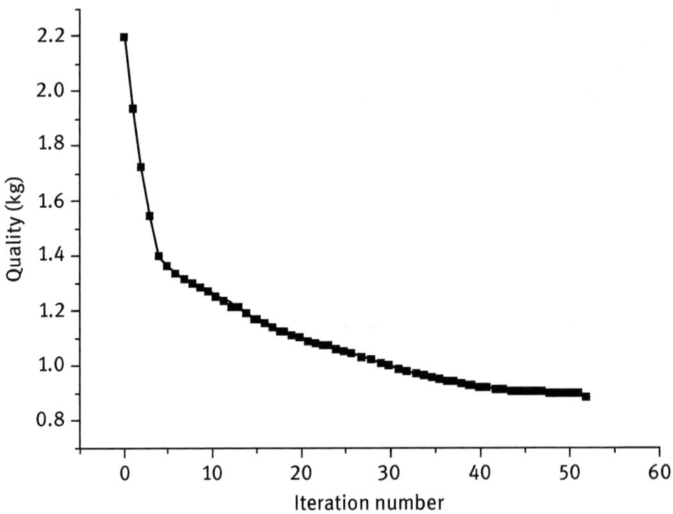

Figure 3.45: Iterative history of quality.

Table 3.8: Comparison of topological results with different constraints.

Constraint type	Initial quality (kg)	Optimized quality (kg)	Iteration number
Displacement	2.19	0.564	42
Buckling	2.19	0.80	49
Displacement + buckling	2.19	0.89	52

6. Example of MultiWorking Conditions

Example 1

As illustrated in Figure 3.46, the base structure is a plane elastic body with a size $600 \times 260 \times 2.0$ mm^3. The elastic modulus of structure material is E = 68,890 MPa, Poisson's ratio is $\gamma = 0.3$ and the material density is $\rho = 2.7 \times 10^{-6}$ kg/mm^3. The right-hand and left-hand underpart corner points of the structure are clasped. A quadrilateral element with four nodes is used and the total number of elements is 60×26. Condition 1: a concentrated force P1 = 15,000 N acts on the upper boundary of the midpoint. Condition 2: two concentrated forces P2 = 7,500 act on the upper boundary point whose distance is 220 mm to the midpoint. The base structure's quality is 0.84 kg. The first-order buckling force of the base structure is 112 N, upper critical buckling is $\bar{\lambda}_2$ = 60 N, second-order buckling critical force is 792 N and the upper critical force of buckling is $\bar{\lambda}_2$ = 400 N. The upper limit of the critical point displacement is 2.0 mm in working condition 1, and the upper limit of the displacement of the working condition 2 is 1.2 mm. The convergence precision is 0.001 (Figure 3.47 to Figure 3.59).

This mainly studies the topology optimization of continuum structures under the combined effects of buckling constraints and displacement constraints under single and multiworking conditions, as well as multimode and multiload conditions. As can be seen from the graph, the topological structure with buckling constraints is "robust," but higher-order constraint plays a major role when considering the multi-order buckling constraints (table 3.9).

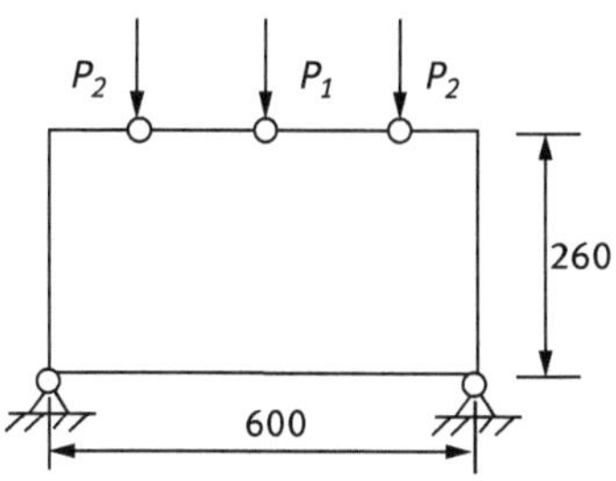

Figure 3.46: Base structure.

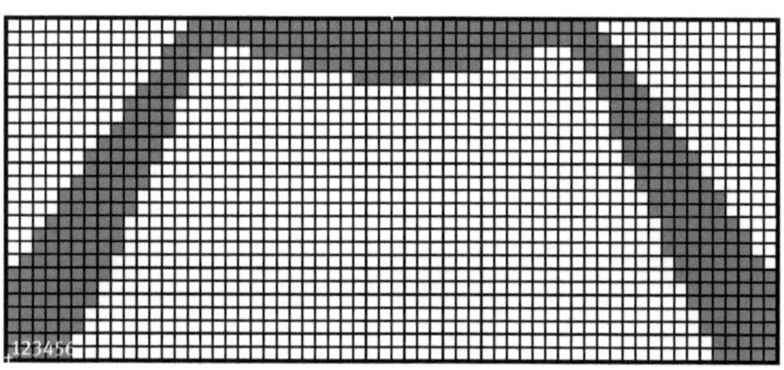

Figure 3.47: Optimal topological configuration of load case 1 with buckling constraints.

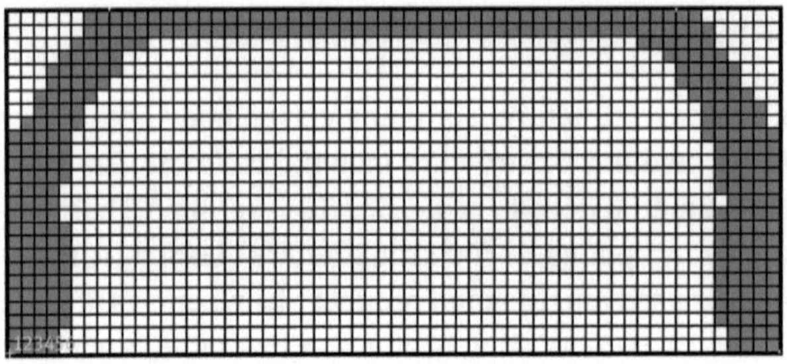

Figure 3.48: Optimal topological configuration of load case 2 with buckling constraints.

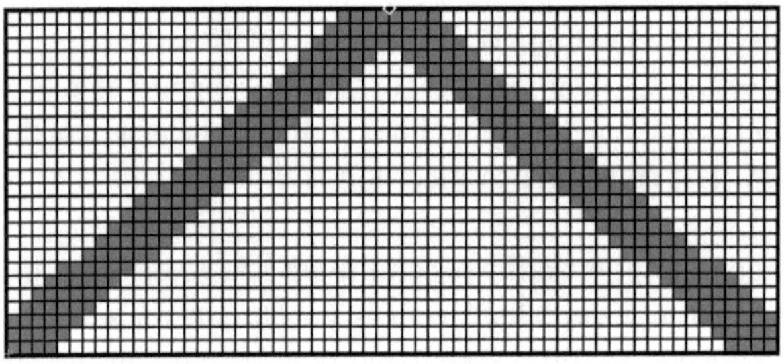

Figure 3.49: Topological configuration of load case 1 with displacement constraints.

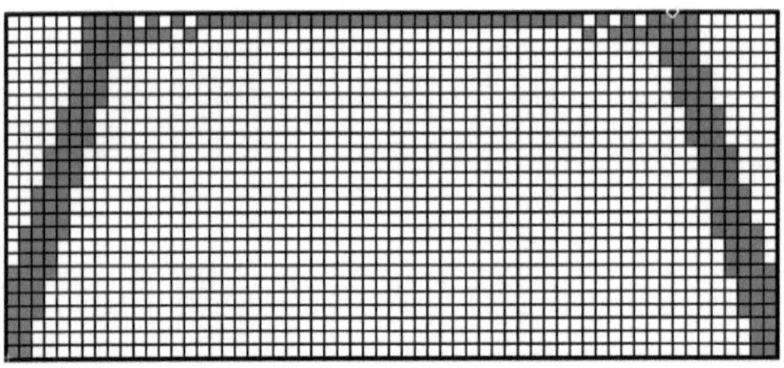

Figure 3.50: Topological configuration of load case 2 with displacement constraints.

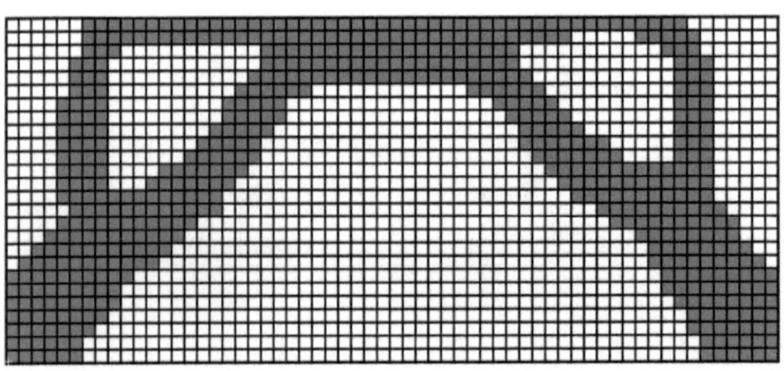

Figure 3.51: Topological configuration of multiple load cases with buckling constraints.

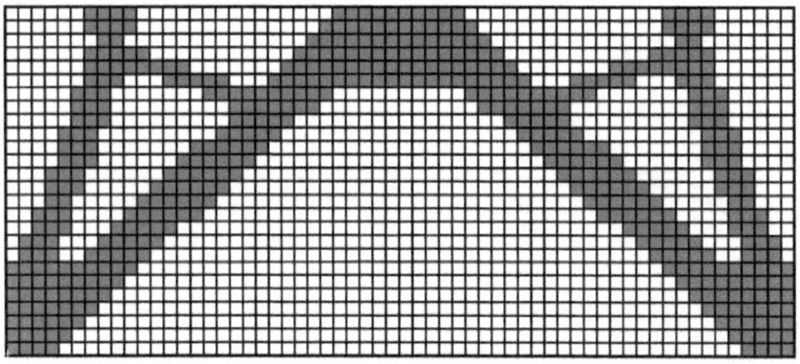

Figure 3.52: Topological configuration of multiple load cases with displacement constraints.

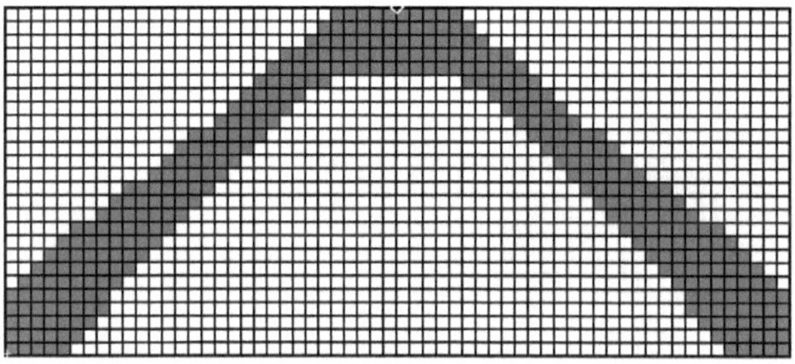

Figure 3.53: Topological configuration of load case 1 with buckling and displacement constraints.

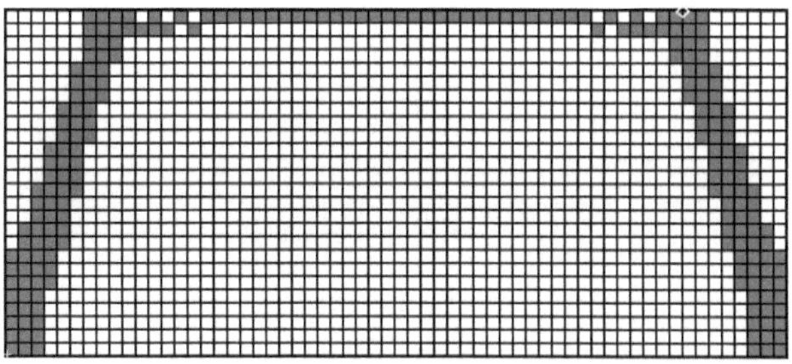

Figure 3.54: Topological configuration of load case 2 with buckling and displacement constraints.

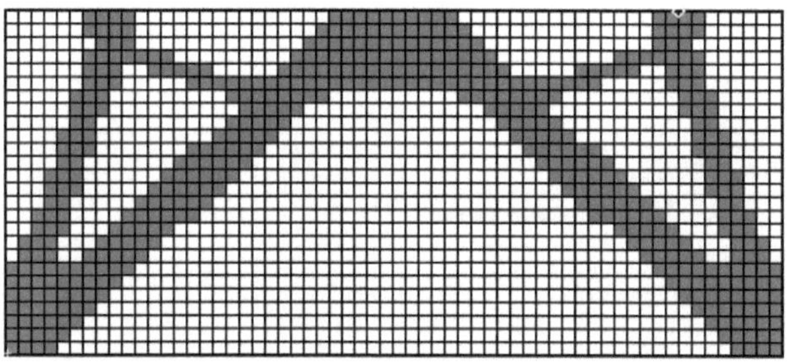

Figure 3.55: First-order mode topological configuration of multiple load cases with buckling and displacement constraints.

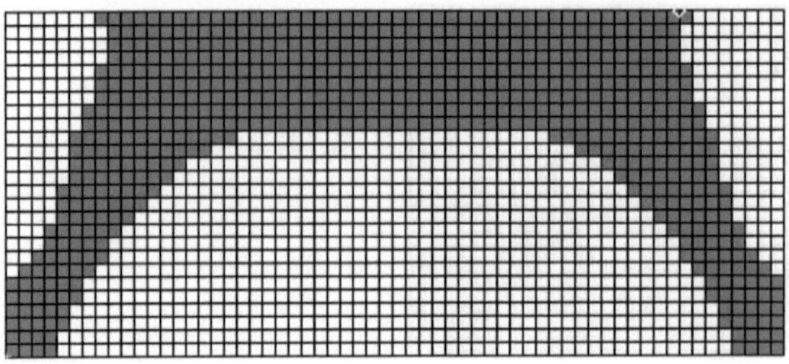

Figure 3.56: Second-order mode topological configuration with multiple load cases $\bar{\lambda}_2 = 400$ N.

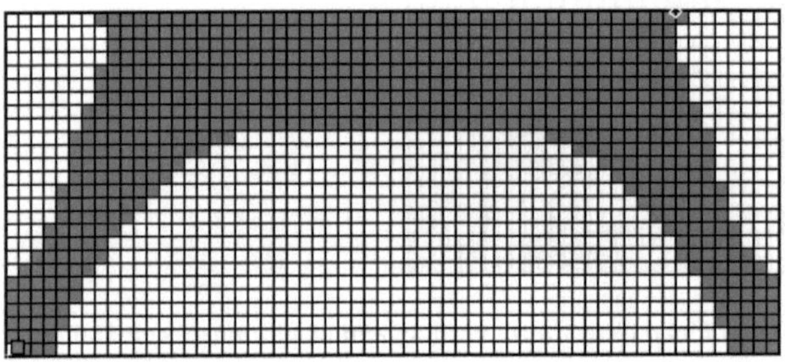

Figure 3.57: First- and second-order mode topological configuration with multiple load cases $\bar{\lambda}_1 = 60$ N, $\bar{\lambda}_2 = 400$ N.

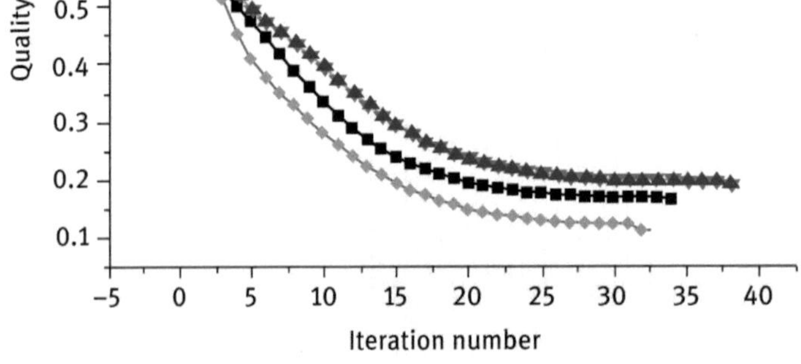

Figure 3.58: Iterative history of single load case quality.

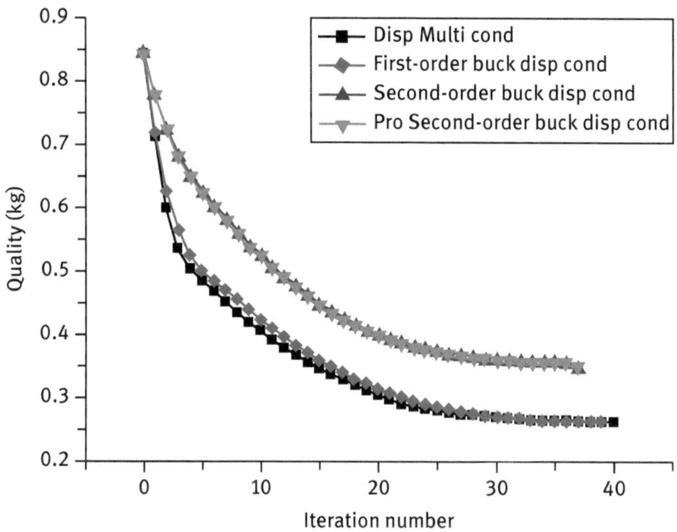

Figure 3.59: Different iterative history of multiple load cases quality.

Table 3.9: Comparison of topological results with different load cases.

constrain type	Initial quality (kg)	Optimized quality (kg)	Iteration number
Single working condition 1 + displacement	0.84	0.16	34
Single working condition 2 + displacement	0.84	0.11	32
Multiload condition + displacement	0.84	0.26	40
Working condition 1 + displacement + buckling	0.84	0.19	38
Working condition 2 + displacement + buckling	0.84	0.116	33
First-order buckling + displacement + multiload condition	0.84	0.258	39
Second-order buckling + displacement + multiload condition	0.84	0.347	38
The first- and second-order buckling + displacement + multiload condition	0.84	0.347	38

Example 2

As illustrated in Figure 3.60, the base structure is a plane elastic body with a size 120 × 60 × 6.0 mm^3. The elastic modulus of structure material is E = 210 GPa, Poisson's ratio is y = 0.25 and the material density is $\rho = 7.8 \times 10^{-6}$ kg/mm^3. Quadrilateral element with four nodes is used and the total number of elements is 60 × 30. There are two concentrated forces P = 2,000 acting on the upper and lower boundary corners; the left boundary has all the fixed constraints.

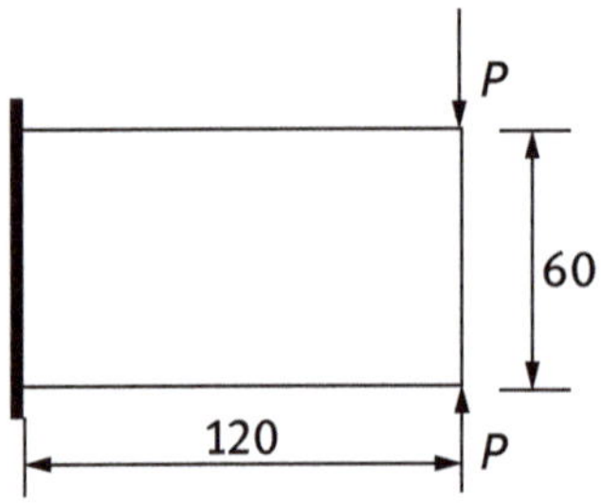

Figure 3.60: Base structure.

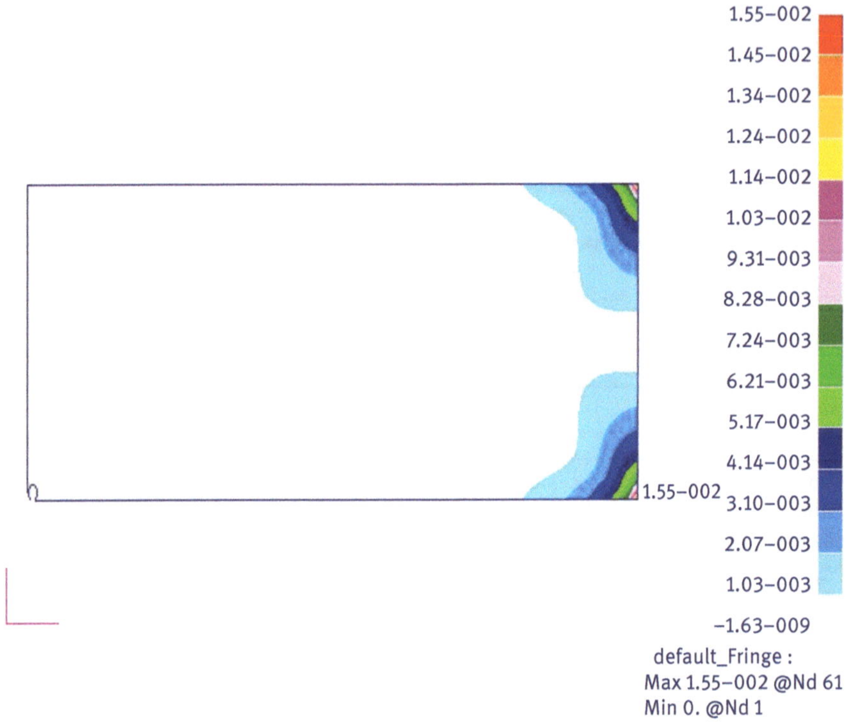

Figure 3.61: Displacement fringe result of the base structure with multiple load cases.

Condition 1: *P* acting on the upper corner point. Condition 2: *P* acting on the lower corner point. The base structure's quality is 0.337 kg. The first-order buckling critical force of the base structure is 67,066 N, and the single condition upper limit of the buckling critical force is 30,000 N. The key point of displacement is action point, and the displacement of the key point is 0.2 mm (Figure 3.61 to Figure 3.71). As can be seen from the graph, the topological structure of multiworking conditions is not a superposition of the single working condition structure (table 3.10).

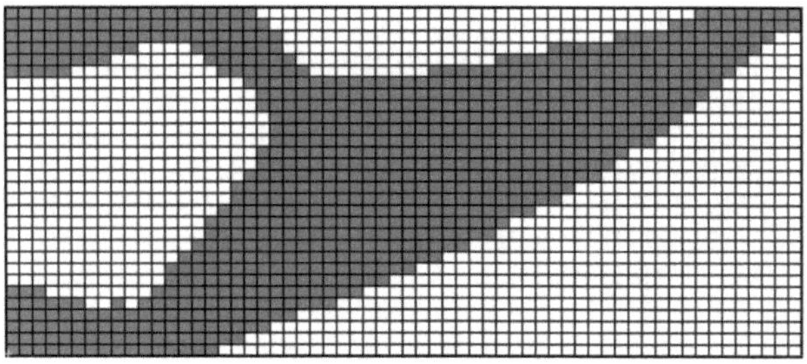

Figure 3.62: Optimal topological configuration of load case 1 with buckling constraints.

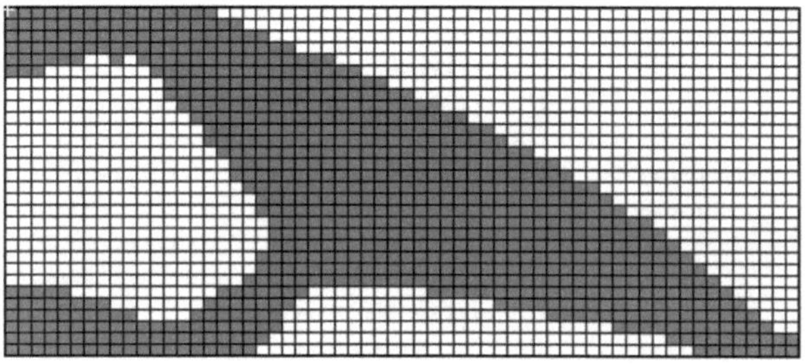

Figure 3.63: Optimal topological configuration of load case 2 with buckling constraints.

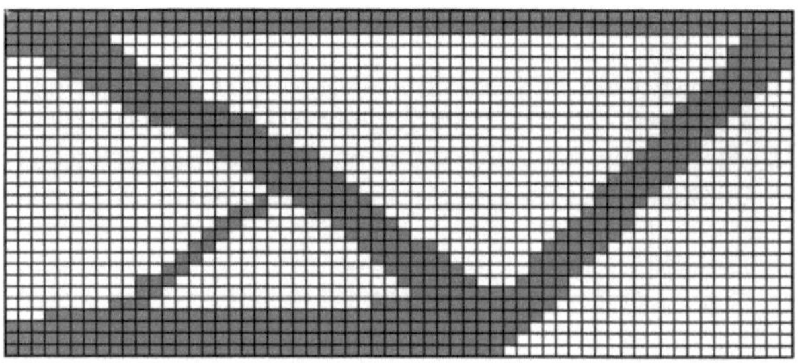

Figure 3.64: Topological configuration of load case 1 with single displacement constraints.

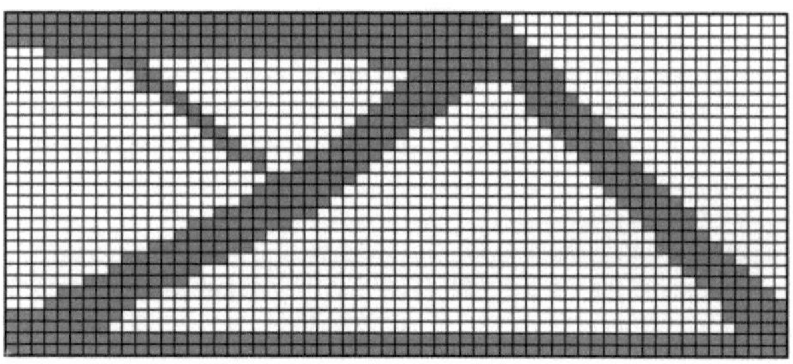

Figure 3.65: Topological configuration of load case 2 with single displacement constraints.

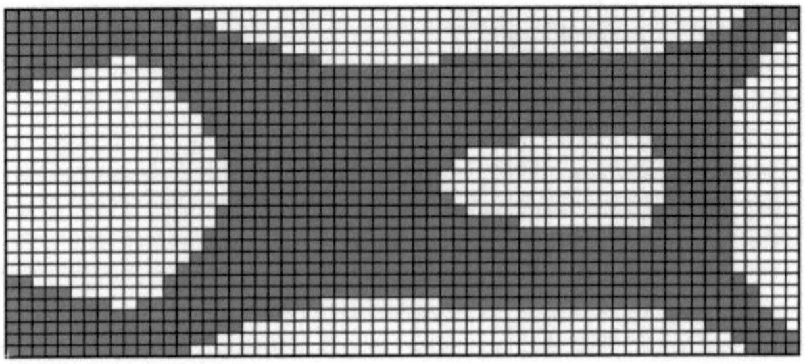

Figure 3.66: Topological configuration of multiple load cases with buckling constraints.

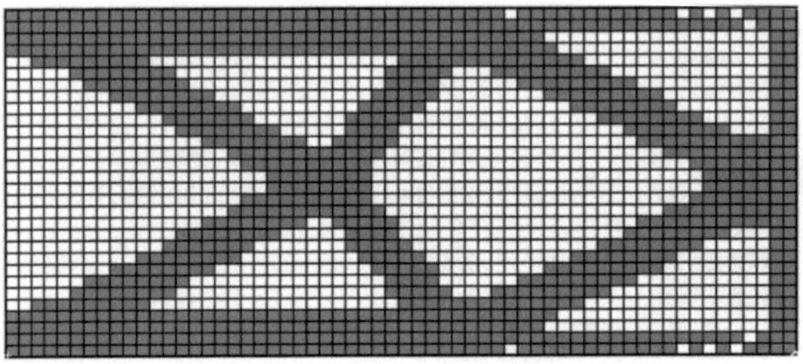

Figure 3.67: Topological configuration of multiple load cases with displacement constraints.

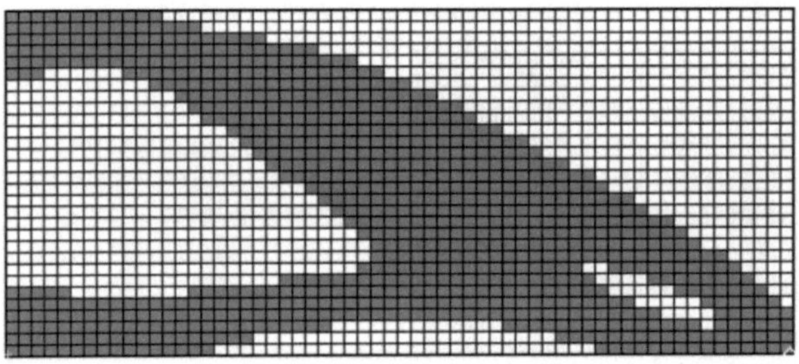

Figure 3.68: Topological configuration of single load case with buckling and displacement constraints.

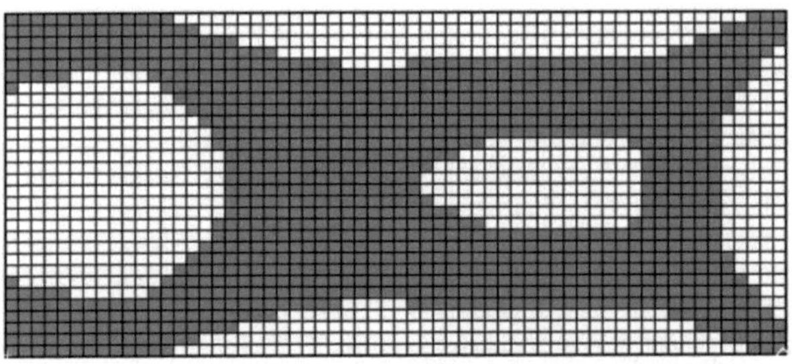

Figure 3.69: Topological configuration of multiple load cases with buckling and displacement constraints.

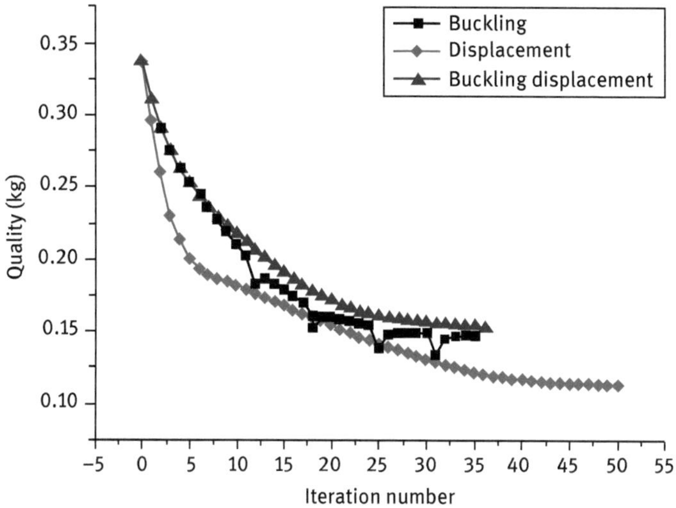

Figure 3.70: Iterative history of single load case quality.

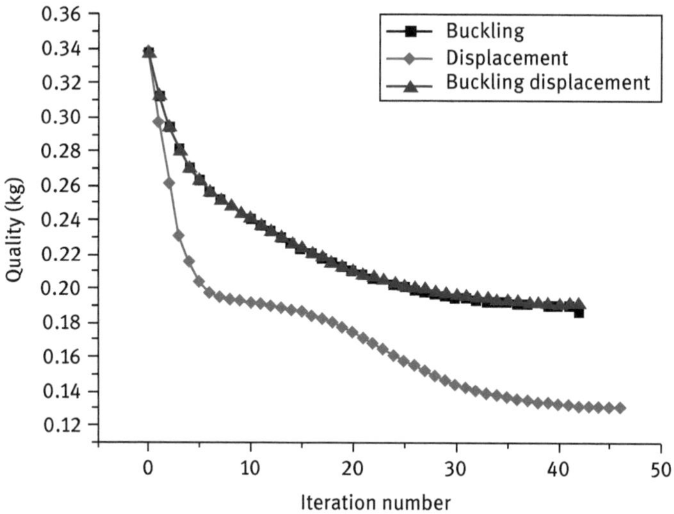

Figure 3.71: Different iterative history of multiple load cases quality.

3.6.2 Numerical Example of Three-dimensional Solid Structure

Example 1

As illustrated in Figure 3.72, the base structure is a 3D elastic body with a size $100 \times 100 \times 10$ mm^3. The elastic modulus of structure material is $E = 68.89$ GPa, Poisson's ratio is $y = 0.30$ and the material density is $\rho = 2.7 \times 10^{-6}$ kg/mm^3. A concentrated force

Table 3.10: Comparison of topological results with different load cases.

Constrain type	Initial quality (kg)	Optimized quality (kg)	Iteration number
Single condition + buckling	0.337	0.1464	35
Multiload condition + buckling	0.337	0.18511	42
Single working condition + displacement	0.337	0.11268	50
Multiload condition + displacement	0.337	0.1301	46
Single condition + buckling + displacement	0.337	0.15276	36
Multiload condition + buckling + displacement	0.337	0.19034	42

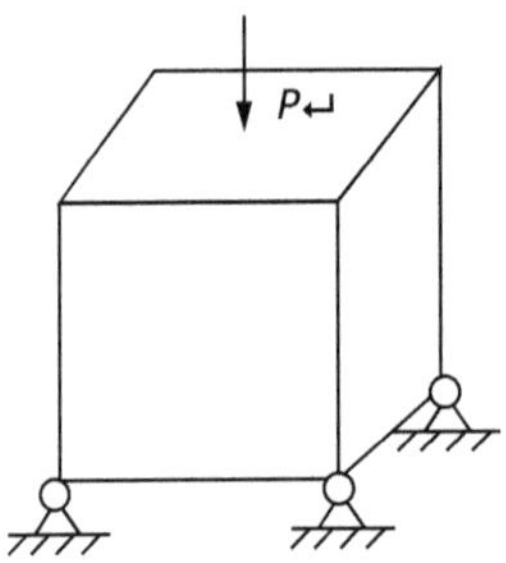

Figure 3.72: Base structure.

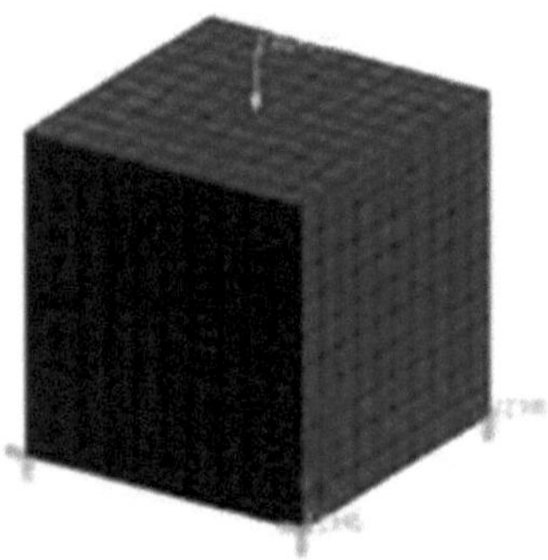

Figure 3.73: Finite Element model.

P = 20,000 N along the up-to-down (-y) direction is located in the upper surface midpoints of the structure. A fixed support is arranged at the four corners of the cube. A solid element with eight nodes is used and the total number of elements is 10 × 10 × 10. The base structure's quality is 2.7 kg (Figure 3.73 to Figure 3.81).

The center point of displacement is 0.104 mm, the buckling factor is 255.06 and the critical force is 5,101,200 N. In the topology optimization of the structure, the force acting point is chosen as the control point of displacement constraint. The upper limits of the displacement and of the buckling critical force are 0.25 mm and 3,000,000 N, respectively. After 15 times iterations, it is in convergence, the quality of the structure is 1.188 kg, the first-order buckling critical force is 2.51×10^6 N and the displacement of the topological structure is 0.253 mm (table 3.11).

Figure 3.74: Optimal topological configuration with displacement constraints.

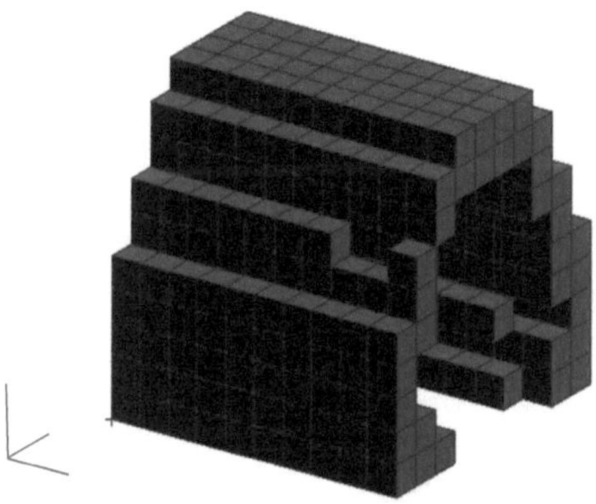

Figure 3.75: Optimal topological configuration with buckling constraints.

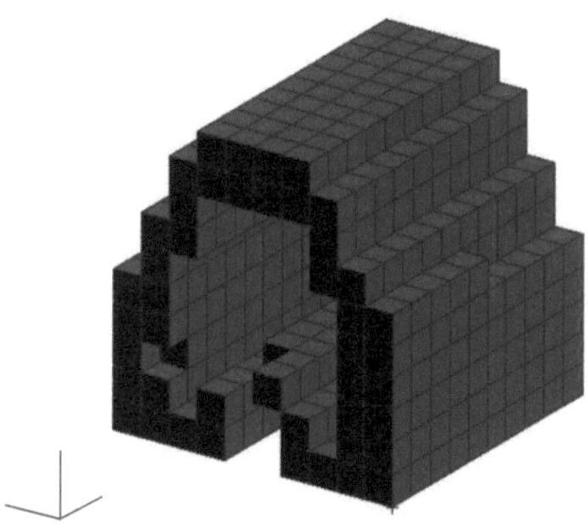

Figure 3.76: Optimal topological configuration with buckling and displacement constraints.

As can be seen from the graph, there is a significant difference between the topological structure of the displacement constraint and the buckling constraint. In the topology of a single constraint, the structure of the bottom is disconnected. In the topological structure of the displacement constraint and the buckling constraint, the

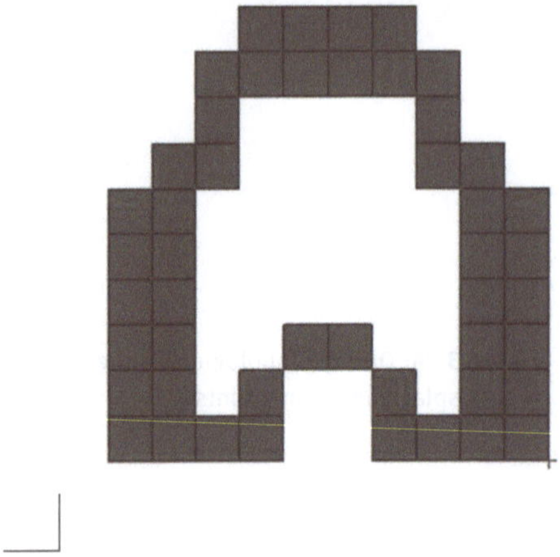

Figure 3.77: Topological configuration's elevation drawing with buckling and displacement constraints.

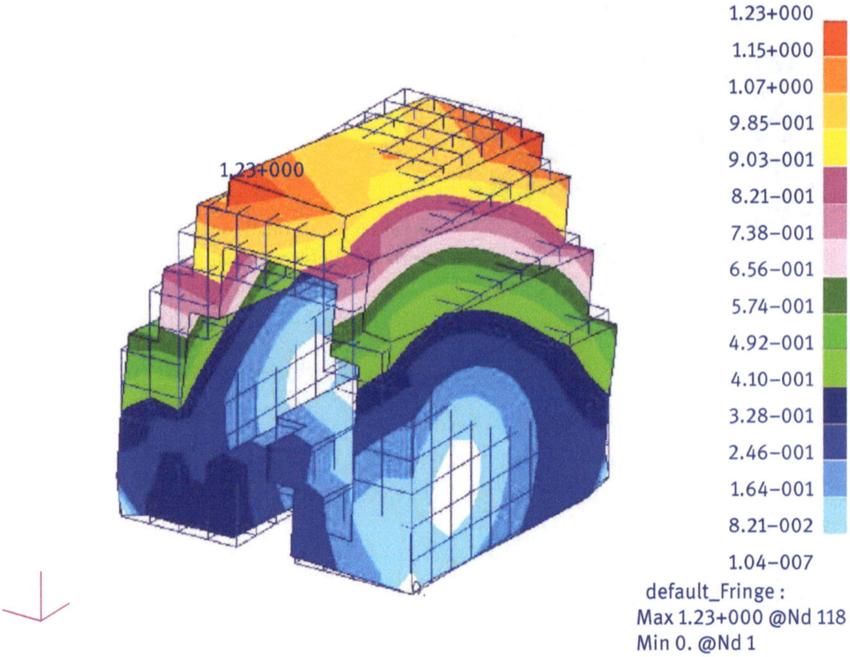

Figure 3.78: First-order buckling mode shape of topological configuration.

bottom structure is strengthened, and the structure is changed into continuous structure. The stability of the structure is strengthened.

Example 2

As illustrated in Figure 3.82, the base structure is a 3D elastic body with a size $120 \times 45 \times 8$ mm^3. The elastic modulus of structure material is $E = 70 \times 10^3$ MPa, Poisson's ratio is

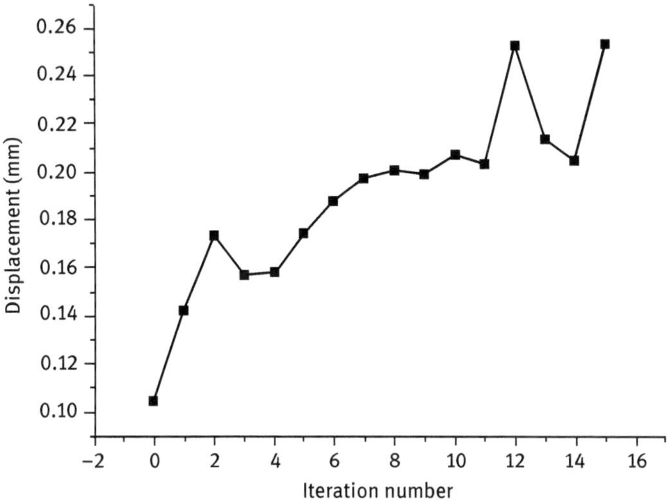

Figure 3.79: Iterative history of displacement constraints.

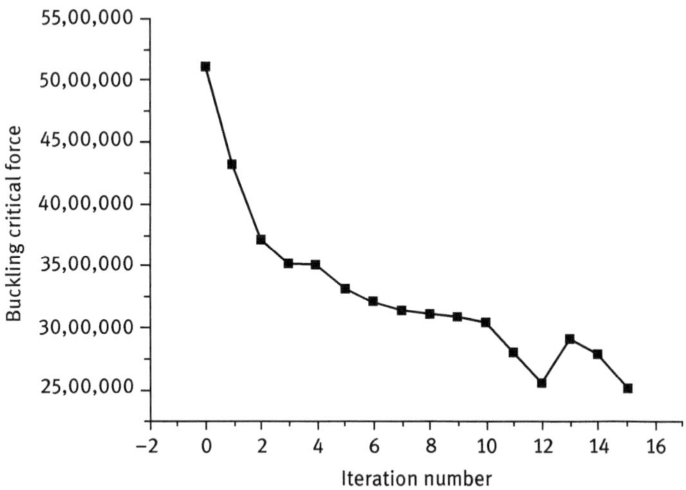

Figure 3.80: Iterative history of buckling critical load.

y = 0.30 and the material density is $\rho = 2.7 \times 10^{-6}$ kg/mm^3. A concentrated force P = 12,500 N along with the up-to-down (-y) direction is located in middle five points of the structure upper surface and in the lower left corner and lower right corner of the fixed support. A solid element with eight nodes is used and the total number of elements is 40 × 15 × 4. The base structure's quality is 0.11664 kg (Figure 3.83 to Figure 3.89).

The center point of displacement is 0.197 mm, the buckling factor is 2.7551 and the critical force is 34,438.75 N. In the topology optimization of structure, the force

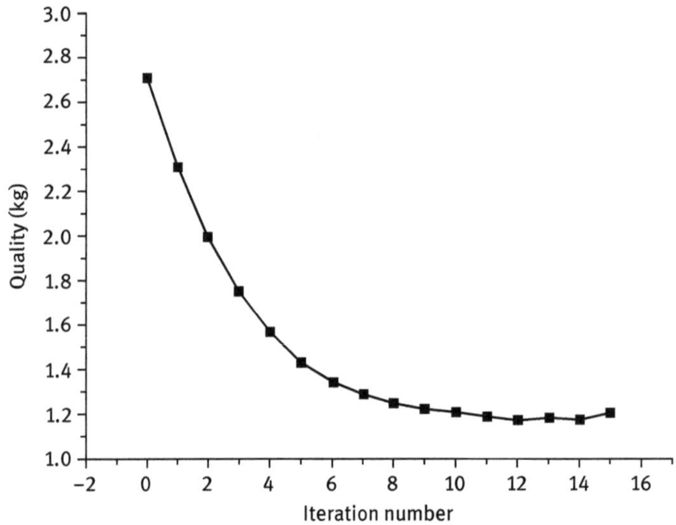

Figure 3.81: Iterative history of quality.

Table 3.11: Comparison of topological results with different constraints.

Constrain type	Initial quality (kg)	Optimized quality (kg)	Iteration number
Displacement	2.7	0.2344	36
Buckling	2.7	1.107	65
Displacement + buckling	2.7	1.188	15

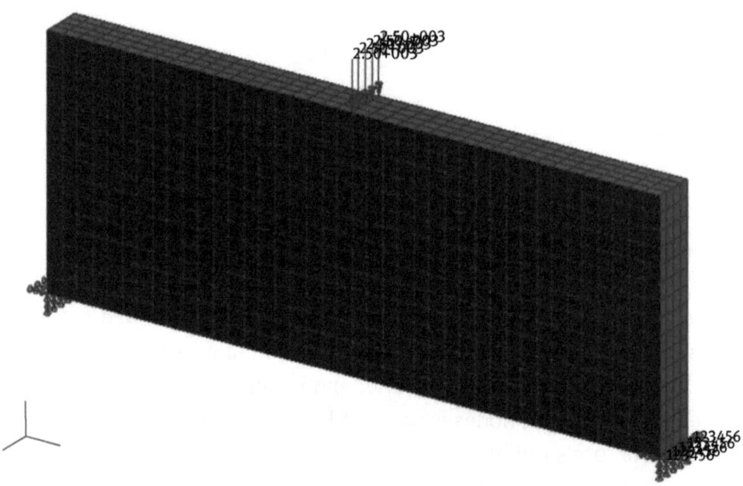

Figure 3.82: Base structure.

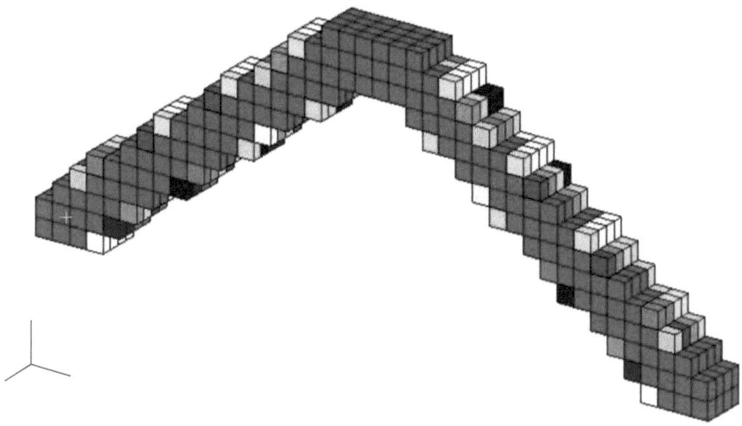

Figure 3.83: Optimal topological configuration with displacement constraints.

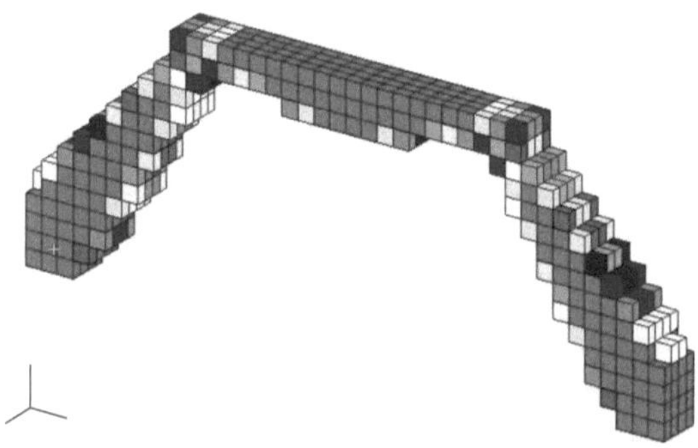

Figure 3.84: Optimal topological configuration with buckling constraints.

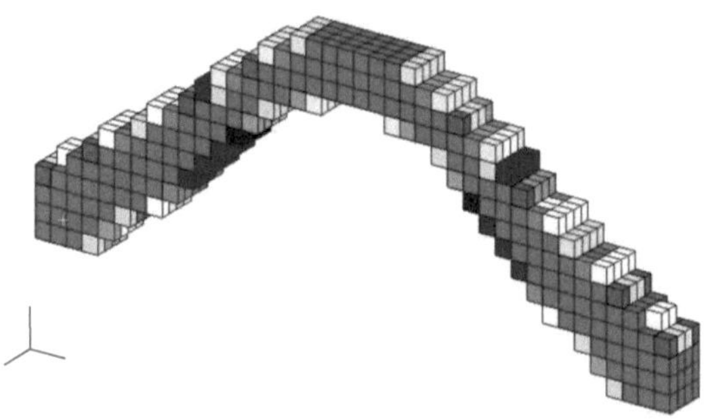

Figure 3.85: Optimal topological configuration with buckling and displacement constraints.

acting point is chosen as the control point of the displacement constraint. The displacement upper limit is 0.4 mm. The upper limit of the buckling critical force is 21,000 N. After 34 times iterations, it is in convergence. After optimization, the

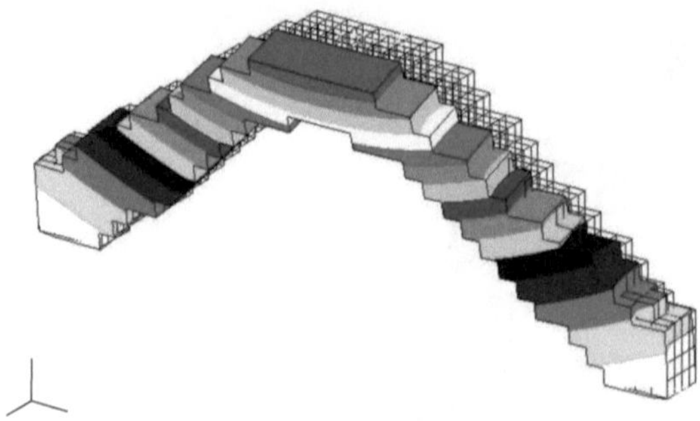

Figure 3.86: First-order buckling mode shape of topological configuration.

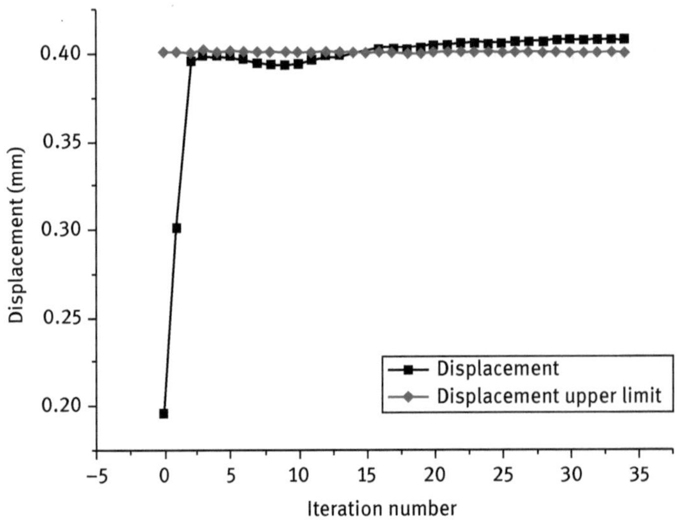

Figure 3.87: Iterative history of displacement constraints.

quality of the structure is 0.03632 kg, and the first-order buckling critical force is 21,102.5 N (table 3.12).

As can be seen from the graph, the topological structure of displacement constraint is relatively uniform. The top level of the topological structure under buckling constraints is relatively long, and the two ends of the horizontal section have a finer topology. The bottom of the topology of buckling constraints and displacement constraints is enhanced. Thus, strengthening the structure can improve the ability of the structure to resist buckling.

Example 3

As illustrated in Figure 3.90, the base structure is a 3D elastic body with a size 1,000 mm × 200 mm × 60 mm. The elastic modulus of structure material is $E = 210$

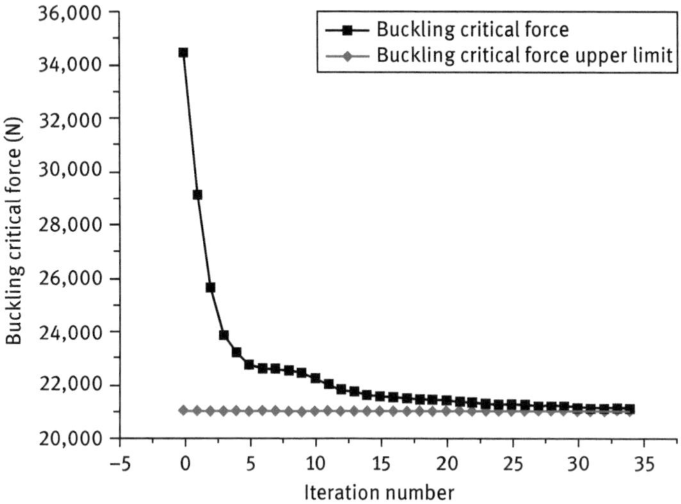

Figure 3.88: Iterative history of buckling critical load.

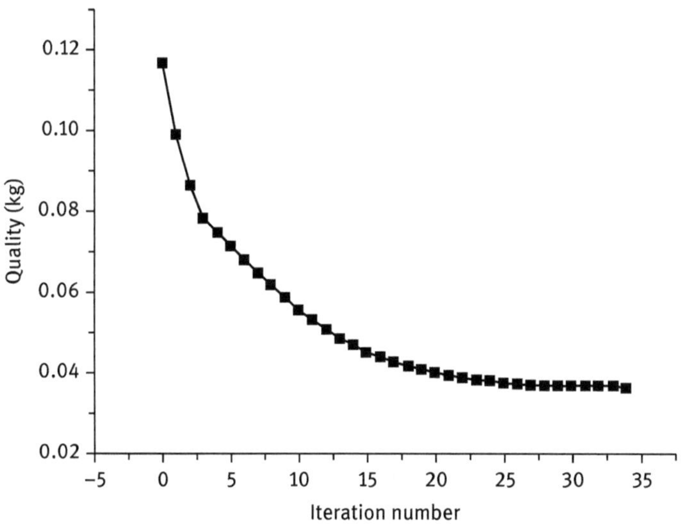

Figure 3.89: Iterative history of quality.

Table 3.12: Comparison of topological results with different constraints.

Constrain type	Initial quality (kg)	Optimized quality (kg)	Iteration number
Displacement	0.11664	0.0316	51
Buckling	0.11664	0.0332	37
Displacement + buckling	0.11664	0.0363	34

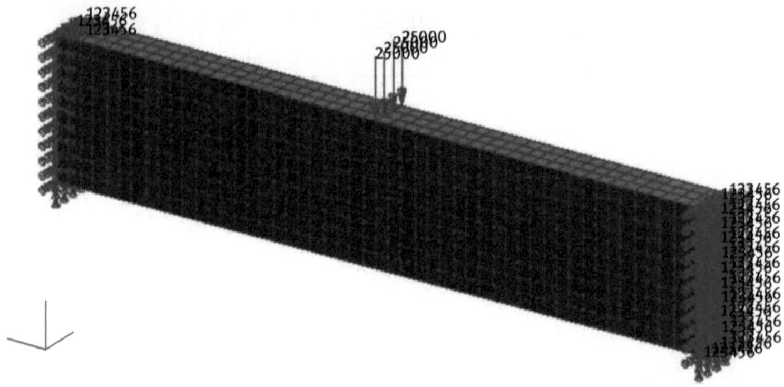

Figure 3.90: Base structure.

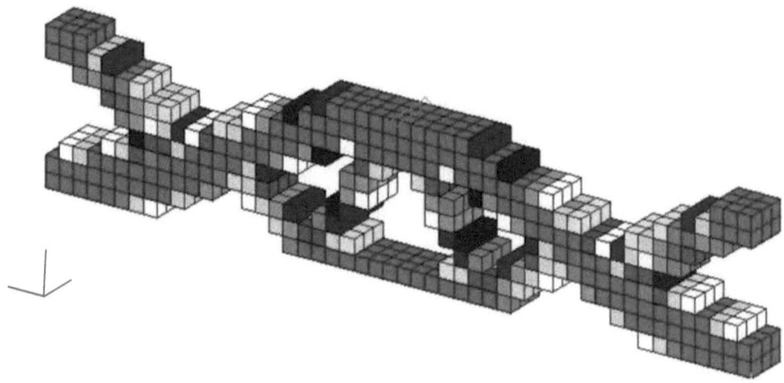

Figure 3.91: Optimal topological configuration with displacement constraints.

GPa, Poisson's ratio is $y = 0.30$ and the material density is $\rho = 7.8 \times 10^{-6}$ kg/mm^3. A concentrated force P = 100 kN along the up-to-down (-y) direction is located in upper surface's middle four points of the structure, and in the lower left corner and lower right corner of the fixed support. A solid element with eight nodes is used and the total number of elements is 50 × 10 × 3. The base structure's quality is 93.6 kg (Figure 3.91 to 3.97).

The initial displacement of the four nodes in the upper boundary is −0.11103 mm, -0.09807 mm, −0.11103 mm and −0.09807 mm, respectively. The upper limit of displacement is 0.225 mm. The buckling factor is 259.0961 and the critical force is 2.591×10^7 N. The upper limit of the buckling critical force is 1.29×10^7 N. After 31 times iterations, it is convergence. quality of the structure is 57.72 kg (table 3.13).

Example 4

As illustrated in Figure 3.98, the base structure is a 3D elastic body with a size 960 mm × 960 mm × 2,880 mm. The elastic modulus of structure material is E = 210 GPa, Poisson's

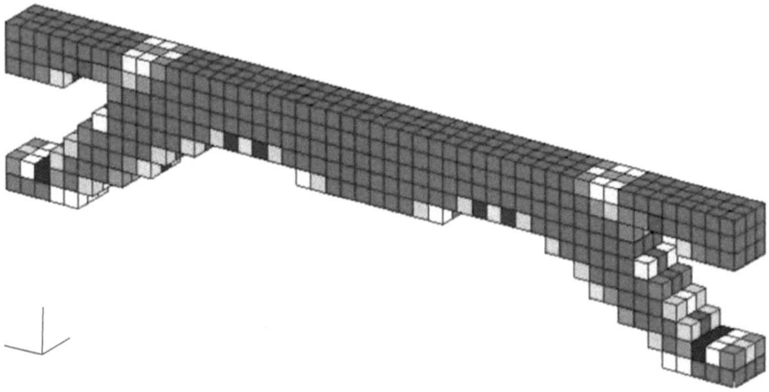

Figure 3.92: Optimal topological configuration with buckling constraints.

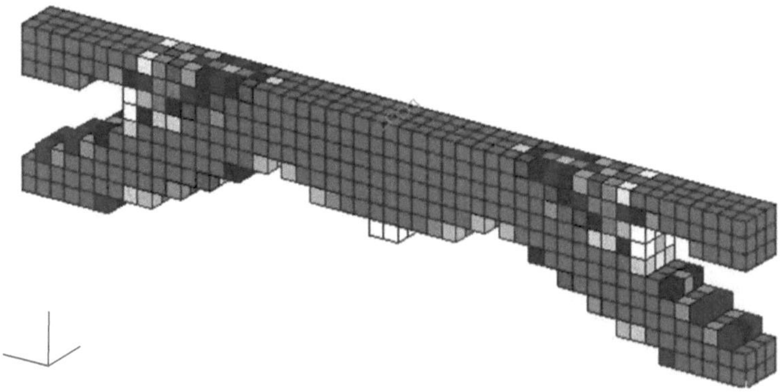

Figure 3.93: Optimal topological configuration with buckling and displacement constraints.

ratio is $\gamma = 0.30$ and the material density is $\rho = 7.8 \times 10^{-6}$ kg/mm^3. A concentrated force $P = 1.8 \times 10^7$ N along the up-to-down (-y) direction is located in upper surface's core points of the structure. To prevent the phenomenon of stress concentration, the force is distributed on the nine nodes in the middle of the above side.

The bottom four corners have fixed support. A solid element with eight nodes is used and the total number of elements is $8 \times 8 \times 36$. The base structure's quality is 20,701.2 kg.

The displacement upper limit is 3.1 mm. The buckling factor is 42.582 and the critical force is 7.665×10^8 N. The upper limit of the buckling critical force is 5.5×10^8 N. After 41 times iterations, it is in convergence. After optimization, the quality of the structure is 10,405.2 kg (table 3.14).

It can be seen from the figure 3.99 to figure 3.105 that the topological structure under the displacement constraint is different from the top of the topological structure under the buckling constraints. The topological structure of the bottom is

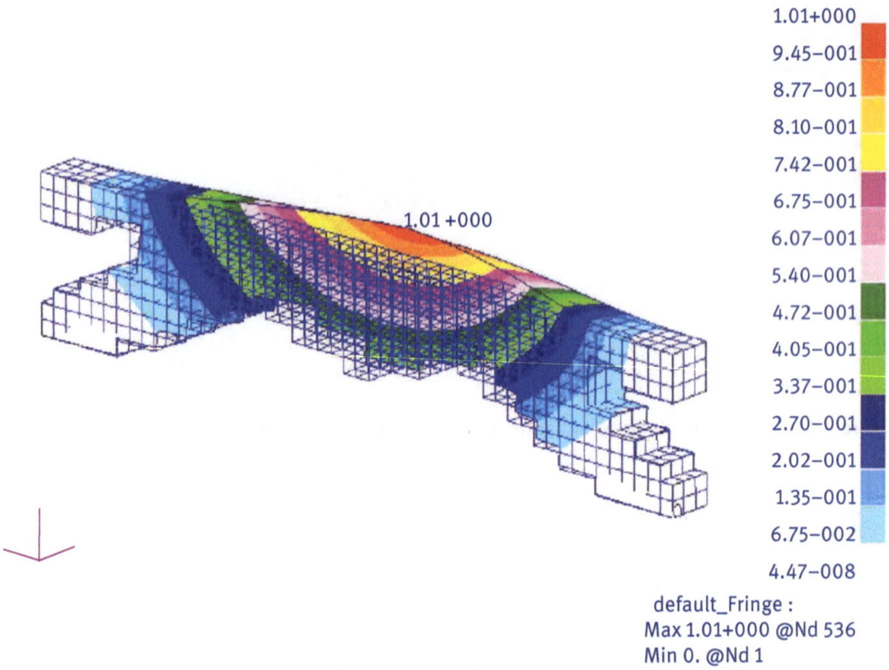

Figure 3.94: First-order buckling mode shape of topological configuration.

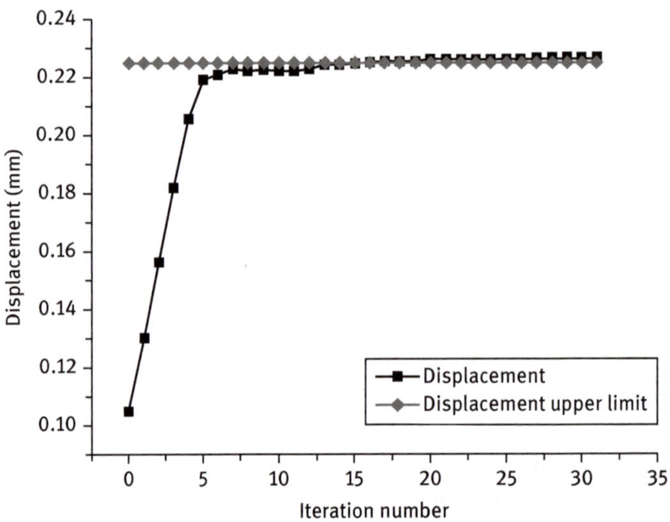

Figure 3.95: Iterative history of displacement constraints.

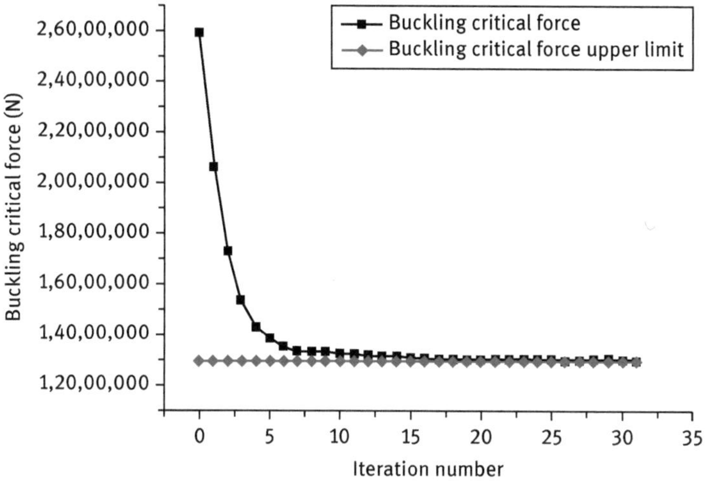

Figure 3.96: Iterative history of buckling critical load.

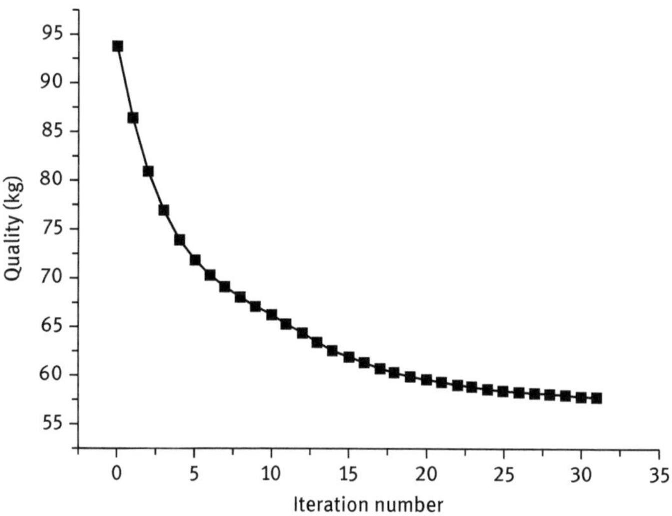

Figure 3.97: Iterative history of quality.

Table 3.13: Comparison of topological results with different constraints.

Constrain type	Initial quality (kg)	Optimized quality (kg)	Iteration number
Displacement	93.6	52.65	47
Buckling	93.6	49.14	32
Displacement + buckling	93.6	57.72	31

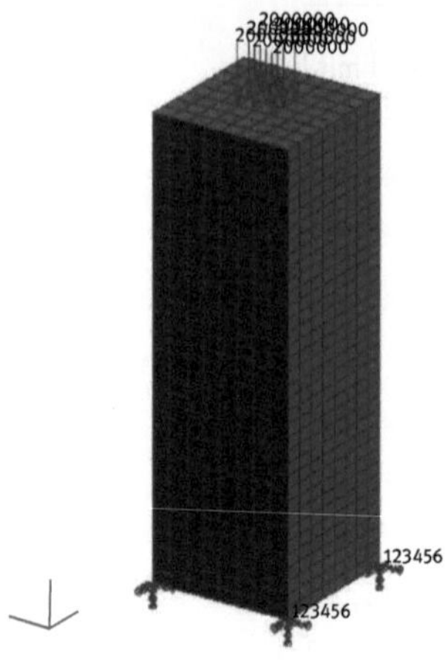

Figure 3.98: Base structure.

Table 3.14: Comparison of topological results with different constraints.

Constrain type	Initial quality (kg)	Optimized quality (kg)	Iteration number
Displacement	20,701.2	5,205	19
Buckling	20,701.2	10,686	33
Displacement + buckling	20,701.2	10405.2	41

Figure 3.99: Optimal topological configuration with displacement constraints.

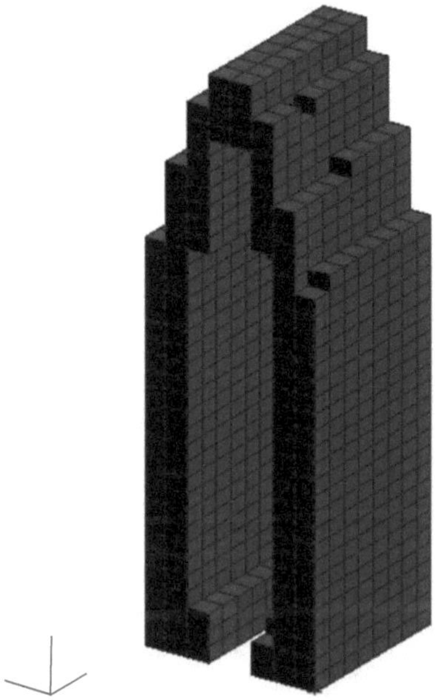

Figure 3.100: Optimal topological configuration with buckling constraints.

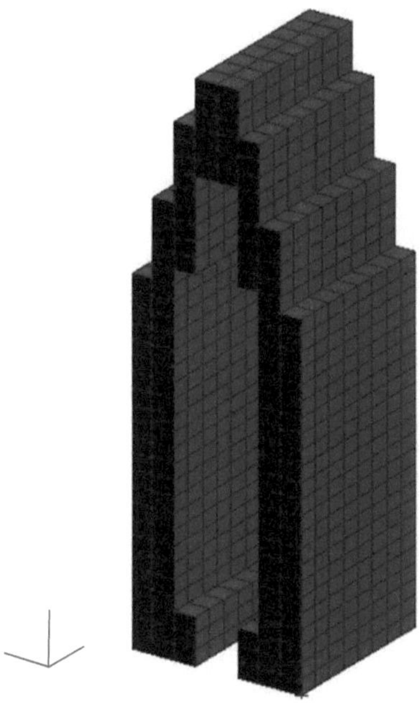

Figure 3.101: Topological configuration with buckling and displacement constraints.

basically the same. The topological structure with the joint action of buckling constraints and displacement constraints is the same as the topological structure of the displacement constraints, but the top thickness is strengthened, and the ability to resist buckling is enhanced.

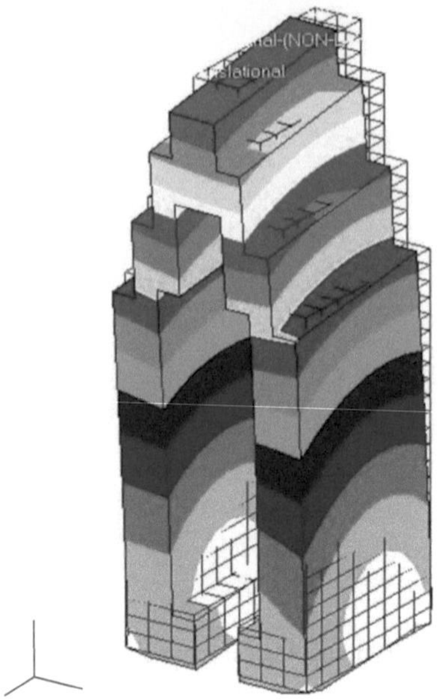

Figure 3.102: First-order buckling mode shape of topological configuration.

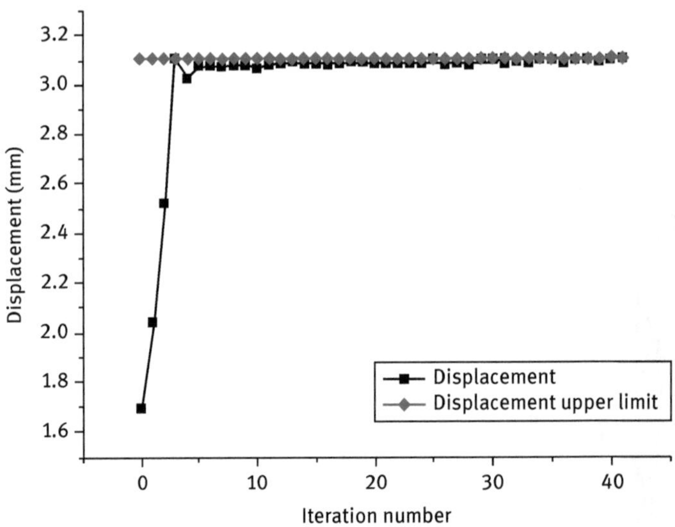

Figure 3.103: Iterative history of displacement constraints.

Example 5

As illustrated in Figure 3.106, the base structure is a 3D elastic column with a section radius 20 mm and a height 600 mm. The elastic modulus of structure material is E = 210 GPa, Poisson's ratio is y = 0.30 and the material density is ρ = 7.8×10^{-6} kg/mm^3. A concentrated force P = 20,000 N along the up-to-down

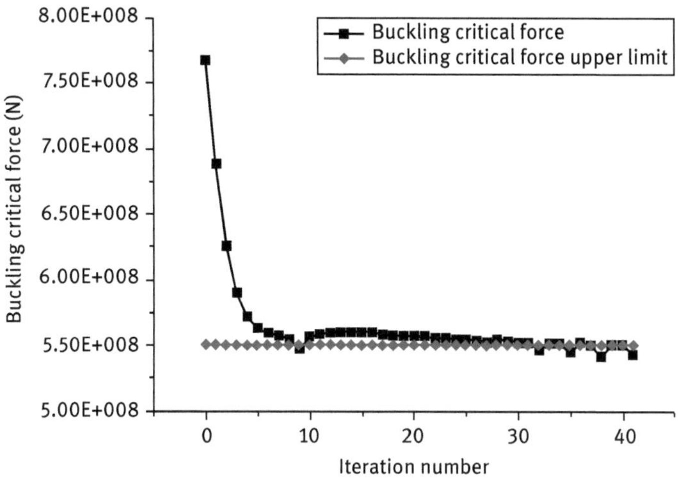

Figure 3.104: Iterative history of buckling critical load.

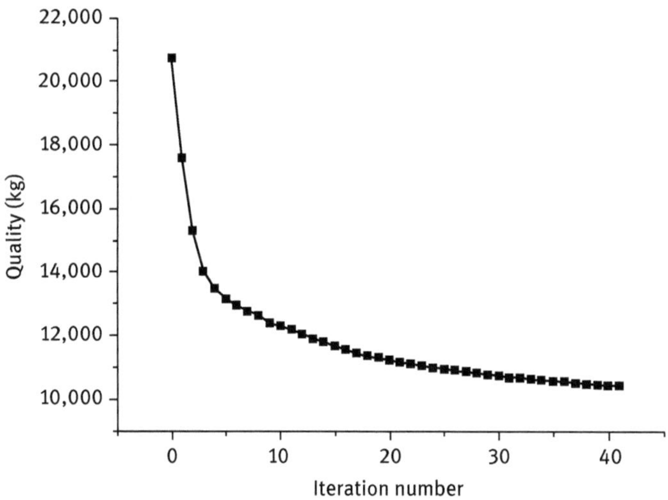

Figure 3.105: Iterative history of quality.

(-y) direction is located in the upper surface core points of the structure. The four ends of the two diameters of the lower boundary are fixed and supported. A solid element with eight nodes is used and the total number of elements is 2,560. The base structure's quality is 5.733 kg.

The initial displacement of the force acting point of the structure is 0.217 mm. The displacement upper limit is 0.45 mm. The buckling factor is 5.97 and the critical

Figure 3.106: Base structure.

force is 134,390 N. The upper limit of the buckling critical force is 70,000 N. After 8 iterations, it is in convergence. After optimization, the quality of the structure is 4.423 kg (Figure 3.107 to Figure 3.113).

The optimal topology model of the first-order buckling constraint of the circular cylinder is given, and the deformation of the buckling mode is also given. As can be seen from the displacement constraint topology graph, the cross-sectional area of the bar will be reduced with an increase in the displacement constraint. In the topological graph of buckling constraints, the bottom of the support is not removed, the displacement is small, the top is reduced and the deformation is increased gradually (table 3.15).

Figure 3.107: Optimal topological configuration with displacement constraints.

Figure 3.108: Optimal topological configuration with buckling constraints.

Figure 3.109: Topological configuration with buckling and displacement constraints.

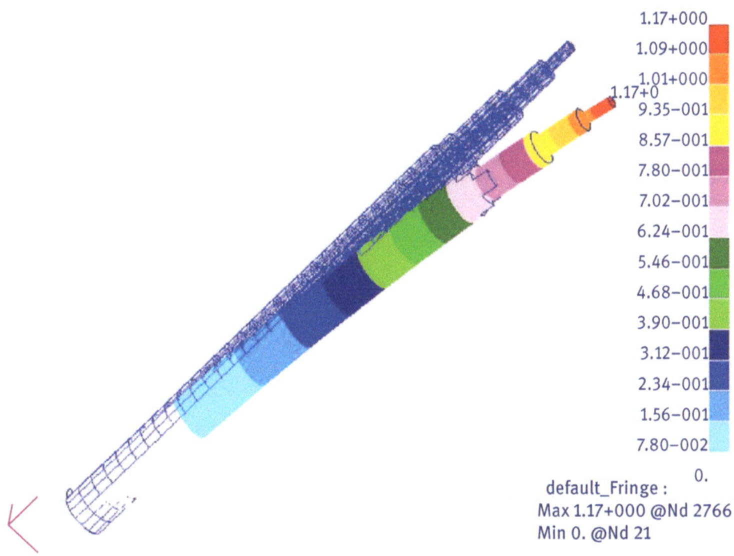

Figure 3.110: First-order buckling mode shape of topological configuration.

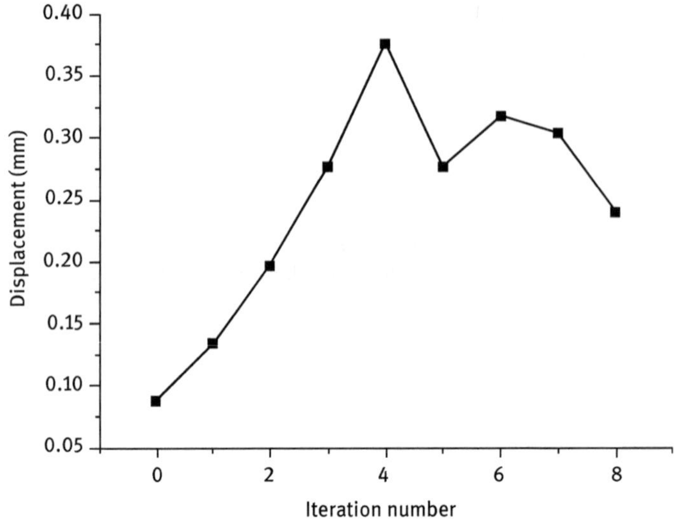

Figure 3.111: Iterative history of displacement constraints.

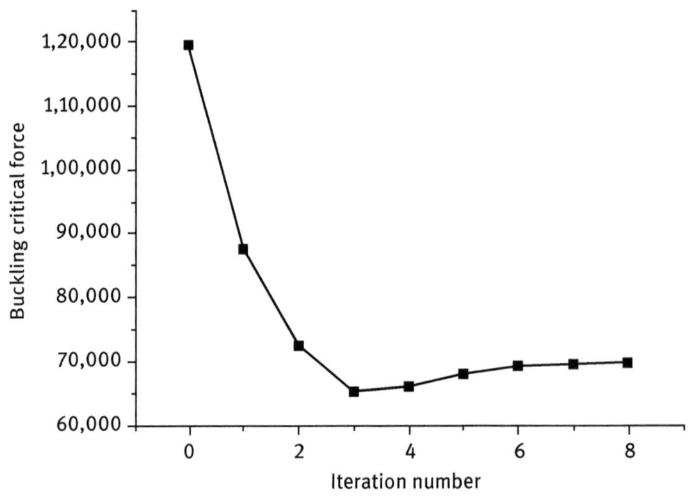

Figure 3.112: Iterative history of buckling critical load.

3.7 Summary

In this chapter, according to the ICM method, the topology optimization model for the continuum structure was constructed. The model had the minimized quality as the objective function was subjected to the buckling constraints and key point displacement constraints. This is mainly because the structural design not only considers the buckling constraints, but also considers the problem of structural stiffness, and the stiffness

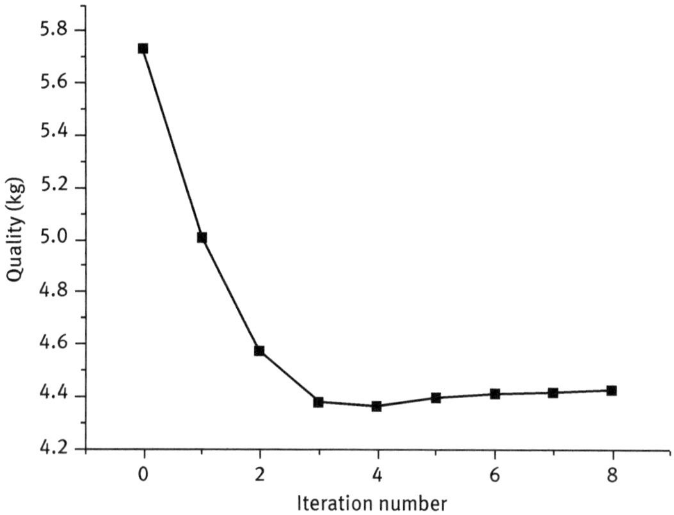

Figure 3.113: Iterative history of quality.

Table 3.15: Comparison of topological results with different constraints.

constrain type	Initial quality (kg)	Optimized quality (kg)	Iteration number
Displacement	5.733	2.556	23
Buckling	5.733	4.368	10
Displacement + buckling	5.733	4.423	8

constraint problem is called the structural displacement constraint problem. That is called the minimum quality structure under the displacement constraints condition.

The concept of "displacement constraint ratio" was defined. The "quality ratio" and "displacement constraints ratio" function curves of the topology optimization structure were established, and the "displacement–quality constraint ratio" curve was fitted by the least squares method. The "quality–buckling constraint ratio" is compared with that of the "quality–displacement constraint ratio," which provides a functional relationship for the selection of buckling constraints and displacement constraints.

To solve the optimization model, the objective function and the buckling critical force constraints are approximated, and the displacement constraints are transformed into explicit constraints of the design variables by using Moore's theorem and virtual work principle. The reasonable topology and clear topology structure are obtained by solving the dual programming. Finally, the method is applied to 2D and 3D continuum structure topology optimization problem with single load and multi-load operating conditions. The effectiveness and feasibility of this method are indicated by the results of a number of examples.

4 Topological Optimization for Continuous Structure with Buckling and Stress Constraints

4.1 Introduction

There are two kinds of constraints' state quantities in the structural optimum. These are global state quantity and local state quantity. The former mainly includes displacement, frequency, and global stability. The latter mainly includes stress and local stability. At present, the majority of studies on topological optimization of the continuum structure are focused on the global state quantity (such as compliance, displacement and frequency) which is a constraint or an object function [13]. However, the local state quantity is seldom. In addition, the topological optimization of the continuum structure with buckling constraints and globalization stress constraints is seldom either. In engineering, the structure strength is an aspect that must be considered, and it is also the basic requirement of the structure. The stress is a basic parameter in structural design, and the stability of the structure is also an important aspect. Therefore, it is significant to study the topology optimization with buckling constraints and stress constraints.

In the topological optimization of the continuum structure with stress constraints, each design variable corresponds to one element. Therefore, the number of constraints is large. The analysis quantity of the sensitivity is large. The model is inefficient. At present, the stress constraints are solved by the full stress method, that is to say the zero- or one-order approximation method, but the analysis quantity of the sensitivity is not decreased. The study by Gengdong Cheng and Wang Jian et al. reported the topology optimization of a plane elastic body and a thin plate structure with the modified full stress method. The stress constraint problem was dealt with the bidirectional evolutionary structural optimization method (BESO) by Guan. The low stress element is removed, and the high stress element is added, which is essentially based on the full stress method. The studies by Jianhua Rong et al. [6] reported that the structure is optimized by using the BESO method which is based on the stress sensitivity analysis.

How to solve the above problems? In order to reduce the number of constraints of the model, the local stress constraints are transformed into the global stress constraints by an innovate idea. Yunkang Sui [74–80] proposed a stress global strategy that the global state variables (such as flexibility, displacement, and frequency) as the constraint or objective function have advantages. Using the fourth strength theory, the local stress constraints are transformed into the whole strain energy constraints of the structure.

In this chapter, the topology optimization of continuum structures with minimum quality object function and stress buckling constraints was established, based

https://doi.org/10.1515/9783110462272-004

on the ICM method. Using the idea of global change, the local stress constraint is transformed into the global strain energy constraint problem. Using the Rayleigh quotient, the Taylor expansion and the filter function, the stability constraint will be bound to approximate explicit methods, to avoid the sensitivity of the calculation. Finally, the optimization model is transformed into dual programming, and the number of design variables is reduced, the solving scale of the model is reduced and the computational efficiency is improved by using two sequential quadratic programming.

4.2 Deformation Energy

The structure is subjected to unidirectional tension or compression, the relation of stress σ and strain ε is linear. The relationship between deformation and outer force is equal in number, and the unit volume deformation energy formula can be obtained by

$$e = \frac{1}{2}\sigma\varepsilon \tag{4.1}$$

In the three-direction stress states, the deformation energy of the unit volume of the elastic body is equal to that of the external force, so the energy density of the stress state is

$$e = \frac{1}{2}\sigma_1\varepsilon_1 + \frac{1}{2}\sigma_2\varepsilon_2 + \frac{1}{2}\sigma_3\varepsilon_3 \tag{4.2}$$

The three principal stresses of the cube element with equal edges are not equal, and the corresponding principal strain is also not equal. Therefore, the deformation of the three edges of the cube element is different: it will be changed into a rectangular cube. The deformation of the element body: one aspect is the change of volume, that is, the volume of the element body becomes larger or smaller, on the other hand, the change of the shape is changed from the cube to the rectangular body.

Therefore, the deformation energy density is composed of two parts:

(1) The volume change of the storage energy density is e_v. The body's volume change is still a cube, so it is called the "volume change energy density."

(2) Keeping volume constant and changing shape, the cube is changed into cuboid and the stored energy density is e_f. It is called "shape change energy density."

Thus, we obtained the following:

$$e = e_v + e_f \tag{4.3}$$

4.3 The Theoretical Background of Deformation Energy

The fourth strength theory assumes that shape change energy density is the main factor that caused the yield stress, no matter what—as long as the shape change energy density e_f reaches a limit value related to material properties—the material will yield.

According to the previous formula, the size of the shape change energy density can be obtained as follows:

$$e_f = (1+\boldsymbol{\mu})(\boldsymbol{\sigma}_s^2)/(3E) \tag{4.4}$$

The shape change energy density under any stress condition can be expressed as

$$e_f = \frac{1+\mu}{6E}[(\sigma_1-\sigma_2)^2+(\sigma_2-\sigma_3)^2+(\sigma_3-\sigma_1)^2] \tag{4.5}$$

According to the above formula, we can obtain the expression of the yield limit:

$$\sqrt{\frac{1}{2}[(\sigma_1-\sigma_2)^2+(\sigma_2-\sigma_3)^2+(\sigma_3-\sigma_1)^2]} = \sigma_s \tag{4.6}$$

The yield limit is divided by the safety factor n_s, which is greater than 1; the allowable stress $\bar{\sigma}$ can be obtained as follows:

$$\bar{\sigma} = \frac{\sigma_s}{n_s} \tag{4.7}$$

According to the industry standard, in the case of static load, the safety factor of the plastic material is generally taken as $n_s = 1.2$–2.5. In this chapter, all safety factors are $n_s = 2.5$ in order to ensure that the structure has enough safety. Therefore, the fourth strength theory can be expressed as

$$\sqrt{\frac{1}{2}[(\sigma_1-\sigma_2)^2+(\sigma_2-\sigma_3)^2+(\sigma_3-\sigma_1)^2]} \le \bar{\sigma} \tag{4.8}$$

4.4 Method of Globalization of Stress Constraints and Strain Revision Coefficient

4.4.1 Global Stress Theory

Using the globalization method of stress constraints and the von Mises yield criterion in mechanics of materials, the local stress constraints were transformed into the whole structure strain energy constraint. According to the fourth strength theory, the change in the element shape is the main representation of the state variable in the yield of the material:

$$e_f = (1+\boldsymbol{\mu})(\boldsymbol{\sigma}_s^2)/(3E) \tag{4.9}$$

Here, e_f is the element shape change energy density, E is the elasticity modulus of the material, μ is the Poisson's ratio, and σ_s is the yield limit of the material.

It can be expressed as the element stress constraint is changed to the element shape change energy density:

$$e_f \le (1+\boldsymbol{\mu})(\boldsymbol{\sigma}_s^2)/(E) \tag{4.10}$$

By formula (4.3), it is known that the element strain energy is larger than the element shape change energy:

$$e \ge e_f \tag{4.11}$$

The element strain energy can be replaced by the element shape change energy. Therefore, the element strain energy can be expressed as

$$e \le (1+\boldsymbol{\mu})(\boldsymbol{\sigma}_s^2)/(3E) \tag{4.12}$$

where e is the element strain energy.

So the formula of the entire continuum structure is established:

$$\sum_{i=1}^{N} e_i \le \sum_{i=1}^{N} (1+\boldsymbol{\mu})(\boldsymbol{\sigma}_s^2) V_i/(3E) \tag{4.13}$$

Here, i is the element number and V_i is the element volume. If in formula (4.13) yield stress σ_s considers the safety coefficient of the structure, using allowable stress $\bar{\sigma}$ instead, then the structure is safe:

$$\sum_{i=1}^{N} e_i \le \sum_{i=1}^{N} (1+\mu)(\bar{\sigma}^2) V_i/(3E) \tag{4.14}$$

Take formula (4.7) into formula (4.14):

$$\sum_{i=1}^{N} e_i \le \sum_{i=1}^{N} \frac{1}{n_s^2} (1+\mu)(\sigma_s^2) V_i/(3E) \tag{4.15}$$

The yield stress σ_s will also instead by allowable stress $\bar{\sigma}$:

$$\sum_{i=1}^{N} e_i \le \sum_{i=1}^{N} \frac{1}{n_s^2} (1+\mu)(\bar{\sigma}^2) V_i/(3E) \tag{4.16}$$

The above formula can be regarded as the inequality of the global change of the stress constraint.

It can be seen from the above formula that the left-hand side of the inequality is the sum of the strain energy of all the elements, that is the structural strain energy. The right-hand side of the inequality is the allowable strain energy of the structure. The local stress constraints of the structure can be transformed into the global strain energy constraints, which can greatly reduce the number of constraints and greatly improve the efficiency of the model. Therefore, the optimization model of the continuum structure is improved by the global strategy of stress, which can solve the problem of continuum structure topology optimization with stress constraints.

4.4.2 Determination of the Strain Energy Correction Coefficient

The local constraints can be replaced with global constraints, so the local stress constraints are transformed into approximate the global structural strain energy constraints, thus greatly reducing the number of constraints, reducing the scale of the problem, and making the problem easier to solve. However, formula (4.16) is bound to be satisfied, and the strain energy of formula (4.12) of each element in the structure may not be fully satisfied. The numerical results show that the elements' von and Mises stress will usually be great which applied in support and near the load, because it is caused by the local stress concentration. The strain energy of these elements is so great and will not meet the formula (4.12). Our goal is that the number of elements which stress stress difference is minimize and fully meet the formula (4-12) (4-16) 's constraint requirements. The appropriate choice of safety factor can achieve this goal.

In eq. (16), $1/n_s^2 = \gamma$, where γ is the modified coefficient of structural strain energy. The numerical examples show that the reasonable topological structures were obtained when the safety factor is 2.5. Then, the modified coefficient of structural strain energy is $\gamma = 0.16$.

4.5 Model Establishment and Solution

4.5.1 Continuum Topology Optimization Model

As in the second chapter, the buckling problem eigenvalue equation is

$$(\boldsymbol{K} + \lambda_j \boldsymbol{G})\boldsymbol{u}_j = \boldsymbol{0} \quad (j = 1, \cdots, J) \tag{4.17}$$

The symbols $\boldsymbol{K}$, $\boldsymbol{G}$, λ_j and $\boldsymbol{u}_j$ denote the stiffness matrix, the geometric stiffness matrix, the j-th eigenvalue, and the corresponding eigenvector, respectively.

The relation between the critical load and the outer load can be rewritten as

$$\lambda_j = \zeta_j P \tag{4.18}$$

where ζ_j is the buckling factor and P is the outer load. The buckling critical force constraint of the structure can be expressed as

$$\lambda_j(f_k(t_i), f_g(t_i)) \le \bar{\lambda}_j \tag{4.19}$$

The filter function is introduced into the expression of strain energy, and the strain energy of the element can be expressed as

$$\begin{aligned} e_i &= \tfrac{1}{2}\boldsymbol{u}_i^{\mathrm{T}}\mathbf{k}_i\boldsymbol{u}_i = \tfrac{1}{2}\boldsymbol{u}_i^{\mathrm{T}} f_k(t_i)\mathbf{k}_i^0\boldsymbol{u}_i = \tfrac{1}{2}\boldsymbol{u}_i^{\mathrm{T}}\left(t_i^{\ b}\right)^{(k)}\mathbf{k}_i^0\boldsymbol{u} \\ &= \tfrac{1}{2}\boldsymbol{u}_i^{\mathrm{T}}\frac{\left(t_i^{\ b}\right)^{(k)}\mathbf{k}_i}{f_k(t_i)}\boldsymbol{u}_i = \frac{\left(t_i^{\ b}\right)^{(k)}}{t_i^{\ b}}\tfrac{1}{2}\boldsymbol{u}_i^{\mathrm{T}}\mathbf{k}_i\boldsymbol{u}_i \end{aligned} \tag{4.20}$$

The strain energy of the element can be changed into

$$e_i = \frac{\left(t_i^{\ b}\right)^{(k)}}{t_i^{\ b}}\frac{1}{2}\boldsymbol{u}_i^{\mathrm{T}}\mathbf{k}_i\boldsymbol{u}_i = \frac{\left(t_i^{\ b}\right)^{(k)}}{t_i^{\ b}}e_i^{(k)} \tag{4.21}$$

where $\left(t_i^{\ b}\right)^{(k)}$ is the value of a topological variable at the k-th iteration and $e_i^{(k)}$ is the strain energy of element i at the k-th iteration.

The strain energy constraints of the structure can be rewritten as

$$Q_j = \sum_{i=1}^{N} e_{ij} \le \sum_{i=1}^{N}\frac{1}{n_s^2}(1+\mu)(\bar{\sigma}^2)V_i/(3E) = \bar{Q}_j \tag{4.22}$$

where Q_j is the strain energy to the j-th buckling mode of the structure and $\bar{Q}_j$ is the upper limit strain energy of the j-th buckling mode of the structure.

Finally, the topology optimization model for the continuum structure is constructed, in which the minimized quality as the objective function has been subjected to the buckling and stress constraints:

$$\begin{cases} \text{find} & \boldsymbol{t} \in (t_1, \cdots, t_N)^{\mathrm{T}} \\ \min & W = \sum_{i=1}^{N} f_w(t_i) w_i^0 \\ \text{s.t.} & \lambda_{jl}(f_k(t_i), f_g(t_i)) \le \bar{\lambda}_{jl} \\ & Q_{jl}(f_k(t_i)) \le \bar{Q}_{jl} \\ & 0 \le t_i \le 1 \qquad (i = 1, \ldots, N; j = 1, \ldots, J; l = 1, \ldots, L) \end{cases} \tag{4.23}$$

where W is the total quality of the structure; w_i and N are the quality of the i-th element and the total number of elements, respectively; $\bar{\lambda}_j$ is the upper limit of the critical load; J is the total number of the buckling mode; Q_{jl} is the l-th load case strain energy of the j-th buckling mode and $\bar{Q}_{jl}$ is the l-th load case strain energy of the upper limit j-th buckling mode; N is the total number of units; and L is the total number of working conditions.

4.5.2 Explicit Approximate Function of the Buckling Constraint

The buckling constraint's explicit approximate function is the same as provided in the second chapter; here a statement is no longer needed.

Uniform expression formula of buckling constraints can be obtained:

$$\sum_{i=1}^{N} d_{ijl} x_i \le h_{jl} \tag{4.24}$$

If we define $A_{ijl}^{(k)} = (U_{ijl}^{(k)} + \lambda_{jl} V_{ijl}^{(k)}) / V_{\Sigma jl}$, the left-hand side of inequality (4.24) is equal to the following:

$$\sum_{i=1}^{N} d_{ijl} x = \sum_{i=1}^{N} x_i (U_{ijl}^{(k)} + \lambda_{jl} V_{ijl}^{(k)}) / V_{\Sigma jl} x_i^{(k)} = \sum_{i=1}^{N} A_{ijl}^{(k)} x_i / x_i^{(k)} = \sum_{i=1}^{N} (t_i^{(k)})^b A_{ijl}^{(k)} x_i \tag{4.25}$$

If we define $d_{ijl} = (t_i^{(k)})^b A_{ijl}^{(k)}$, the left-hand side of inequality (4.24) is rewritten as

$$h_{jl} = \bar{\lambda}_{jl} - \lambda_{jl}(\mathbf{x}^{(k)}) + \sum_{i=1}^{N} (U_{ijl}^{(k)} + \lambda_{jl} V_{ijl}^{(k)}) / V_{\Sigma ijl} = \bar{\lambda}_{jl} - \lambda_{jl}(\mathbf{x}^{(k)}) + \sum_{i=1}^{N} A_{ijl}^{(k)} \tag{4.26}$$

where $U_{ijl} = \boldsymbol{u}_{jl}^{\mathrm{T}} \boldsymbol{k}_i \boldsymbol{u}_{jl}/2$ is the mode strain energy for the i-th element of the j-th mode under the l-th condition. $V_{ijl} = \boldsymbol{u}_{jl}^{\mathrm{T}} \boldsymbol{g}_i \boldsymbol{u}_{jl}/2$ is the mode geometric strain energy for the i-th element of the j-th mode under the l-th condition. $V_{\Sigma ijl} = \boldsymbol{u}_{jl}^{\mathrm{T}} \boldsymbol{g}_i \boldsymbol{u}_{jl}/2$ is the total mode geometric strain energy for the structure of the j-th mode under the l-th condition.

4.5.3 Explicit Approximate Function of the Stress Constraint

According to eq. (4.21) and eq. (4.22), the strain energy constraints are rewritten as

$$\sum_{i=1}^{N}\frac{\left(t_i{}^b\right)^{(k)}}{t_i{}^b}e_i^{(k)} \le \sum_{i=1}^{N}\gamma(1+\mu)(\bar{\sigma}^2)V_i/(3E) \tag{4.27}$$

The left- and right-hand sides of the inequation are rewritten as $\sum_{i=1}^{N}(1+\mu)(\bar{\sigma}^2)V_i/(3E)=\sum_{i}^{N}\overline{e_i}=q$, $(t_i^b)^{(k)}e_i^{(k)}=s_i$. Therefore, the strain energy constraint becomes

$$\sum_{i=1}^{N}\frac{s_i}{t_i{}^b}\le\gamma q \tag{4.28}$$

4.5.4 Explicit Approximate Function of the Objective Function

The processing of the objective function is basically the same as provided in the third chapter; here a statement is no longer needed. The object function is transformed into an explicit expression by second-order Taylor expansion:

$$W=\sum_{i=1}^{N}\{[w_i^0\beta(\beta+1)/2x_i^{(k)\beta+2}]x_i^2+[-w_i^0\beta(\beta+2)/x_i^{(k)\beta+1}]x_i\} \tag{4.29}$$

where $x_i^{(k)}$ is the initial design variable of the i-th element at the k-th iteration.

We define $A_i=w_i^0\beta(\beta+1)/2x_i^{(k)\beta+2}$ and $B_i=-w_i^0\beta(\beta+2)/x_i^{(k)\beta+1}$.

The objective function can be written as $W=\sum_{i=1}^{N}\left(A_ix_i^2+B_ix_i\right)$.

The topology optimization model with minimum quality as objective for continuum structure subjected to buckling constraints can be rewritten as

$$\begin{cases}\text{find} & \boldsymbol{x}\in E^N\\ \min & W=\sum_{i=1}^{N}(A_ix_i^2+B_ix_i)\\ \text{s.t} & \sum_{i=1}^{N}d_{ijl}x_i\le h_{jl}\quad (j=1,\dots,J;l=1,\dots,L)\\ & \sum_{i=1}^{N}s_{ijl}x_i\le\gamma q_{jl}\\ & 1\le x_i\le\bar{x}_i\qquad (i=1,\dots,N)\end{cases} \tag{4.30}$$

4.5.5 Solution of the Topological Optimization Model

According to the relation between the topology and the design variables, the topology optimization model is transformed into the model with design variables. In order to reduce the number of variables and the model's scale, the optimization model is translated into a dual programming:

$$\begin{cases} \text{find} & \chi \\ \max & \Phi(\chi) \\ \text{s.t.} & \chi \geq 0 \end{cases} \tag{4.31}$$

In equation $\chi = \begin{bmatrix} \mu \\ \eta \end{bmatrix}$, $\Phi(\boldsymbol{\mu}, \boldsymbol{\eta}) = \min\limits_{1 \leq x_i \leq \overline{x_i}} L(\boldsymbol{x}, \boldsymbol{\mu}, \boldsymbol{\eta})$,

$$\begin{aligned} L(\boldsymbol{x}, \boldsymbol{\mu}, \boldsymbol{\eta}) = & \sum_{i=1}^{N} (A_i x_i^2 + B_i x_i) + \sum_{j=1}^{J} \sum_{l=1}^{L} \mu_{jl} \left(\sum_{i=1}^{N} d_{ijl} x_i - h_{jl} \right) \\ & + \sum_{j=1}^{J} \sum_{l=1}^{L} \eta_j \left(\sum_{i=1}^{N} s_{ijl} x_i - \gamma q_{jl} \right) \end{aligned} \tag{4.32}$$

According to the Kutta condition, the derivative of an augmented Lagrangian function with respect to the design variable can be written as follows:

$$\begin{aligned} \partial L(\boldsymbol{x}, \boldsymbol{\mu}, \boldsymbol{\eta}) / \partial x_i = & 2A_i x_i^* + B_i + \sum_{j=1}^{J} \sum_{l=1}^{L} \boldsymbol{\mu}_{jl} d_{ijl} + \sum_{j=1}^{J} \sum_{l=1}^{L} \boldsymbol{\eta}_{jl} S_{ijl} \\ & \begin{cases} \leq 0 & (x_i^* = \bar{x}_i) \\ = 0 & (1 < x_i^* < \bar{x}_i) \\ \geq 0 & (x_i^* = 1) \end{cases} \end{aligned} \tag{4.33}$$

Making $I_a = \{ i | 1 < x_i^* < \bar{x}_i \}$ the active variable set, the equation is established:

$$2A_i x_i^* + B_i + \sum_{j=1}^{J} \sum_{l=1}^{L} \boldsymbol{\mu}_{jl} d_{ijl} + \sum_{j=1}^{J} \sum_{l=1}^{L} \boldsymbol{\eta}_{jl} s_{ijl} \equiv 0 \tag{4.33}$$

For, $\mu_{ml}, \mu_{jm}, \eta_{nl}, \eta_{jn}$, respectively, we obtain the following derivations:

$$\begin{aligned} & \partial x_i^* / \partial \mu_{ml} = -d_{iml} / 2A_i, \ \partial x_i^* / \partial \eta_{nl} = -s_{inl} / 2A_i \\ & \partial x_i^* / \partial \mu_{jm} = -d_{ijm} / 2A_i, \ \partial x_i^* / \partial \eta_{jn} = -s_{inl} / 2A_i \end{aligned} \tag{4.34}$$

Because it is difficult to obtain the explicit expression of $\Phi(\chi)$, it will be transformed into a two-order Taylor expansion approximation. According to the dual theory, the objective function of formula (4.32) for the design of a variable's one- and two-order derivative is:

$$\partial\Phi/\partial\mu_{jl} = \sum_{i=1}^{N} d_{ijl}x_i^* - h_{jl},\ \partial\Phi/\partial\eta_{jl} = \sum_{i=1}^{N} s_{ijl}x_i^* - \gamma q_{jl} \tag{4.35}$$

$$\begin{aligned} \partial^2\Phi/(\partial\mu_{jl}\partial\mu_{ml}) &= \sum_{i=1}^{N} d_{ijl}\partial x_i^*/\partial\mu_{ml},\ \partial^2\Phi/\left(\partial\eta_{jl}\partial\eta_{nl}\right) = \sum_{i=1}^{N} s_{ijl}\partial x_i^*/\partial\eta_{nl} \\ \partial^2\Phi/\partial\mu_{jl}\partial\mu_{jm} &= \sum_{i=1}^{N} d_{ijl}\partial x_i^*/\partial\mu_{jm},\ \partial^2\Phi/\left(\partial\eta_{jl}\partial\eta_{jn}\right) = \sum_{i=1}^{N} s_{ijl}\partial x_i^*/\partial\eta_{jn} \end{aligned} \tag{4.36}$$

Formula (4.34) was taken into formula (4.36) to obtain the following:

$$\begin{aligned} \partial^2\Phi/(\partial\mu_{jl}\partial\mu_{ml}) &= -\sum_{i=1}^{N} d_{ijl}d_{iml}/2A_i,\ \partial^2\Phi/\left(\partial\eta_{jl}\partial\eta_{nl}\right) = -\sum_{i=1}^{N} s_{ijl}s_{inl}/2A_i \\ \partial^2\Phi/(\partial\mu_{jl}\partial\mu_{ml}) &= -\sum_{i=1}^{N} d_{ijl}d_{iml}/2A_i,\ \partial^2\Phi/\left(\partial\eta_{jl}\partial\eta_{nl}\right) = -\sum_{i=1}^{N} s_{ijl}s_{inl}/2A_i \end{aligned} \tag{4.37}$$

$$\partial^2\Phi/\left(\partial\mu_{jl}\partial\eta_{jl}\right) = \partial^2\Phi/\left(\partial\eta_{jl}\partial\mu_{jl}\right) = -\sum_{i=1}^{N} d_{ijl}s_{ijl}/2A_i \tag{4.38}$$

We can obtain the second-order programming model:

$$\begin{cases} \text{find} & \chi \\ \min & -\Phi(\chi) = \frac{1}{2}\chi^T\mathbf{D}_\chi + \left(\mathbf{H}^T - \chi^{0^T}\mathbf{D}\right)\chi \\ \text{s.t.} & \chi \geq 0 \end{cases} \tag{4.39}$$

where $\boldsymbol{D} = -\nabla^2\Phi$ and $\boldsymbol{H} = -\nabla\Phi$. The second-order programming model was solved, and we get χ. The active variable set was updated. Accordingly, this process was repeated until

$$\left\|x^{(k+1)} - x^{(k)}\right\| / \left\|x^{(k)}\right\| \leq \varepsilon$$

The optimum values of design variables x^* are solved, and the optimum values of topological variables t^* are accordingly obtained. With the new topological variables, the structure is modified. Then the structural analysis and topological optimization are renewable and begin until the following convergence condition is satisfied:

$$\Delta W = \left|\left(W^{(k+1)} - W^{(k)}\right)/W^{(k+1)}\right| \leq \varepsilon \tag{4.40}$$

where $W^{(k)}$ and $W^{(k+1)}$ are the structural quality of the previous iteration and the current iteration, respectively, and ε is the convergence precision. In the present study, $\varepsilon = 0.001$ is employed.

4.6 Optimization Strategy

4.6.1 Solving the Board Format and Grid Dependence

To solve the problem of the grid phenomenon and the grid dependence, the method of image filtering is used, which was introduced in [62]. It does not filter the sensitivity of the objective function, but filters the displacemente coefficient of the element.

Defining the geometric concentric distance function of elements i and j as $d(i,j)$:

$$d(i,j) = \max(0, r_{\min} - dis(i,j)) \tag{4.41}$$

where $\mathrm{dis}(i,j)$ is the distance between the center of elements i and j. The distance function $d(i,j)$has symmetry on elements i and j.

The element strain energy e_{ij} is filtrated by weight coefficient matrix Φ, the checkerboard and mesh dependent are eliminated in topology optimization, in which

$$e_{ij} = \sum_{k=1}^{N} \Phi_{ik} e_{kl} \tag{4.42}$$

In the quality coefficient matrix Φ elements

$$\Phi_{ij} = \frac{d(i,j)}{\sum_{j=1}^{N} d(i,j)} \tag{4.43}$$

The quality coefficient matrix Φ has the following attributes: $\sum_{j=1}^{N} \Phi_{ij} = 1$. The meaning is: for i element, the sum of all element weight coefficients is 1.

4.6.2 Inversion Criterion of Element Topological Variable

In the iterative process, the topological variables are continuous variables, which guarantee the continuity of the topology structure. In the optimization process, the topological variables are gradually close to 0 or 1. At the last step of the iteration, the topology optimization structure satisfies the convergence precision, and the topology structure is obtained. In the optimal topology, there are still a few non-white nonred regions, namely, the topological variable values between 0 and 1 of the elements. In order to get a clearer topology, we have given a deletion threshold T. When the topology variable is lower than the threshold value, the topology variable is 0. When the topology is higher than the threshold value, the topology variable is 1. The threshold T generally is taken as 0.3–0.7, so that the continuous topological variables were inversed to 0–1 type of discrete topology, and you can get a clear topology.

4.7 Constraint Selection

4.7.1 Constraint Primary Select

In section optimization and shape optimization, the multiple constraints—in order to improve efficiency—reduce the number of structure, and the effective constraint primaries select method is taken, so as to reduce the number of constraints and the cost of computation [59]. In topology optimization with buckling and stress constraints, this strategy can be used to remove unnecessary constraints, reduce the number of calculation models, and improve calculation speed, so that satisfactory result to topology optimization can be quickly obtained.

4.7.2 Selection Criterion of Stress Upper Limit

The upper limit of stress can also affect the optimal topology shape and the quality of the optimal topology. Accurate selection of the upper limit of the stress and the upper limit of the critical buckling force can be quick and effective to obtain a reasonable topology. Similarly, the size of the upper bound can be estimated by the quality of the topology. The function correspondence will give the designer more convenience to apply the topology optimization in the product conceptual design stage.

According to a specific numerical example, the corresponding relationship between the quality ratio and the stress ratio of the topological structure is obtained. In order to ensure the comparability, the numerical example calculated is the same as in the second and third chapters. The specific data of the numerical example is same, the application force is 6,800 N, and the change of force will not affect the structure of the buckling critical force. The maximum stress of the structure is of 369 MPa, and a buckling critical force is $\lambda_1 = 119$ N.

In order to eliminate the influence of the corresponding relationship between the stress constraint and the quality of the topological structure, the concept of "stress constraint ratio" is introduced here. Due to the increasing stress of the structure, the quality of the topology will be reduced. Therefore, the stress constraint ratio is defined as the ratio between the upper limit of the stress constraint and the maximum allowable stress—360 MPa, in this chapter—of the material.

Table 4.1 lists the stress constraints, topological quality and topology optimization of 15 times numerical experiments.

Table 4.1: Stress constraint and quality of optimal structure.

Test number	Stress constraint	Topological quality	Repeated analysis times	Topological graph
1	80	0.9870	30	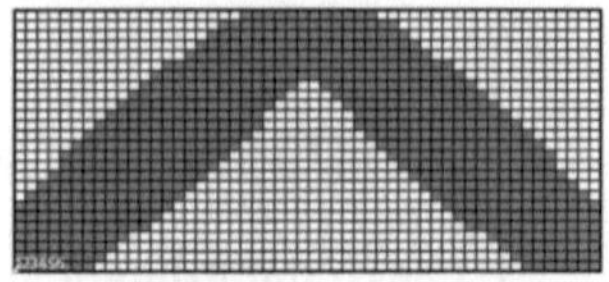
2	90	0.6629	31	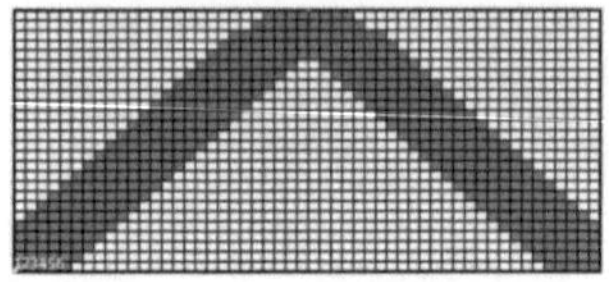
3	110	0.3886	34	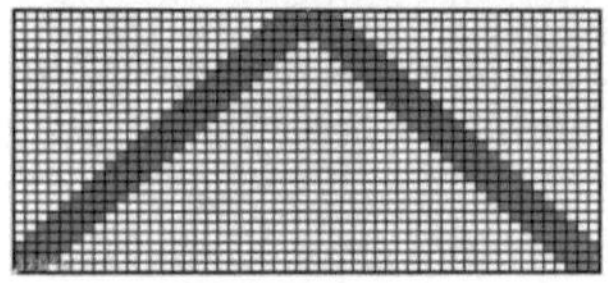
4	130	0.3231	37	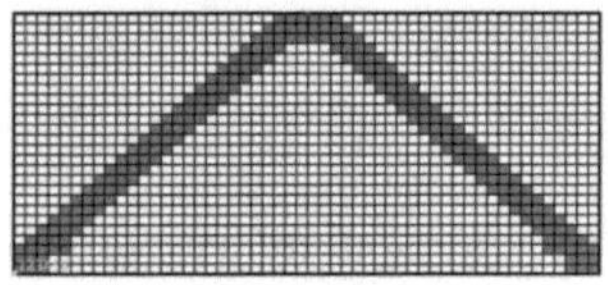
5	150	0.3107	38	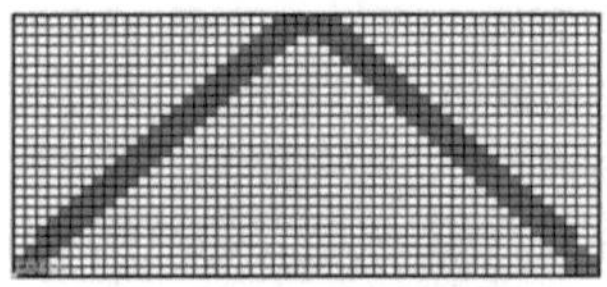
6	170	0.3076	39	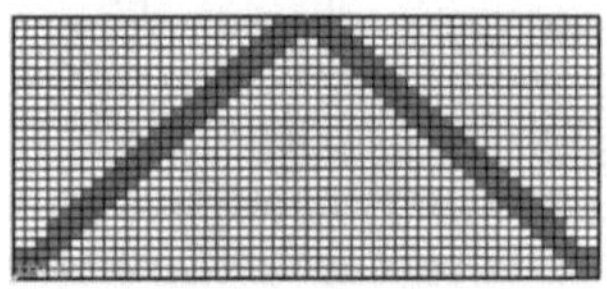
7	190	0.2951	38	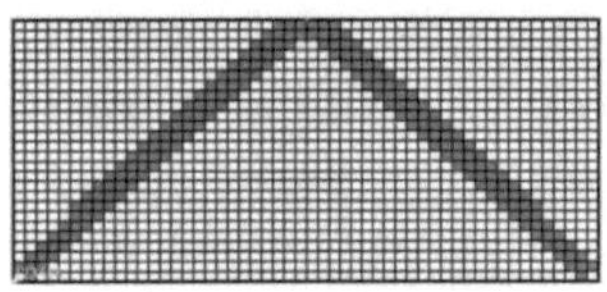
8	210	0.2795	39	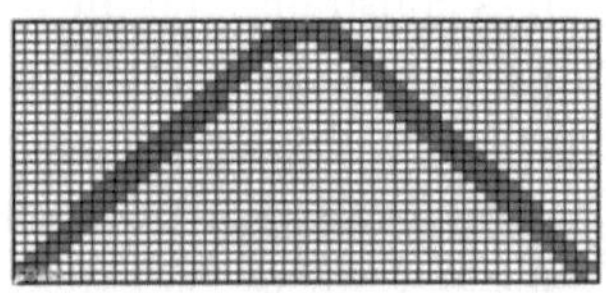

(continued)

Table 4.1: (continued)

Test number	Stress constraint	Topological quality	Repeated analysis times	Topological graph
9	230	0.2639	40	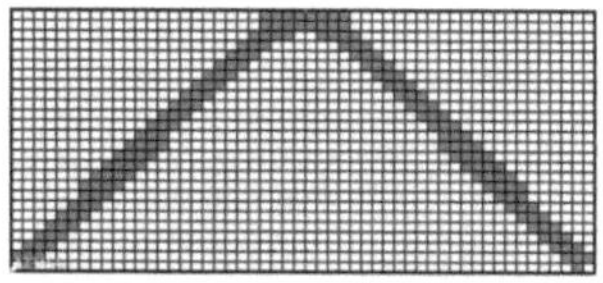
10	250	0.2515	40	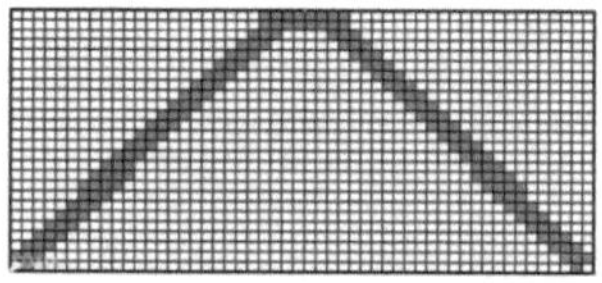
11	270	0.2483	40	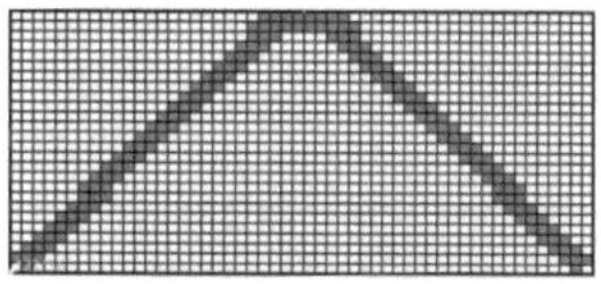
12	290	0.2452	42	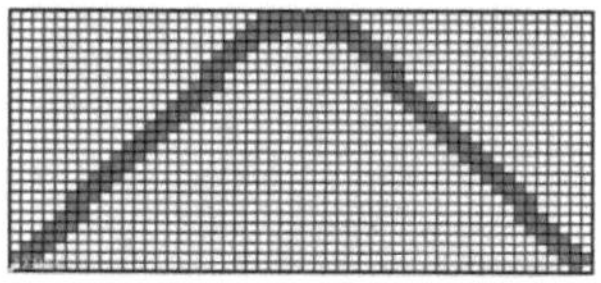
13	310	0.2390	41	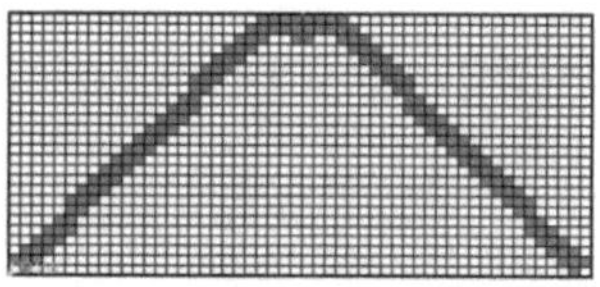
14	330	0.2203	41	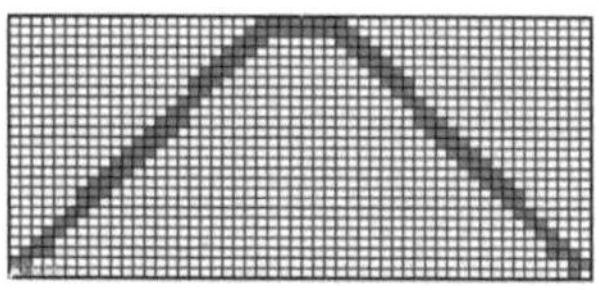
15	350	0.2172	40	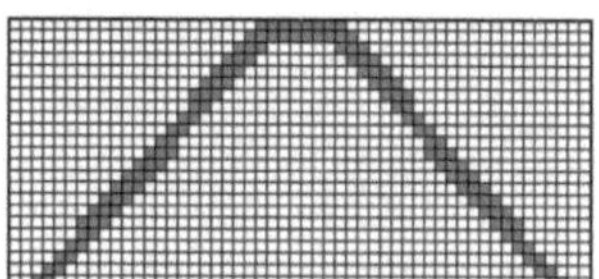

Table 4.2 lists the "quality ratio" and the "stress constraint ratio" of 15 times topology optimization.

The horizontal coordinate x denotes the "quality ratio," and the vertical coordinate y denotes the "stress constraint ratio" in Figure 4.1.

Table 4.2: Ratio of stress constraint and ratio of quality.

Test number	Stress constraint ratio	Quality ratio of topological structure
1	0.2222	0.4680
2	0.2500	0.3143
3	0.3055	0.1843
4	0.3611	0.1532
5	0.4167	0.1473
6	0.4722	0.1459
7	0.5278	0.1399
8	0.5833	0.1325
9	0.6389	0.1251
10	0.6944	0.1193
11	0.7500	0.1177
12	0.8056	0.1163
13	0.8611	0.1133
14	0.9167	0.1045
15	0.9722	0.1030

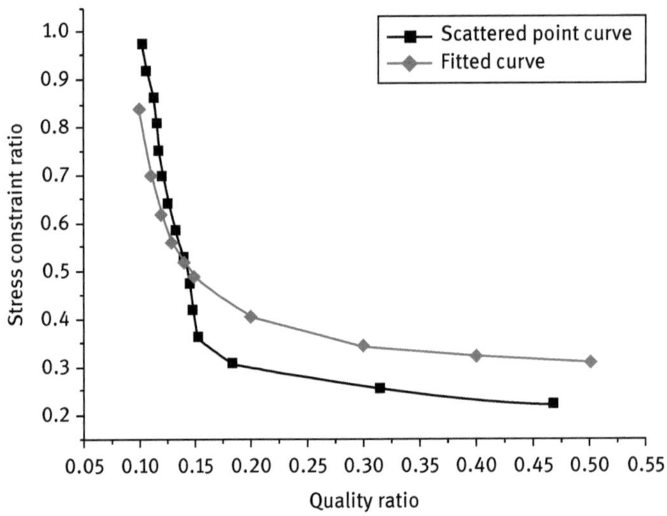

Figure 4.1: Curve of quality ratio and stress constraint ratio.

It can be seen from the graph that the curve is nonlinear. In order to get the function relationship between the quality ratio and the stress constraint ratio, the least squares method is used to fit the curve. The hyperbolic function is selected according to the distribution of the discrete points:

$$\frac{1}{y} = a + \frac{b}{x} \tag{4.44}$$

Table 4.3: Simulation result of the stress constraint curve.

Serial number	x	y	$X=1/x$	$Y=1/y$	X^2	XY
1	1	1	1	1	1	1
2	0.468	0.2222	2.1368	4.5005	4.5659	9.6167
3	0.3143	0.2500	3.1817	4.0000	10.1232	12.7268
4	0.1843	0.3055	5.4259	3.2733	29.4404	17.7606
5	0.1532	0.3611	6.5274	2.7693	42.6070	18.0763
6	0.1473	0.4167	6.7889	2.3998	46.0892	16.2920
7	0.1459	0.4722	6.8540	2.1177	46.9773	14.5147
8	0.1399	0.5278	7.1480	1.8947	51.0940	13.5433
9	0.1325	0.5833	7.5472	1.7144	56.9602	12.9389
10	0.1251	0.6389	7.9936	1.5652	63.8976	12.5116
11	0.1193	0.6944	8.3822	1.4401	70.2613	12.0712
12	0.1177	0.7500	8.4962	1.3333	72.1854	11.3280
13	0.1163	0.8056	8.5985	1.2413	73.9342	10.6733
14	0.1133	0.8611	8.8261	1.1613	77.9000	10.2498
15	0.1045	0.9167	9.5694	1.0909	91.5734	10.4393
16	0.1030	0.9722	9.7087	1.0286	94.2589	9.9864
$\sum$			108.1846	32.5304	832.868	193.7289

Using the data in Table 4.3 to write equations:

$$\begin{bmatrix} \sum_{i=1}^{16} 1 & \sum_{i=1}^{16} X_i \\ \sum_{i=1}^{16} X_i & \sum_{i=1}^{16} X_i^2 \end{bmatrix} \begin{bmatrix} a \\ b \end{bmatrix} = \begin{bmatrix} \sum_{i=1}^{16} Y_i \\ \sum_{i=1}^{16} X_i Y_i \end{bmatrix} \tag{4.45}$$

$$\begin{cases} 16a + 108.1846b = 32.5304 \\ 108.1846a + 832.868b = 193.7289 \end{cases} \tag{4.46}$$

The solution of the equations is $a = 3.7831$ and $b = -0.2588$. Taking them into the hyperbolic equation:

$$y = \frac{x}{3.7831x - 0.2588} \tag{4.47}$$

In order to facilitate comparison, the "quality ratio–buckling constraint ratio" curve and the "quality ratio–stress constraint ratio" curve were drawn in the same coordinate system, as shown in Figure 4.2.

From the graph, we can see that the constraint ratio is in the range of 0.2–0.5, there is little change in quality, and the quality changes in the range of 0.5–1.0, and the structure topology will have a greater change. The influence of different finite

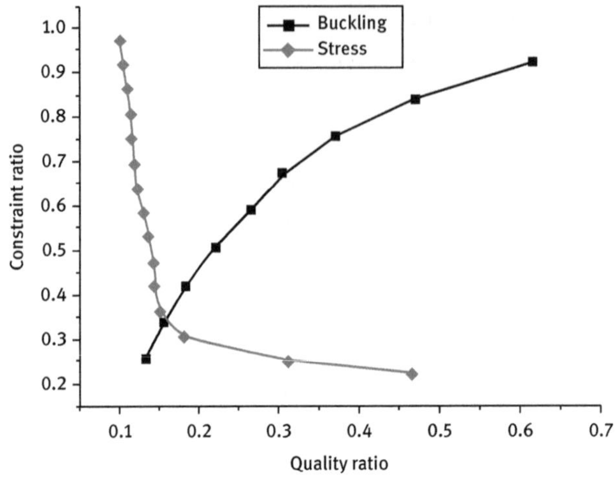

Figure 4.2: Curve of quality ratio and buckling displacement constraint ratio.

element models of the curve can be eliminated by the quality ratio and the constraint ratio. The function relation obtained by this method can be applied to many kinds of topological structures accurately.

4.8 Optimization Algorithms

The optimization algorithms of the stress and buckling constraints are based on the second development platform of MSC.Patran/Nastran software, and the use of the development tool PCL language to design the optimization algorithm program and the user input interface. The specific steps of optimization are as follows:

(1) Finite element model: The geometric model of the continuum structure is established on the MSC.Patran platform, which is divided into finite element mesh, and defines the material properties, boundary conditions, and the properties of the element. Due to the structural topology optimization by buckling constraints, the properties of planar elements are set to shell elements.
(2) Basic optimization parameter input: It includes the convergence accuracy and the radius of the filter, the external force, etc.
(3) Stress constrained optimization parameter input: It includes the name of constraint, stress value, etc.
(4) Buckling constraint optimal parameter input: It includes the sum of the total number of buckling modes, the name of the modal constraint, the numerical value of the buckling constraint and the number of corresponding buckling modes.
(5) Submit MSC.Nastran solution: The linear static finite element analysis of the structure and the finite element analysis of the buckling mode. The linear static

finite element analysis solver is SOL101, and the buckling mode finite element analysis solver is SOL105.

(6) The corresponding finite element analysis results are calculated and extracted.
(7) The effective stress constraint and the buckling constraint are selected.
(8) Generate input files required for the dual quadratic programming.
(9) Call dual quadratic programming to solve the optimization model.
(10) According to the iterative convergence criterion to determine whether to meet the convergence conditions, if not satisfied, go back to step (5); if satisfied, then stop the iteration and record the data, determine the optimal topology and inversion topology variables, get the optimal topology graph, and end the optimization calculation.

The block diagram of the optimization program is shown in Figure 4.3.

4.9 Numerical Simulation and Case Study

4.9.1 Numerical Example of Two-dimensional Plane Continuum

Example 1

As illustrated in Figure 4.4, the base structure is a plane elastic body with size 520 × 260 × 2.0 mm. The elastic modulus of structure material is $E = 68.89 \times 10^3$ MPa, the Poisson's ratio is $y = 0.3$, and the material density is $\rho = 2.7 \times 10^{-6}$ kg/mm^3. The right- and left-hand underpart corner points of the structure are clasped. A concentrated force $P = 7{,}000$ N along the up-to-down (-y) direction is located on the top of structure. A quadrilateral element with four nodes is used and the total number of elements is 52 × 26. The quality of the base structure is 0.73 kg.

The allowable stress of the structure is 380 MPa. The buckling factor is 0.017006. The initial critical load is 119.042 N. The upper limit allowable stress of the structure is 150 MPa. In this example, the upper limit maximal value of the critical load is 75 N. The convergence precision is 0.001, after 49 times of iteration. The quality of the optimized topology is 0.21 kg (Figure 4.5 to Figure 4.11). Table 4.4 lists the comparison of the results of topology optimization with different constraints.

The reliability of the function expression of the "quality stress ratio" is verified after fitting.

The quality of the base structure is 0.73 kg, when the stress constraints alone are taken; After 60 times iteration, it is in convergence, topology structure's quality is 0.11 kg. It is 15% of the base structure quality. The maximum allowable stress of the base structure is 360 MPa, the maximum stress limit is 150 MPa, and the stress ratio is 41.7%. Taking the quality ratio into the quality–stress constraint ratio formula, we obtain the constraint ratio of 48%. The results between the function relation calculation and the numerical calculation have some errors, which are mainly due to the following reasons: the calculation error, fitting error, quality inversion topology and curve

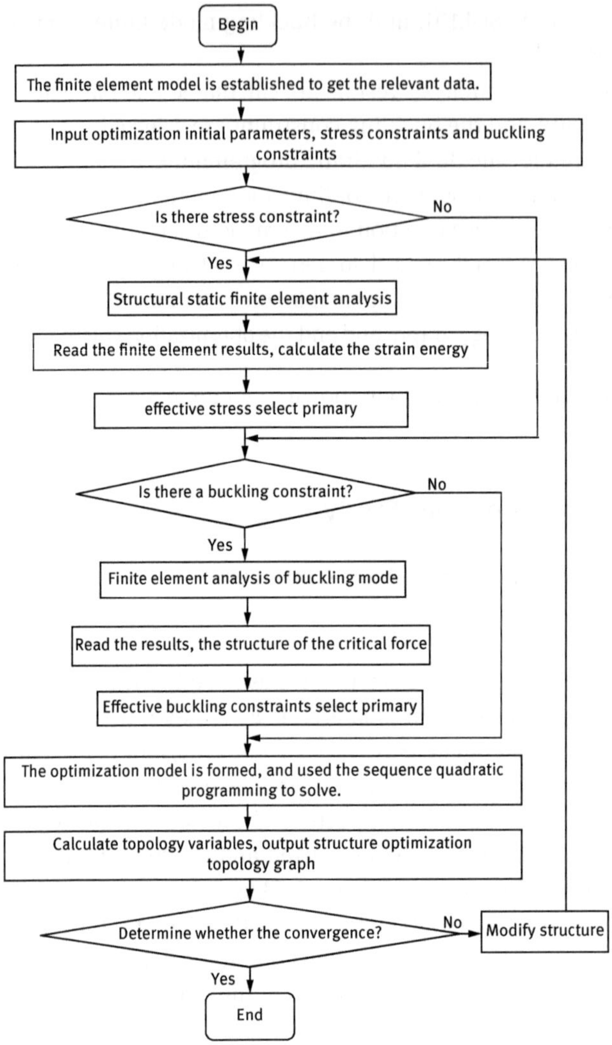

Figure 4.3: Flow chart for the algorithm of topology optimization of continuum structure subjected to buckling constraints and stress constraints.

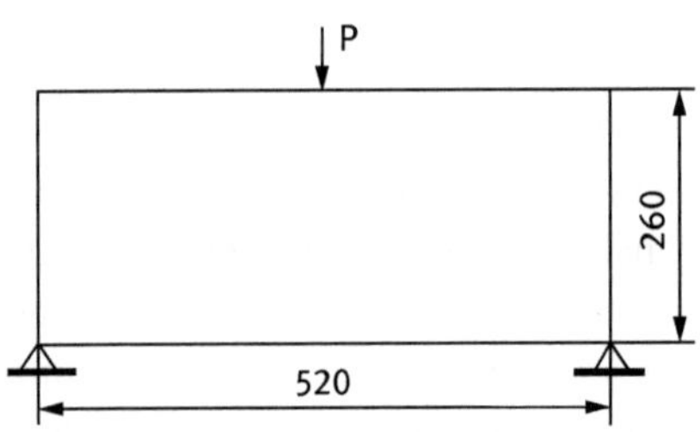

Figure 4.4: Base structure.

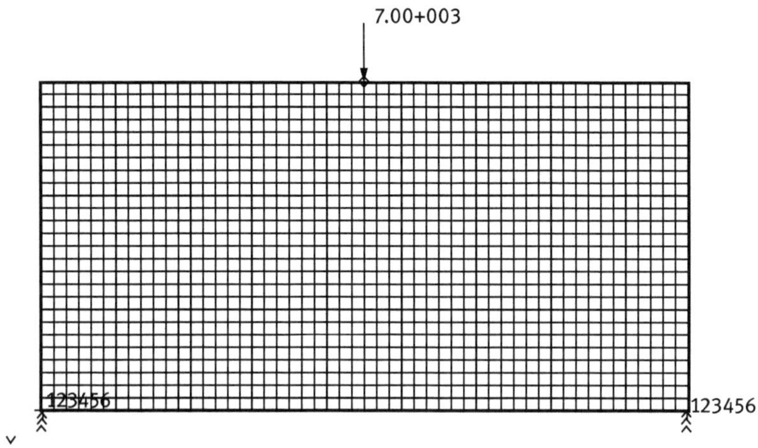

Figure 4.5: Finite element model.

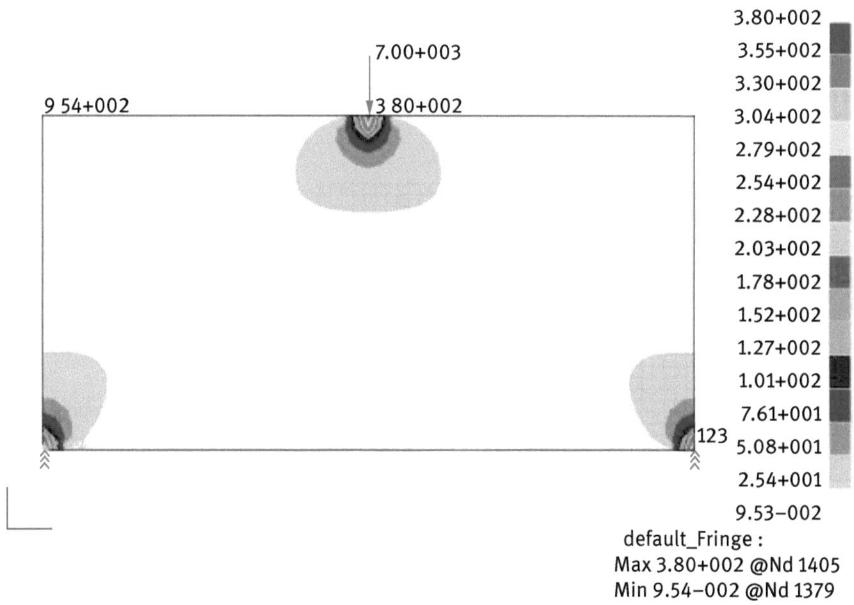

Figure 4.6: Stress fringe result of the base structure.

caused. Although the quality stress ratio has some errors, it is feasible to estimate the topological quality and stress constraints in a conceptual design.

From the topology graph of stress constraints, it can be seen that the topology is relatively slender and more uniform, so the stability of the structure is poor. In the topology of buckling constraints, the topology becomes thicker and the stability is

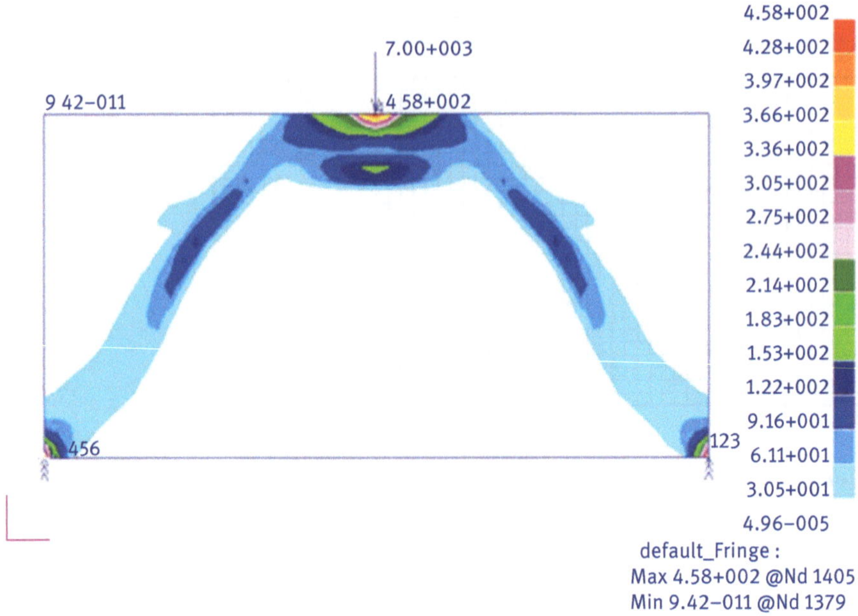

Figure 4.7: Stress distribution of the topology structure.

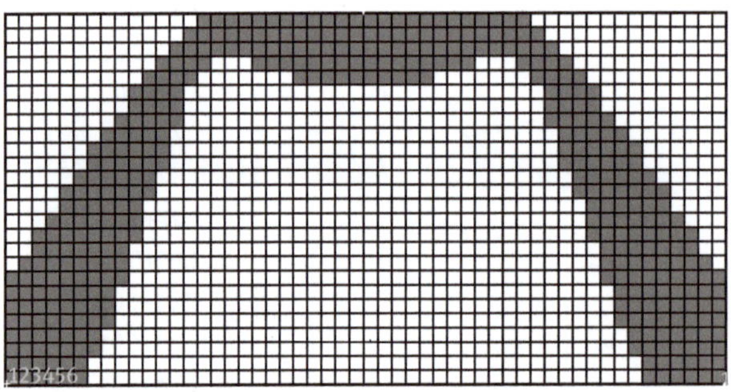

Figure 4.8: Optimal topological configuration with buckling constraint.

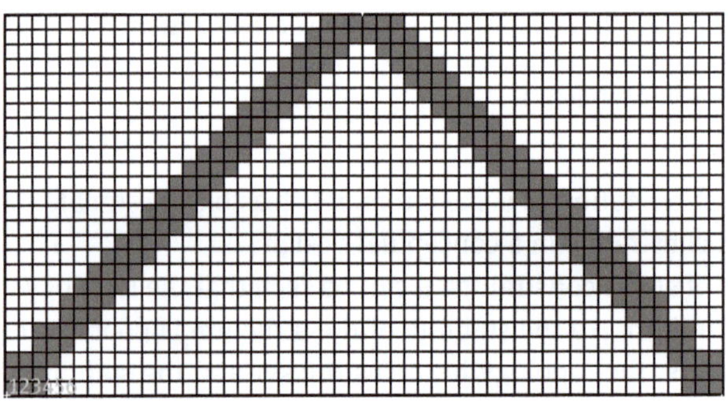

Figure 4.9: Optimal topological configuration with stress constraint.

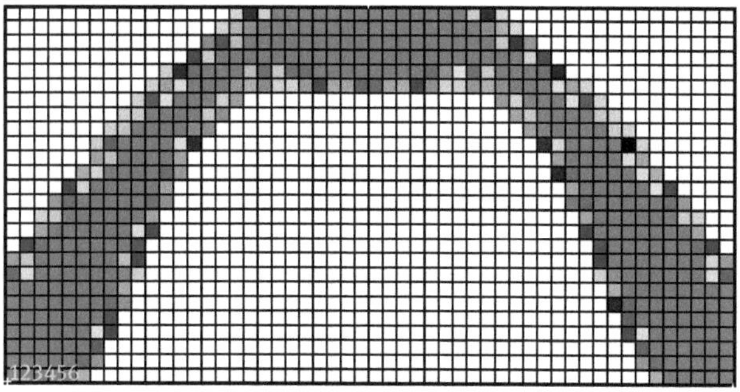

Figure 4.10: Topological configuration with buckling and stress constraints before inversion.

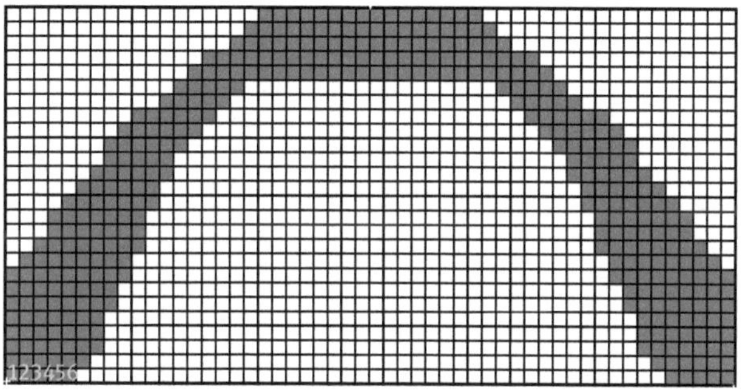

Figure 4.11: Topological configuration with buckling and stress constraints after inversion.

Table 4.4: Comparison of topological results with different constraints.

Constraint type	Initial quality (kg)	Optimized quality (kg)	Iteration number
Stress	0.73	0.11	36
Buckling	0.73	0.22	44
Stress + Buckling	0.73	0.21	60

enhanced. When the stress constraints and buckling constraints are considered, the topology is changed with respect to the single constraint, and is not the simple sum of the two topologies. At the intersection of the horizontal direction and the vertical branch, the topological structure becomes rough, the quality of the whole topology is also improved, and the stability of the whole structure is strengthened.

Example 2

As illustrated in Figure 4.12, the base structure is a plane elastic body with size 160 × 100 × 6.0 mm. The elastic modulus of the structure material is $E = 68.89\times10^3$ MPa, the Poisson's ratio is $y = 0.3$, and the material density is $\rho = 2.7\times10^{-6}$ kg/mm^3. The left boundary points of the structure are clasped. A concentrated force $P = 2{,}500$ N along the up-to-down (-y) direction is located at upper-right corner of the boundary of

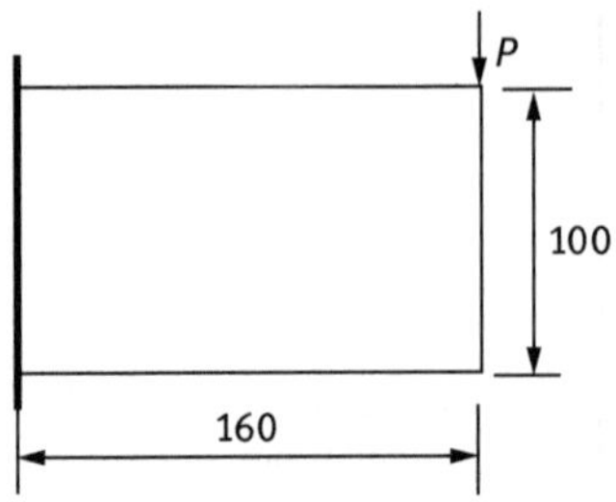

Figure 4.12: Base structure.

Table 4.5: Comparison of topological results with different constraints.

Constraint type	Initial quality (kg)	Optimized quality (kg)	Iteration number
Stress	0.26	0.065	53
Buckling	0.26	0.1002	40
Stress + Buckling	0.26	0.1033	39

structure. A quadrilateral element with four nodes is used and the total number of elements is 48 × 30. The quality of the base structure is 0.26 kg.

The allowable stress of the structure is 384 MPa. The buckling factor is 7.869. The initial critical load is 19,672.5 N. The upper limit allowable stress of the structure is 150 MPa. In this example, the upper limit maximal value of the critical load is 8,000 N. The convergence precision is 0.001, after 39 times of iteration. The quality of the optimized topology is 0.21 kg (Table 4.5).

It can be seen from the graph Figure 4.13 to Figure 4.19 that the topological structure under stress constraints is relatively thin and uniform. The topological structure of the buckling constraint is larger than that of the stress constraint, the thin rod is removed, and the other parts are strengthened. At the same time, the position of force acting point of the structure is strengthened in considering the effect of buckling constraints and displacement constraints.

Example 3

As illustrated in Figure 4.20, the base structure is a plane elastic body with size 520 × 260 × 6.0 mm. The elastic modulus of the structure material is $E = 68.89 \times 10^3$ MPa, the Poisson's ratio is $y = 0.3$, and the material density is $\rho = 2.7 \times 10^{-6}$ kg/mm^3. The right- and left-hand underpart corner points of the structure are clasped. A concentrated force $P = 14{,}100$ N along the up-to-down ($-y$) direction is located in the three nodes in the middle of the lower boundary of structure. A quadrilateral element with four nodes is used and the total number of elements is 52 × 26. The quality of the base structure is 2.2 kg.

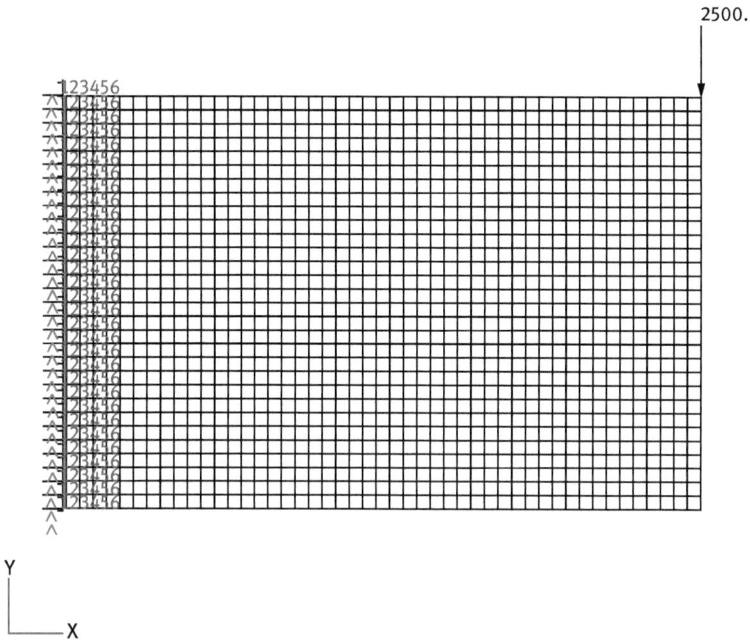

Figure 4.13: Finite element model.

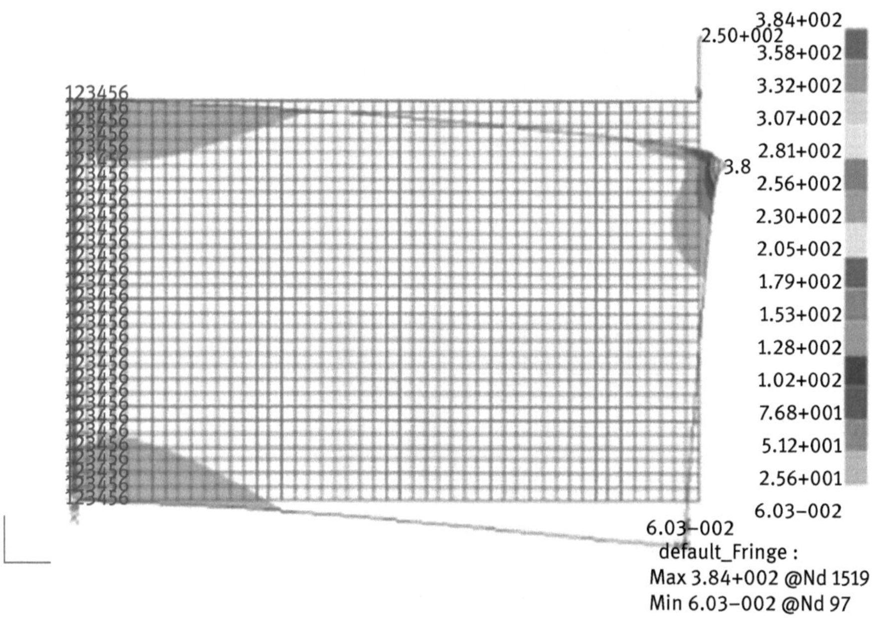

Figure 4.14: Stress fringe result of the base structure.

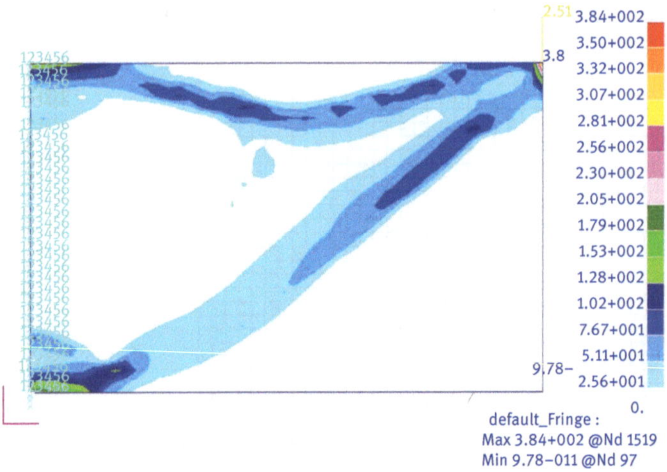

Figure 4.15: Stress distribution of the topology structure.

Figure 4.16: Optimal topological configuration with buckling constraint.

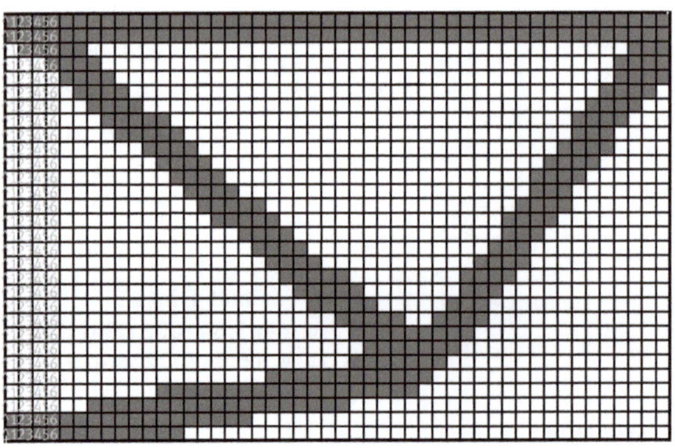

Figure 4.17: Optimal topological configuration with stress constraint.

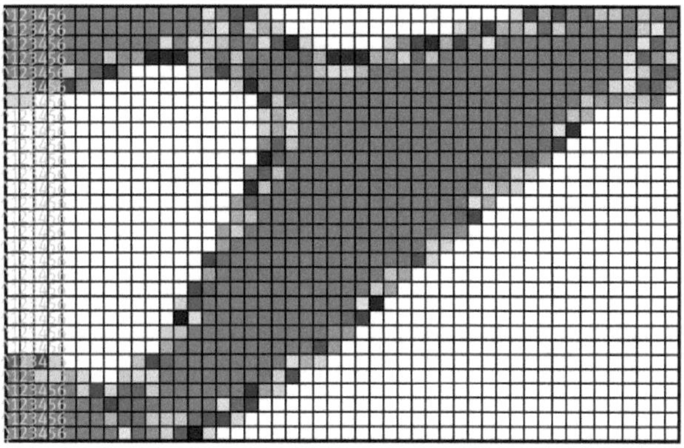

Figure 4.18: Topological configuration with buckling and stress constraints before inversion.

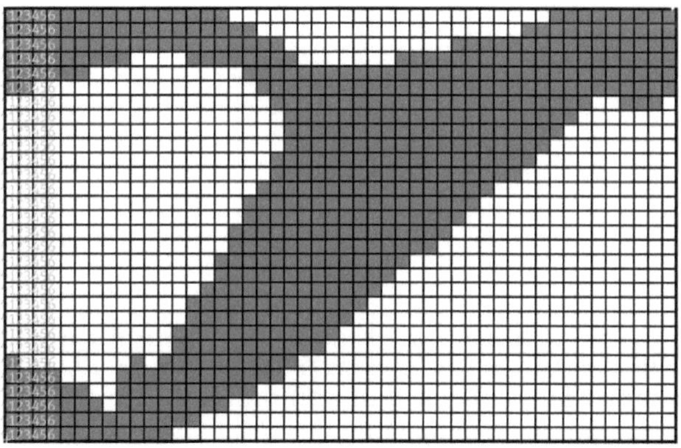

Figure 4.19: Topological configuration with buckling and stress constraints after inversion.

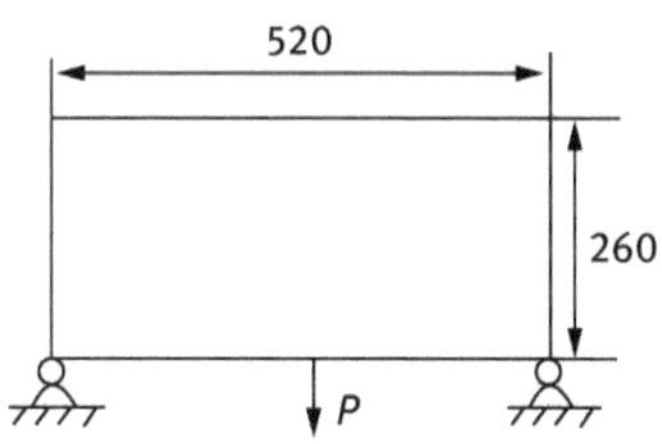

Figure 4.20: Base structure.

The allowable stress of the structure is 360 MPa. The buckling factor is −0.53854. The initial critical load is 7,593.4 N. The upper limit allowable stress of the structure is 150 MPa. In this example, the upper limit maximal value of the critical load is 3,700 N. The convergence precision is 0.001, after 35 times of iteration. The quality of the optimized topology is 0.69 kg (Table 4.6).

Table 4.6: Comparison of topological results with different constraints.

Constraint type	Initial quality (kg)	Optimized quality (kg)	Iteration number
Stress	2.2	0.62	50
Stress+Buckling	2.2	0.69	35

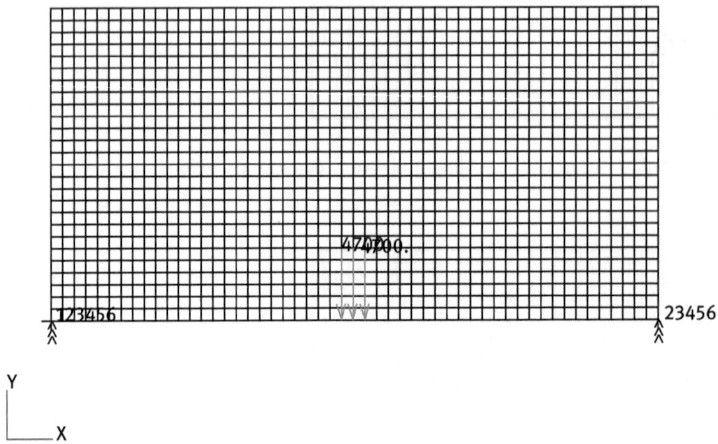

Figure 4.21: Finite element model.

It can be seen from the graph Figure 4.21 to Figure 4.27 that the topological structure under stress constraints is relatively thin and uniform. Because the force is applied to the lower boundary, the buckling factor is negative and the buckling critical force is not convergent. When the buckling constraint and displacement constraint are considered, the beam of the lower part is added in the topology, and all the connections of the beams are strengthened, so that the stability of the whole structure is enhanced.

Example 4

As illustrated in Figure 4.28, the base structure is a plane elastic body with size 2,000 × 400 × 9.0 mm. The elastic modulus of the structure material is $E = 68.89 \times 10^3$ MPa, the Poisson's ratio is $y = 0.3$, and the material density is $\rho = 2.7 \times 10^{-6}$ kg/mm^3. The right- and left-hand underpart corner points of the structure are clasped. A concentrated force $P = 43{,}000$ N along the up-to-down (-y) direction is located at the upper boundary point of structure. A quadrilateral element with four nodes is used and the total number of elements is 100 × 20. The quality of the base structure is 19.44 kg.

The allowable stress of the structure is 366 MPa. The buckling factor is 0.09. The initial critical load is 3,873.31 N. The upper limit allowable stress of the structure is

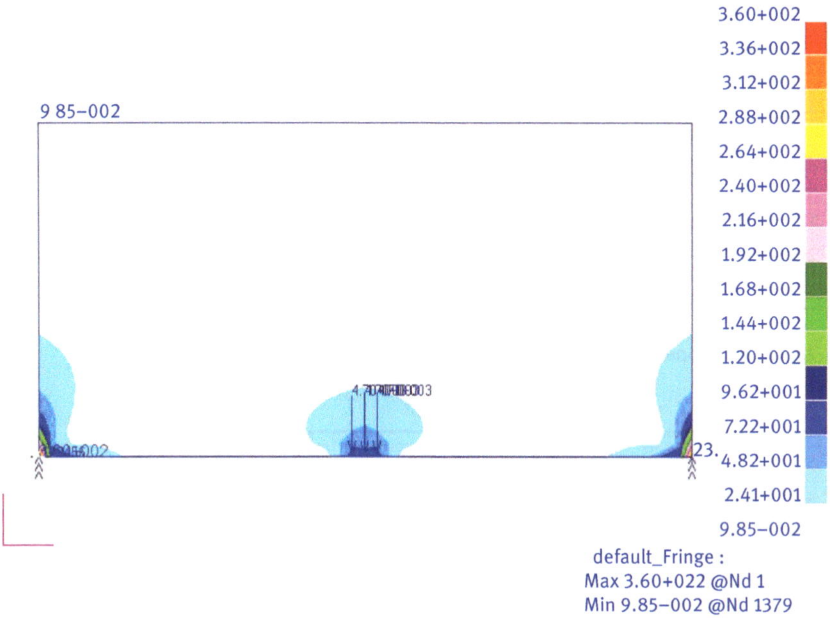

Figure 4.22: Stress fringe result of the base structure.

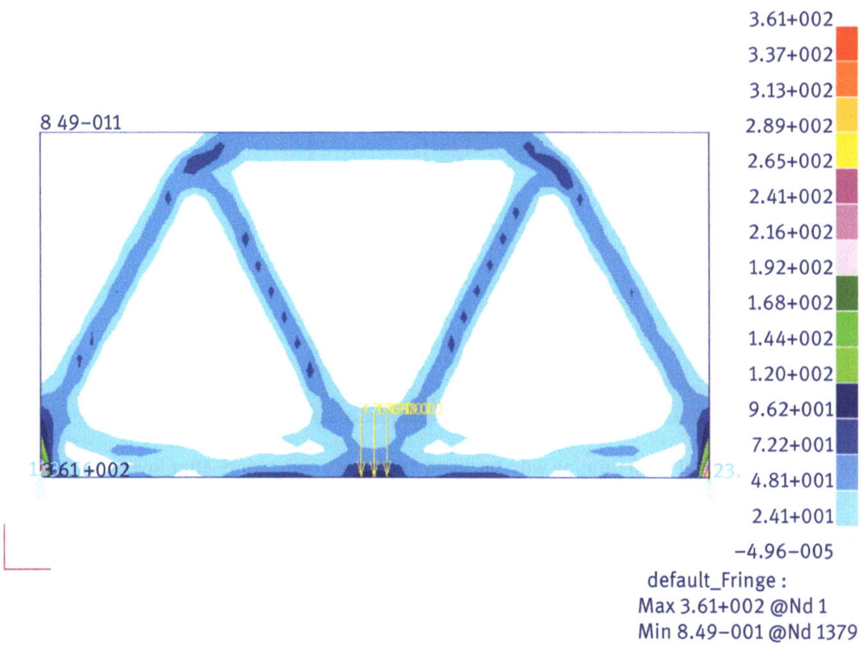

Figure 4.23: Stress distribution of the topology structure.

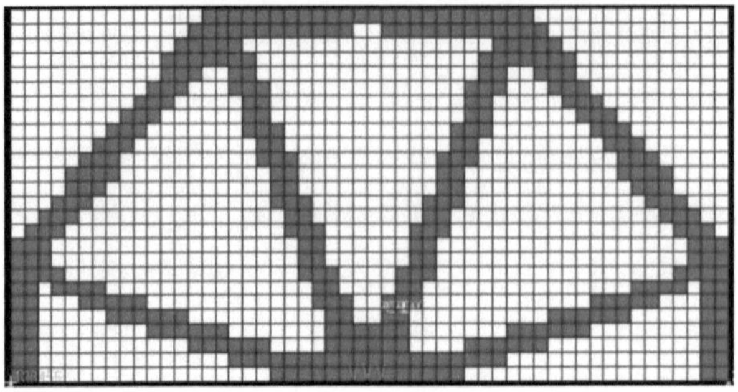

Figure 4.24: Optimal topological configuration with stress constraint.

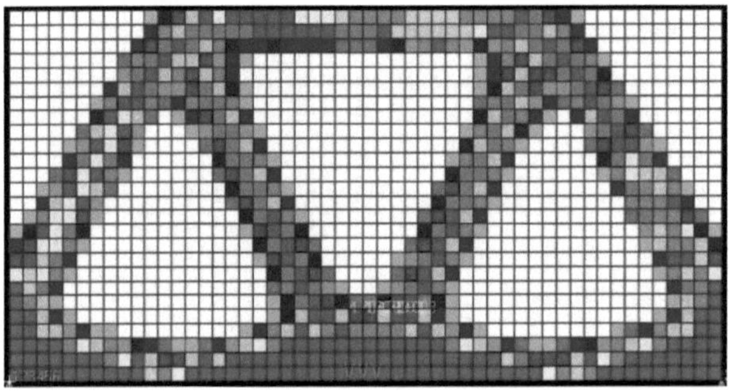

Figure 4.25: Topological configuration with buckling and stress constraints before inversion.

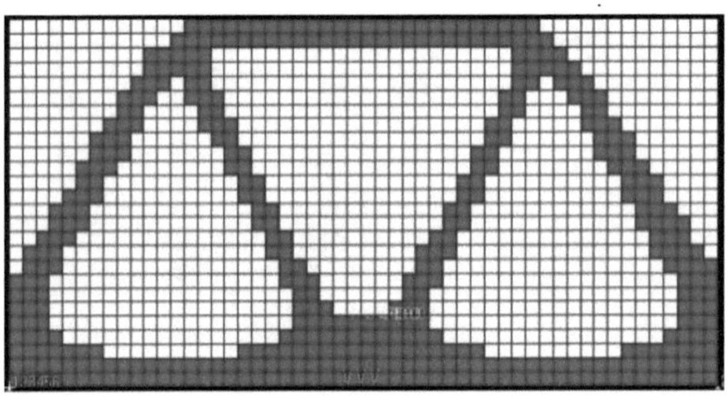

Figure 4.26: Topological configuration with buckling and stress constraints after inversion.

150 MPa. In this example, the upper limit maximal value of the critical load is 2,300 N. The convergence precision is 0.001, after 49 times of iteration. The quality of the optimized topology is 9.15 kg (table 4.7).

It can be seen from the graph Figure 4.29 to Figure 4.35 that the topological structure of the stress constraints in the literature [62] is relatively thin and uniform,

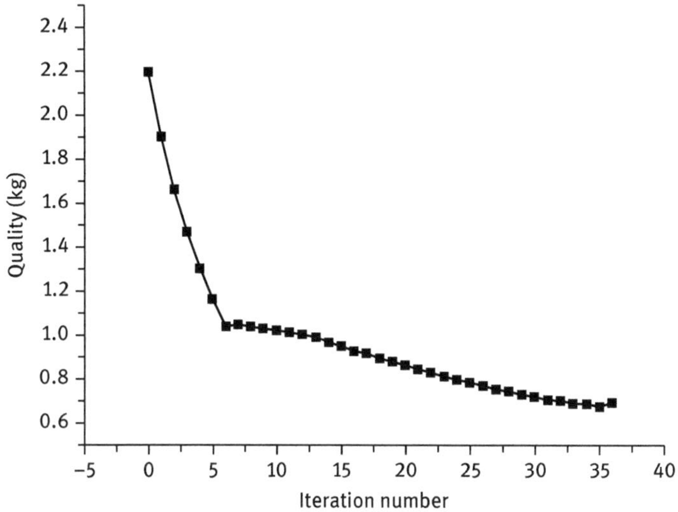

Figure 4.27: Iterative history of quality.

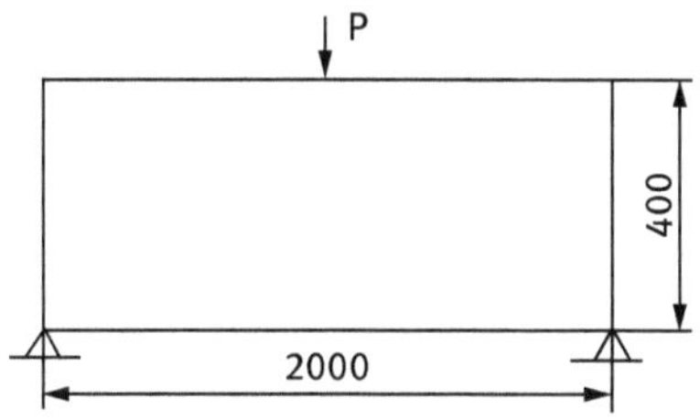

Figure 4.28: Base structure.

Table 4.7: Comparison of topological results with different constraints.

Constraint type	Initial quality (kg)	Optimized quality (kg)	Iteration number
Stress	19.44	8.1	51
Stress + Buckling	19.44	9.15	49

and the upper beam is relatively long. Because the length of the structure is far greater than the width and the structure's main effect is bend, the topological optimization with buckling constraint does not converge. When the buckling constraints and displacement constraints are considered, the length of the upper beam of the topological structure is reduced, and the whole structure becomes compact. The stability of the whole structure is enhanced in the connection of all the beams in the topology.

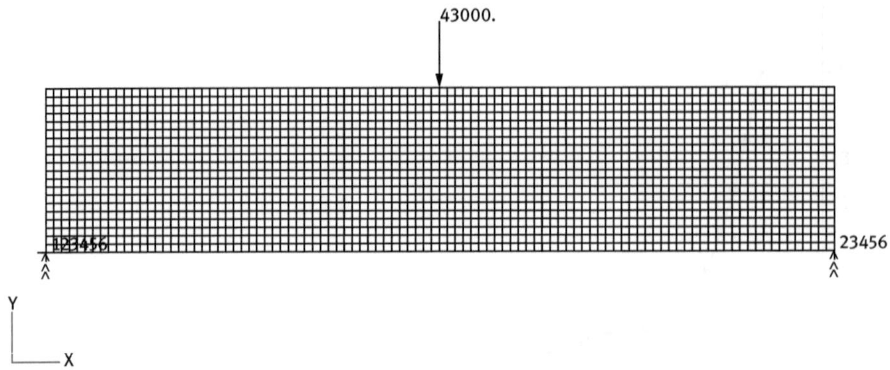

Figure 4.29: Finite element model.

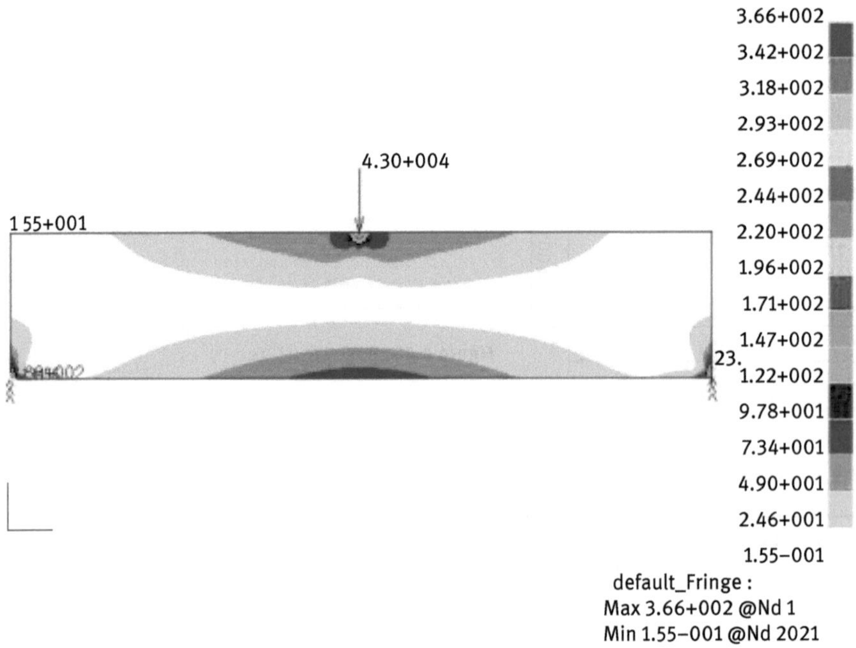

Figure 4.30: Stress fringe result of the base structure.

Example 5

As illustrated in Figure 4.36, the base structure is a plane elastic body with size 100 × 240 × 9.0 mm. The elastic modulus of the structure material is $E = 68.89 \times 10^3$ MPa, the Poisson's ratio is $y = 0.3$, and the material density is $\rho = 2.7 \times 10^{-6}$ kg/mm^3. Fixed constraints are used at the left boundary. A concentrated force $P = 30{,}000$ N along the

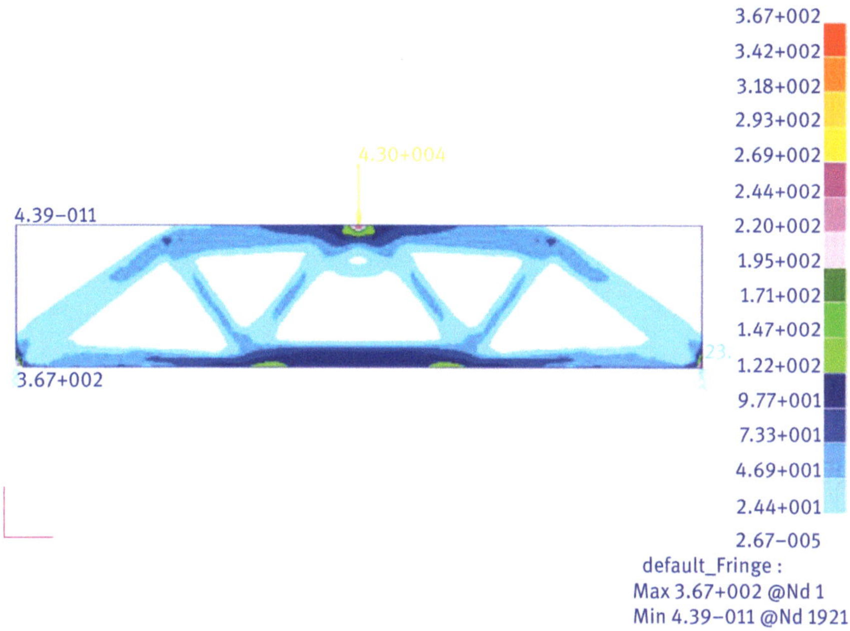

Figure 4.31: Stress distribution of the topology structure.

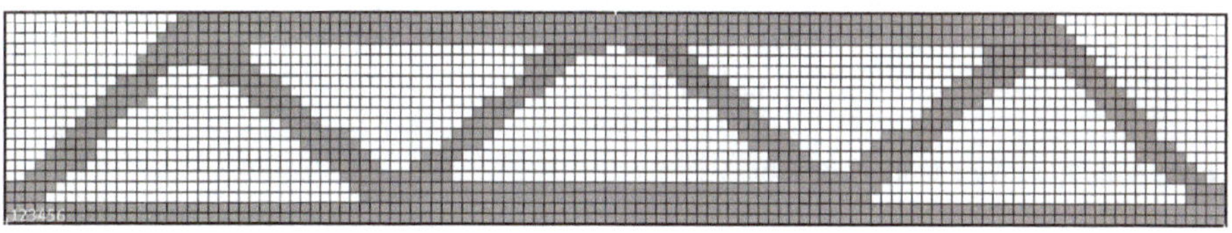

Figure 4.32: Topological configuration with stress constraint in [62].

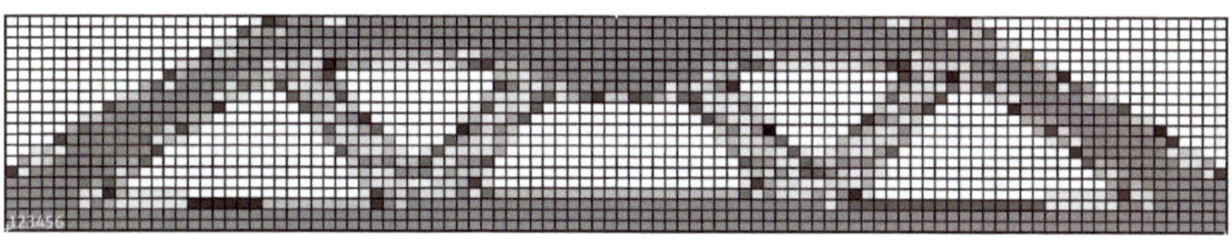

Figure 4.33: Topological configuration with buckling and stress constraints before inversion.

up-to-down (-*y*) direction is located in the right boundary midpoint of structure. A quadrilateral element with four nodes is used and the total number of elements is 20 × 48. The quality of the base structure is 1.685 kg (table 4.8).

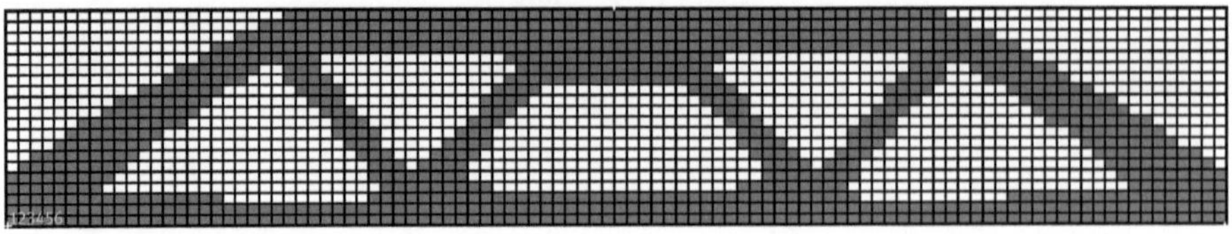

Figure 4.34: Topological configuration with buckling and stress constraints after inversion.

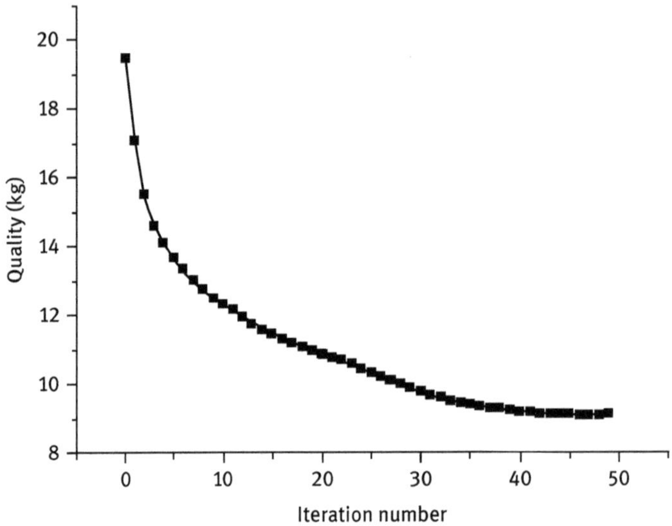

Figure 4.35: Iterative history of quality.

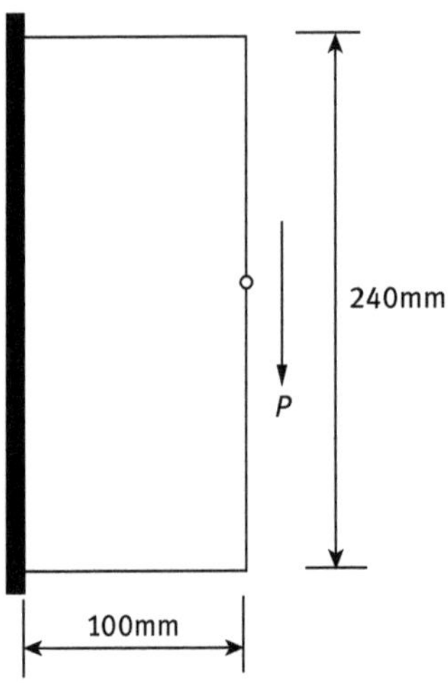

Figure 4.36: Base structure.

Table 4.8: Comparison of topological results with different constraints.

Constraint type	Initial quality (kg)	Optimized quality (kg)	Iteration number
Stress	1.685	0.386	36
Buckling	1.685	0.51	29
Stress + Buckling	1.685	0.49	31

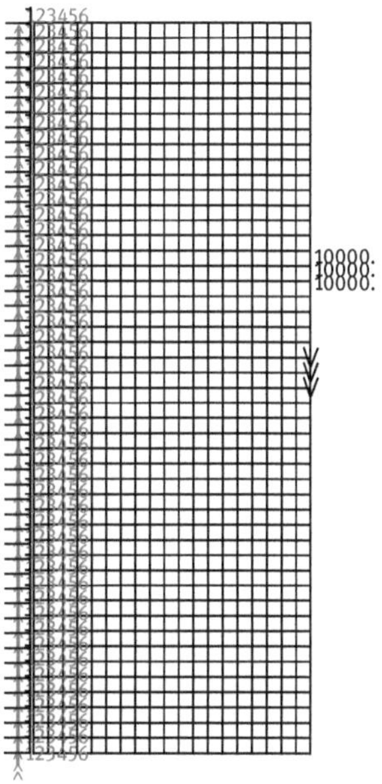

Figure 4.37: Finite element model.

The allowable stress of the structure is 345 MPa. The buckling factor is 18.995. The initial critical load is 569,850 N. The upper limit allowable stress of the structure is 200 MPa. In this example, the upper limit maximal value of the critical load is 200,000 N. The convergence precision is 0.001, after 31 times of iteration. The quality of the optimized topology is 0.49 kg. Compared with the literature [63], the topological structures of the stress and buckling constraints become thick and strong, the distance between the two pillars is reduced, and the stability is increased (Figure 4.37 to Figure 4.43).

6. Examples of multiworking conditions

Example 1

As illustrated in Figure 4.44, the base structure is a plane elastic body with size 600 × 260 × 2.0 mm. The elastic modulus of the structure material is $E = 68.89 \times 10^3$MPa, the Poisson's ratio is $y = 0.3$, and the material density is $\rho = 2.7 \times 10^{-6}$ kg/mm^3. The right-

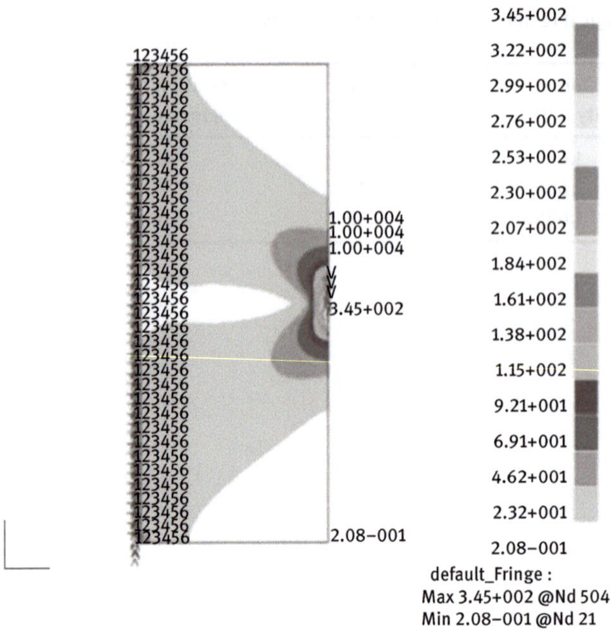

Figure 4.38: Stress fringe result of the base structure.

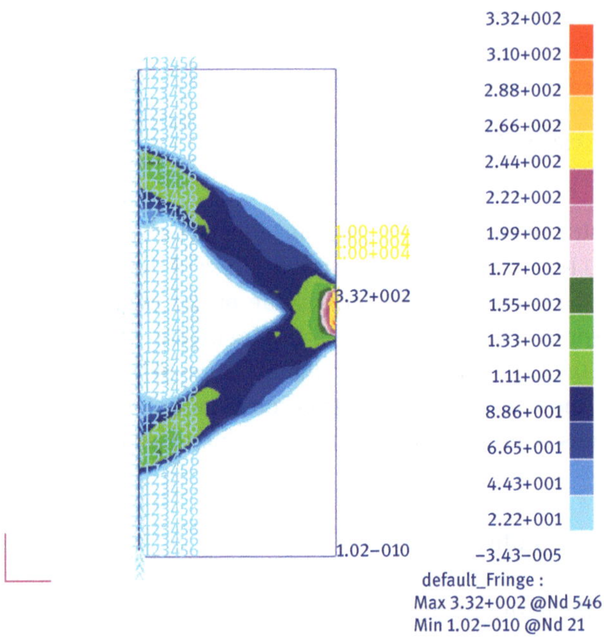

Figure 4.39: Stress distribution of the topology structure.

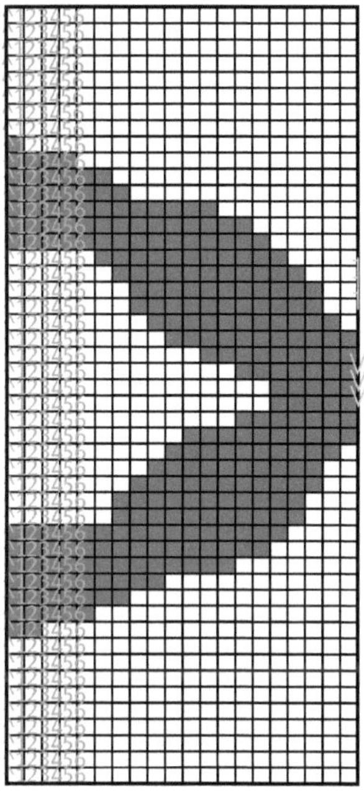

Figure 4.40: Topological configuration with buckling constraint.

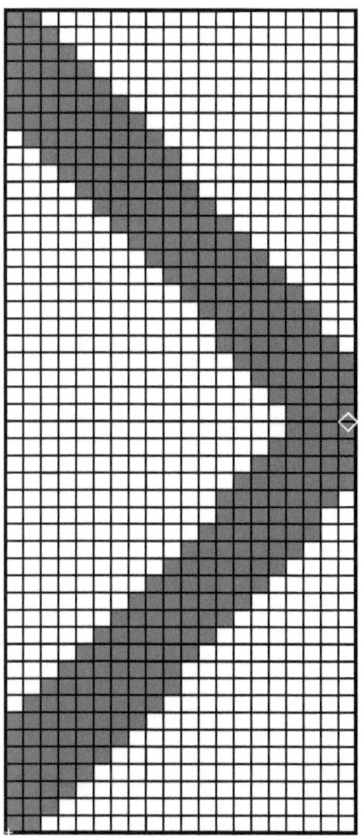

Figure 4.41: Topological configuration with stress constraint in [63].

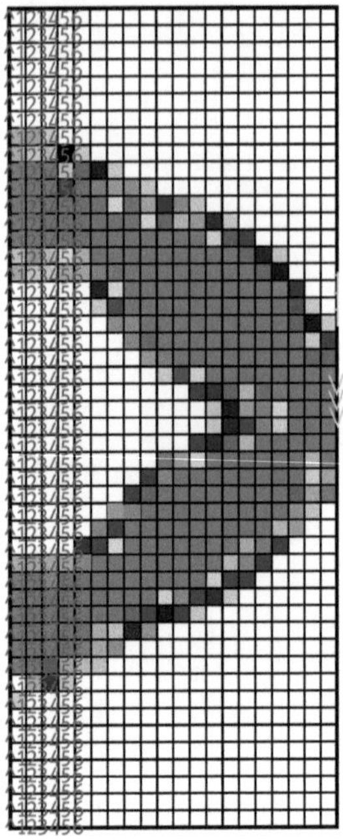

Figure 4.42: Topological configuration with buckling and stress constraints before inversion.

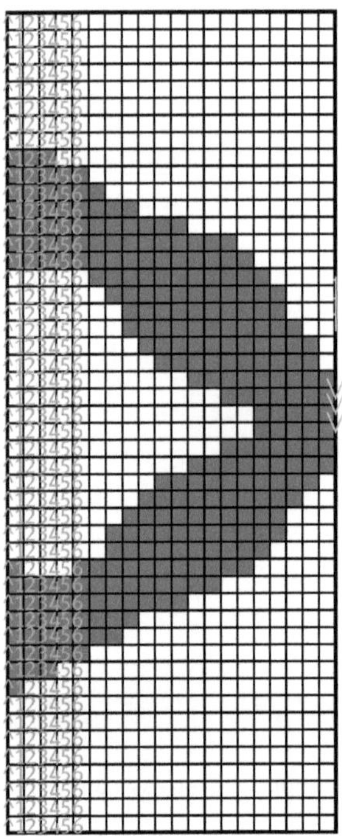

Figure 4.43: Topological configuration with buckling and stress constraints after inversion.

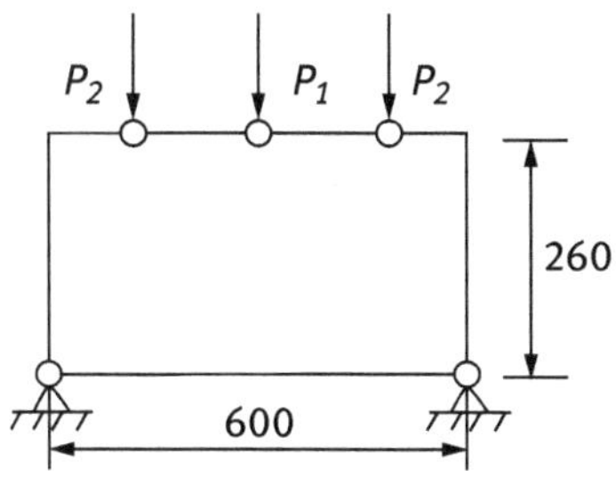

Figure 4.44: Base uretructure.

Table 4.9: Comparison of topological results with different constraints and load cases.

Constraint type	Initial quality (kg)	Optimized quality (kg)	Iteration number
Single-working condition 1 + stress	0.84	0.12	39
Single-working condition 2 + stress	0.84	0.09	30
Working condition 1 + stress + buckling	0.84	0.198	41
Working condition 2 + stress + buckling	0.84	0.117	35
Multiworking condition + stress	0.84	0.19	42
First-order buckling + stress + Multiworking condition	0.84	0.22	42
Second-order buckling + stress + Multiworking condition	0.84	0.367	34
The first- and second-order buckling + stress + Multiworking condition	0.84	0.367	34

and left-hand underpart corner points of the structure are clasped. A quadrilateral element with four nodes is used and the total number of elements is 60 × 26. Condition 1: a concentrated force $P1 = 7{,}000$ N acts on the midpoint of the upper boundary. Condition 2: two concentrated forces $P2 = 3{,}500$ act on the upper boundary point which distance is 220 mm to the midpoint. The quality of the base structure is 0.84 kg. The first-order buckling force of the base structure is 112 N, the upper critical buckling is $\bar{\lambda}_1 = 60$ N, the second-order buckling critical force is 792 N, the upper critical force of buckling is $\bar{\lambda}_2 = 400$ N, and the stress upper limit is 150 MPa. The convergence precision is 0.001 (Table 4.9).

This mainly studies the topology optimization of continuum structures under the combined effects of buckling constraints and stress constraints under single- and multiworking conditions, as well as multimode and multiload conditions. As can be seen from the graph Figure 4.45 to Figure 4.57, the topological structure with buckling constraints is "robust," but higher order constraint plays a major role when considering the multi-order buckling constraints. For example, the first two-order buckling constraints are the same as the topological structure of the multicondition topology and the two-order buckling constraints.

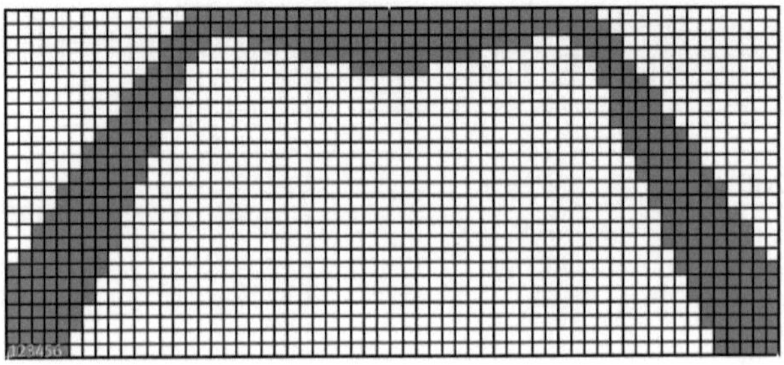

Figure 4.45: Optimal topological configuration of load case 1 with buckling constraints.

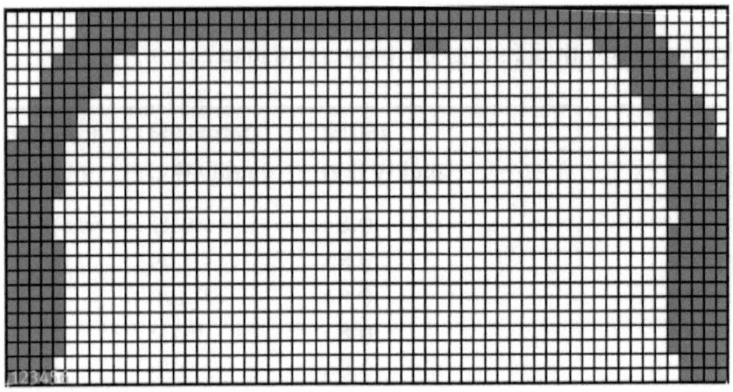

Figure 4.46: Optimal topological configuration of load case 2 with buckling constraints.

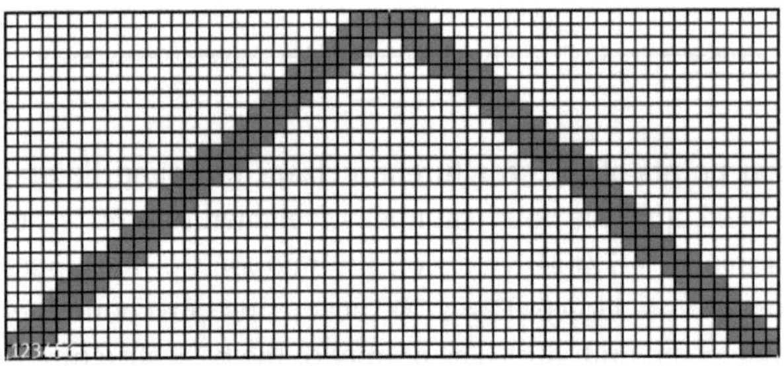

Figure 4.47: Topological configuration of load case 1 with stress constraints.

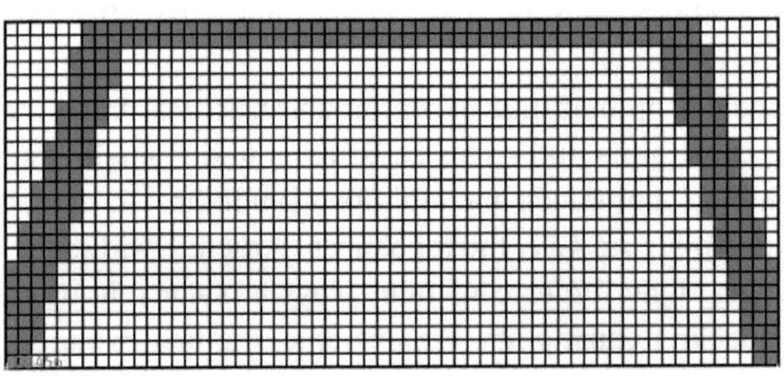

Figure 4.48: Topological configuration of load case 2 with stress constraints.

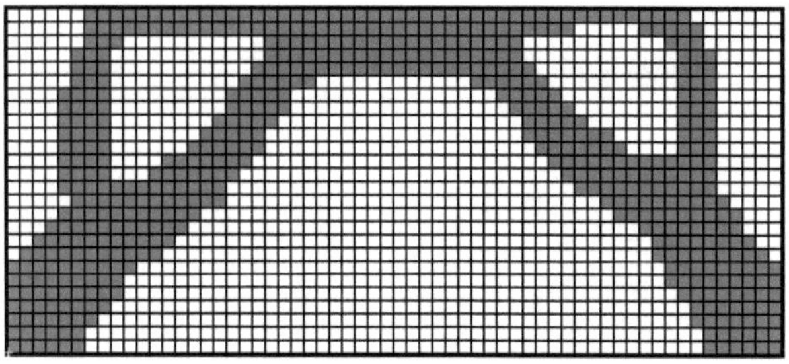

Figure 4.49: Topological configuration of multiple load cases with buckling constraints.

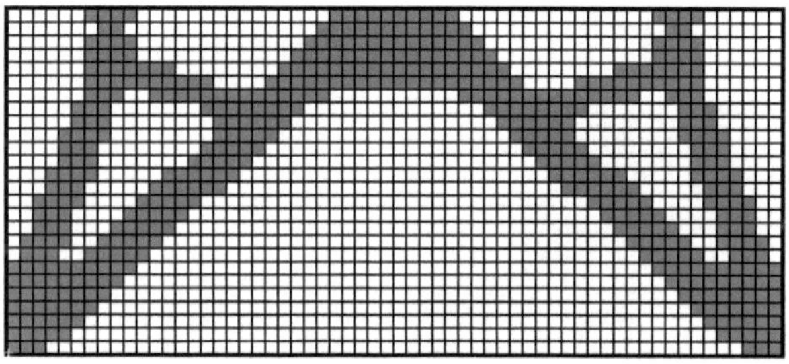

Figure 4.50: Topological configuration of multiple load cases with stress constraints.

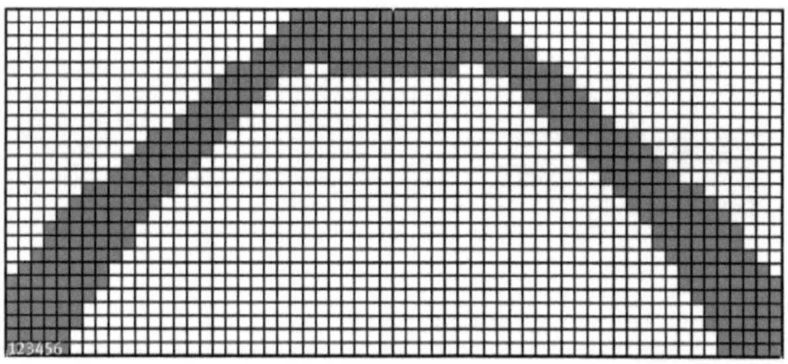

Figure 4.51: Topological configuration of load case 1 with buckling and stress constraints.

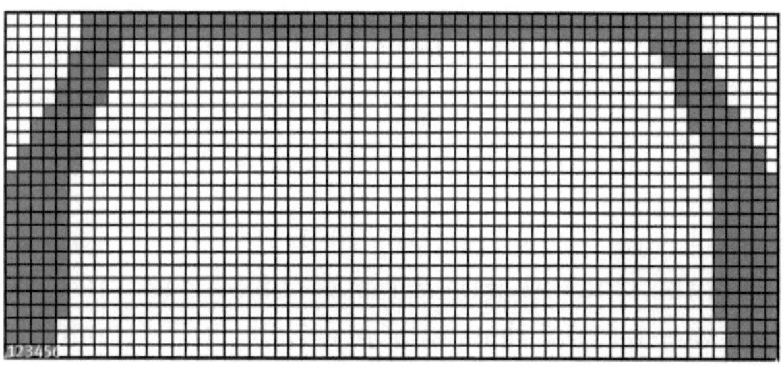

Figure 4.52: Topological configuration of load case 2 with buckling and stress constraints.

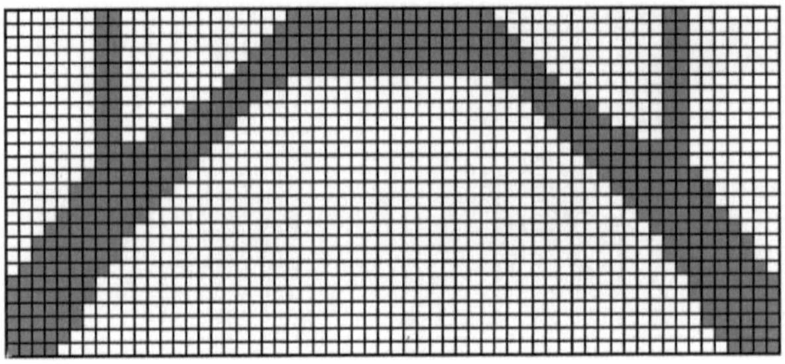

Figure 4.53: First-order mode topological configuration of multiple load cases with buckling and stress constraints.

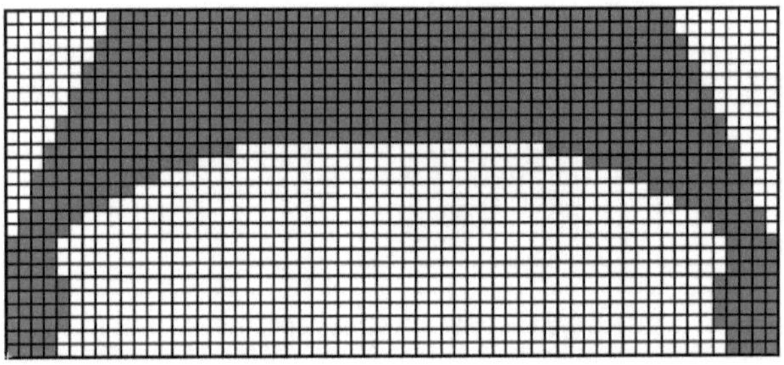

Figure 4.54: Second-order mode topological configuration with multiple load cases $\bar{\lambda}_2 = 400$ N.

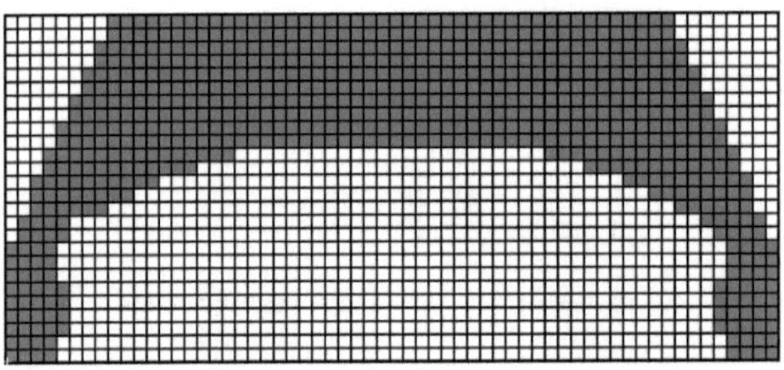

Figure 4.55: First- and second-order mode topological configurations with multiple load cases $\bar{\lambda}_1 = 60$ N and $\bar{\lambda}_2 = 400$ N.

Example 2

As illustrated in Figure 4.58, the base structure is a plane elastic body with size 120 × 60 × 6.0 mm. The elastic modulus of the structure material is $E = 68.89 \times 10^3$ MPa, the Poisson's ratio is $y = 0.25$, and the material density is $\rho = 7.8 \times 10^{-6}$ kg/mm^3. A quadrilateral element with four nodes is used and the total number of elements is 60 × 30. Two concentrated forces $P = 1{,}500$ act on the upper and lower boundary corners;

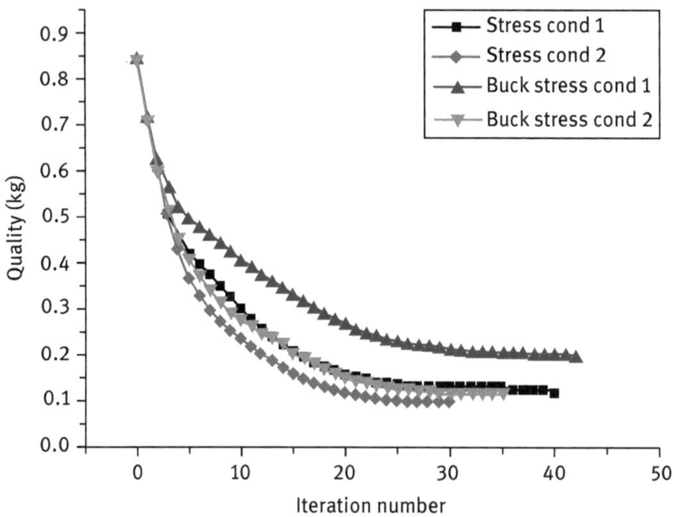

Figure 4.56: Iterative history of single load case quality.

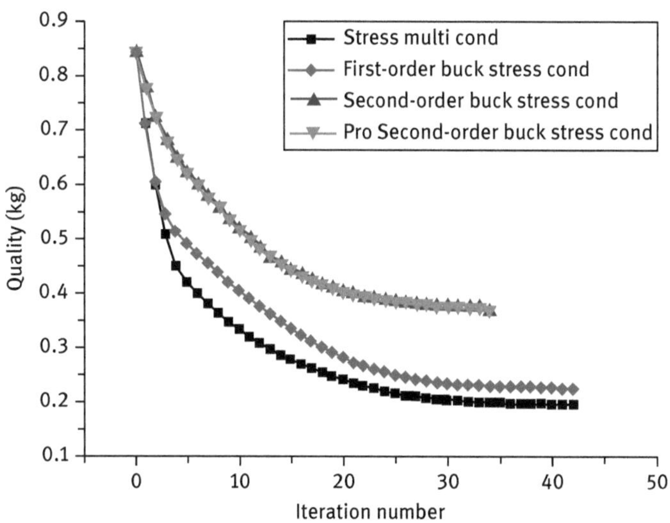

Figure 4.57: Different iterative history of multiple load cases quality.

all the fixed constraints are in the left boundary. Condition 1: *P* acts on the upper corner point. Condition 2: *P* acts on the lower corner point. The quality of the base structure is 0.337 kg (Figure 4.59 to Figure 4.69). The first-order buckling critical force of the base structure is 67,066 N, and the single condition upper limit of the buckling critical force is 30,000 N. The stress upper limit is 150 MPa (table 4.10).

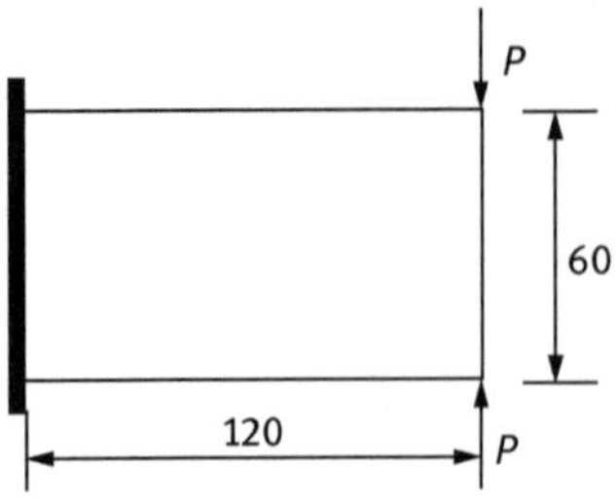

Figure 4.58: Base structure.

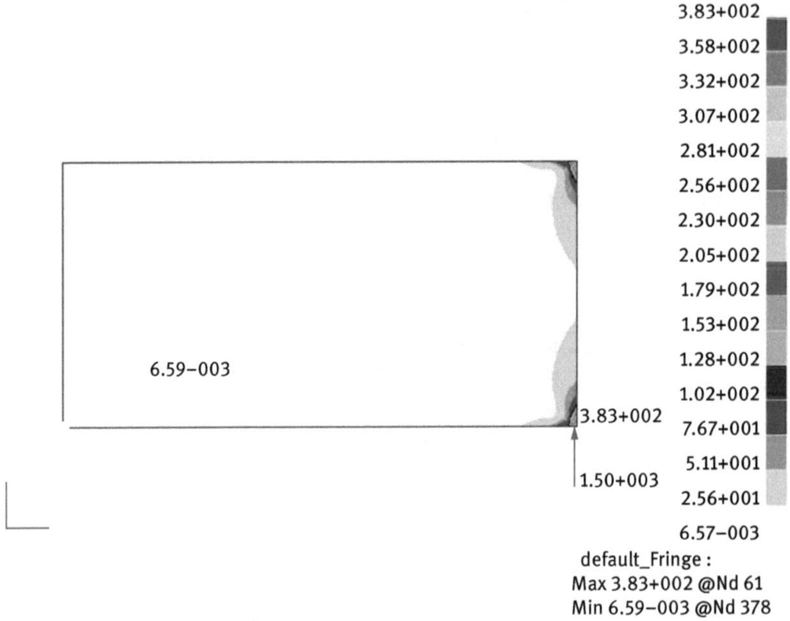

Figure 4.59: Stress fringe result of the base structure with multiple load cases.

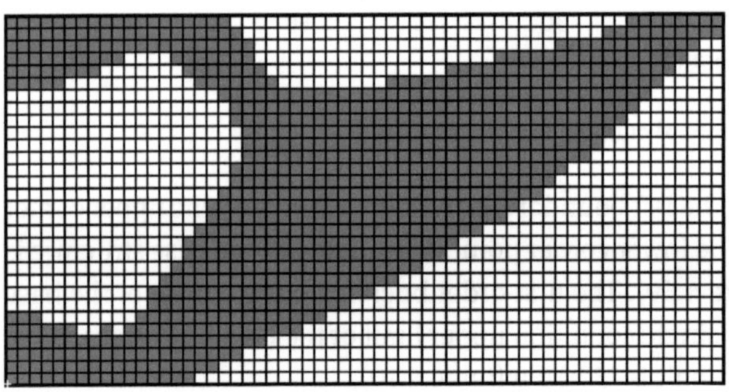

Figure 4.60: Optimal topological configuration of load case 1 with buckling constraints.

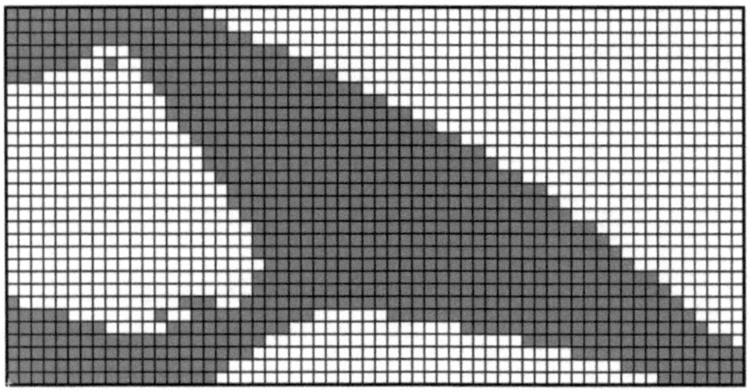

Figure 4.61: Optimal topological configuration of load case 2 with buckling constraints.

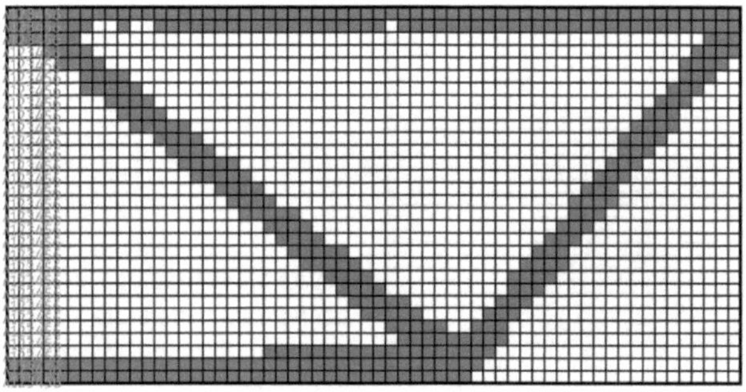

Figure 4.62: Topological configuration of load case 1 with stress constraints.

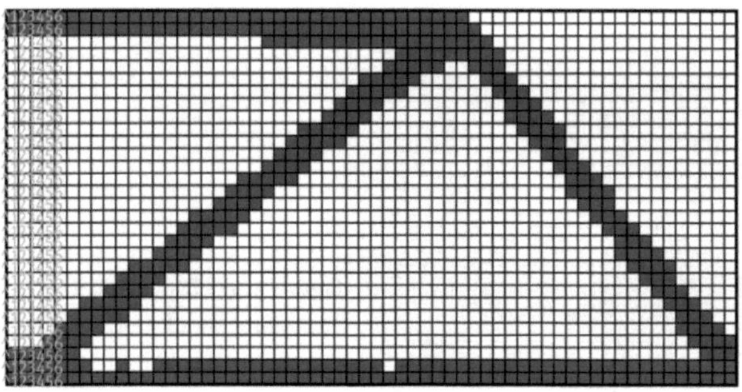

Figure 4.63: Topological configuration of load case 2 with stress constraints.

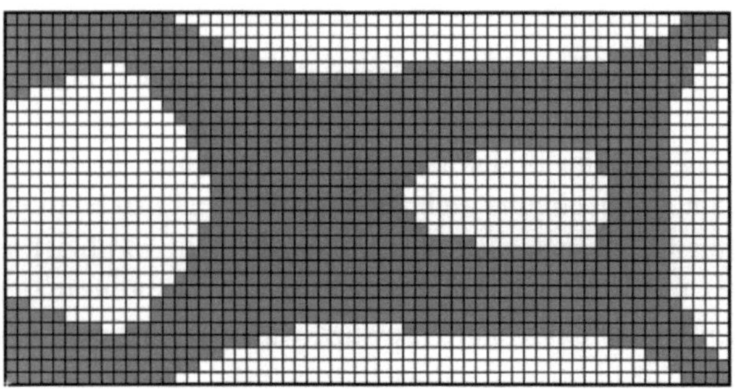

Figure 4.64: Topological configuration of multiple load cases with buckling constraints.

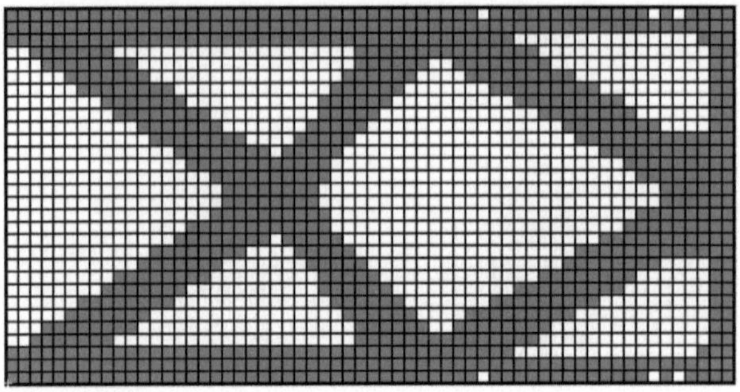

Figure 4.65: Topological configuration of multiple load cases with stress constraints.

Figure 4.66: Topological configuration of single load case with buckling and stress constraints.

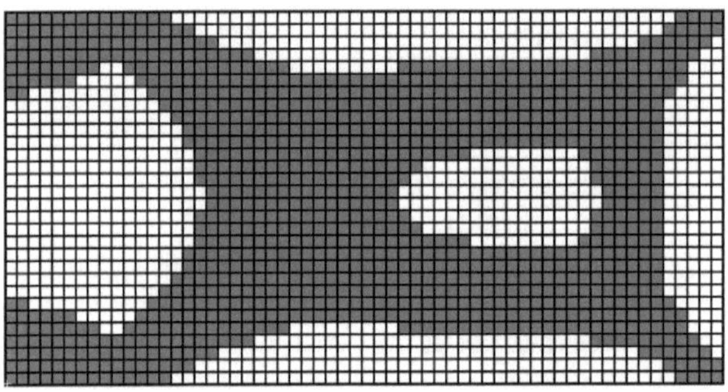

Figure 4.67: Topological configuration of multiple load cases with buckling and stress constraints.

4.9.2 Engineering Application 1: Topology Optimization of the Zhaozhou Bridge

As shown in Figure 4.70, the Zhaozhou Bridge is a famous stone arch bridge of ancient China. It is the most famous cut circular arch and arch plus small arch bridge in the world. It was built during the Sui Dynasty, 1,400 years earlier. The width of the bridge is 9 m, the main arch span is 37.02 m, its height is 7.23 m, and the radius of the circular arc is 27.3 m. The two arch shoulders are provided with two unequal arch spans, respectively, of 3.8 m and 2.8 m, so as to reduce the quality of the deck and is

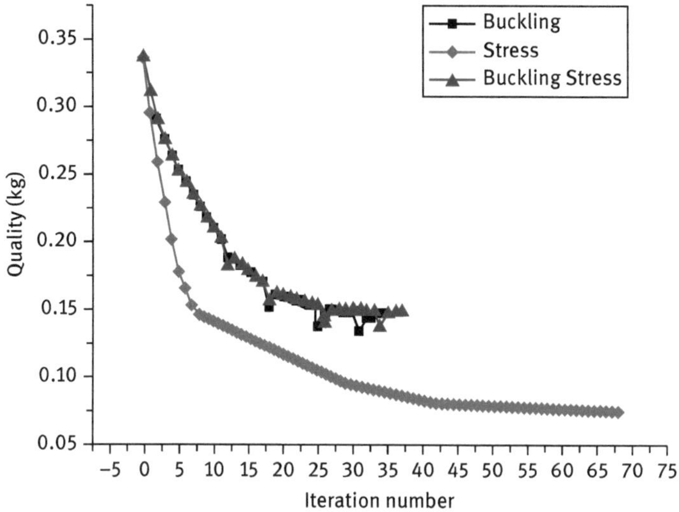

Figure 4.68: Iterative history of single load case quality.

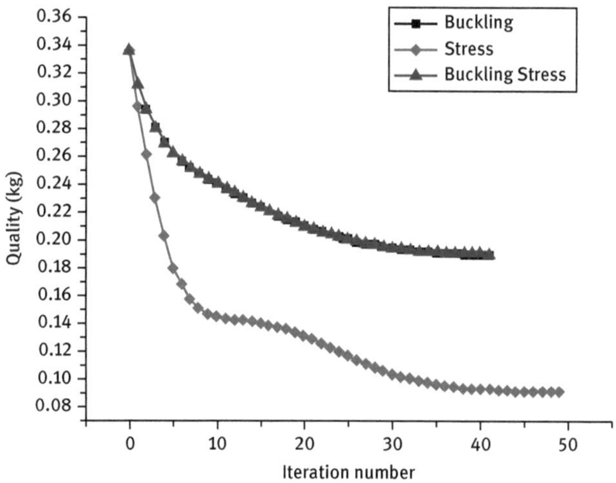

Figure 4.69: Different iterative history of multiple load cases quality.

convenient for flood discharge, while enhancing the appearance of the bridge. The Zhaozhou Bridge shows the diligence and wisdom of the people of ancient China. Whether the structure design of Zhaozhou Bridge can be used to verify by the current scientific theory? Academician Lingxi Qian [4] using elastoplastic theory to calculate and find that the main arch shoulder two "flood discharge hole" is key measures of the structure design. The design ensures the balance in the arch, and also ensures the reasonable distribution of the pressure line.

Table 4.10: Comparison of topological results with different constraints and load cases.

Constraint type	Initial quality (kg)	Optimized quality (kg)	Iteration number
Single working condition + buckling	0.337	0.1464	35
Multiworking conditions + buckling	0.337	0.18511	42
Single working condition + stress	0.337	0.07364	68
Multiworking conditions + stress	0.337	0.08994	49
Single working condition + Buckling + stress	0.337	0.1487	37
Multiworking conditions + Buckling + stress	0.337	0.1894	41

Figure 4.70: Picture of the Zhaozhou Bridge (source: http://image.so.com/v?q).

The rationality of the bridge design is analyzed and verified, based on the structural topology optimization design theory. According to the actual data, the finite element model of the Zhaozhou Bridge is established.

The length of the base structure is 37,020 mm, the height is 6,900 mm, the upper boundary is arc, the intermediate top height of the arc is 7,700 mm, the upper bound of the arc bridge deck is applied uniformly distributed, the bridge floor is the nondesign area, and both sides of the model have fixed support .The structure analysis related parameters are sane as stone: the density of the material is $\rho = 2.8 \times 10^3$ kg/m^3, the elastic modulus is $E = 5.488 \times 10^9$ N/m^2, and the Poisson's ratio is $\nu = 0.18$. The stress upper limit value is 200 MPa, and the buckling critical force limit is 96,000 N.

In order to establish the finite element model similar to the actual Zhaozhou Bridge, the abutment was increased in the model. The slope of two lines and the bridge abutment slope according to Figure 4.70 probably were selected, the uniform force was on the bridge deck, the bridge abutment and the deck were nondesign areas, and Two slope lines intersected by fixed abutment and foundation. They were the symmetry constraint. The structure size is shown in Figure 4.71, and the nondesign area element grid is shown in Figure 4.72 to 4.73.

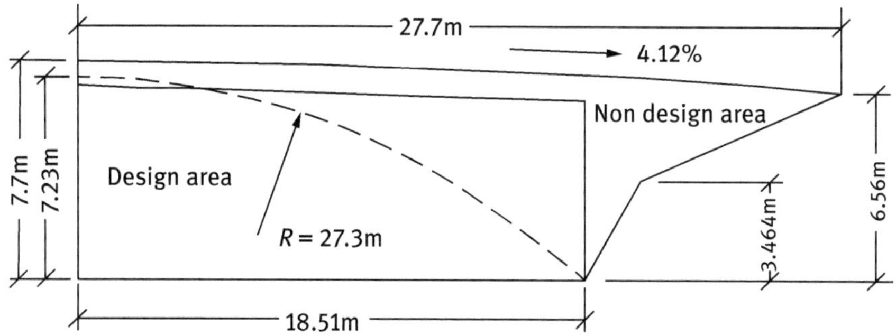

Figure 4.71: Base structure of topology optimization (half-bridge model).

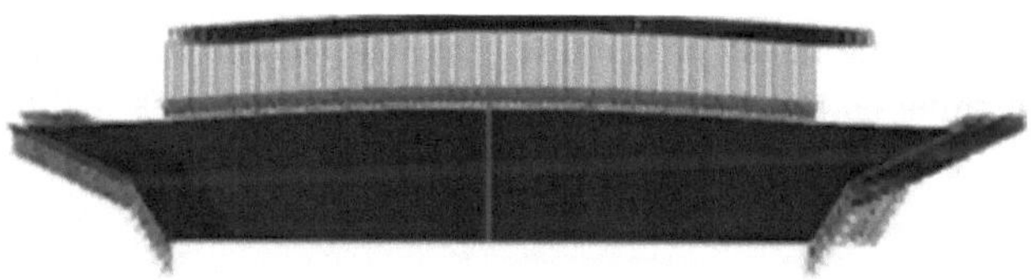

Figure 4.72: Finite element model of a bridge with abutment.

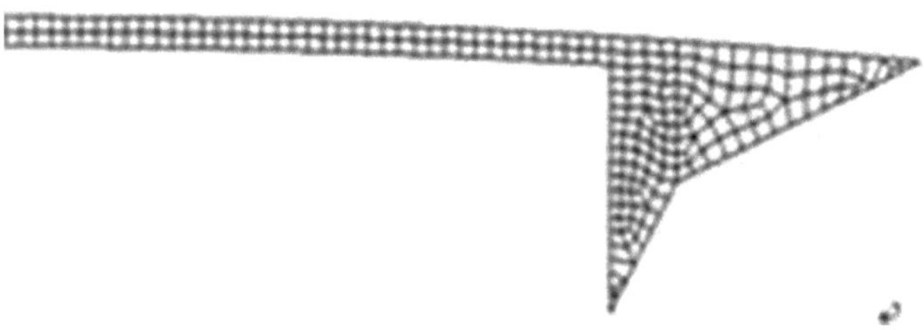

Figure 4.73: Elements of a nondesign region (half-bridge model).

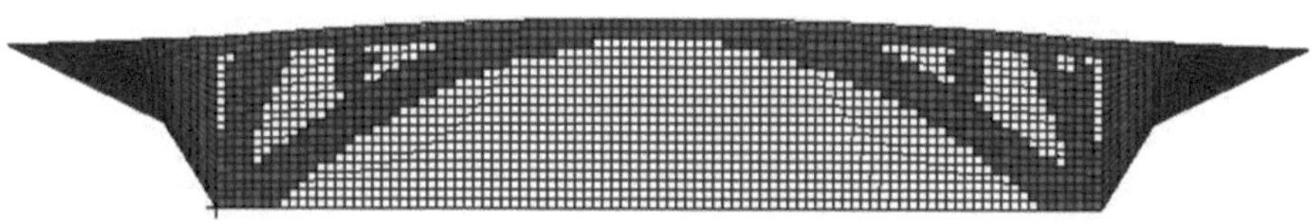

Figure 4.74: Bridge shape obtained by topology optimization with stress constraint.

The stress constraint is transformed into the total strain energy constraint, and the optimized topology graph is obtained as shown in Figure 4.74. It can be seen from the figure that the shape of the arc line of the original bridge is consistent with the main arch. There are three holes on the two shoulders: the size and position of the two big holes are the very same as in the original bridge, and the third hole is very small. Considering the requirements of the construction process, it can be ignored in the actual topology structure. Comparing the abstract topology structure as shown in Figure 4.75 with the original bridge, it can be seen that the shape in two figures is very similar. It shows that the rationality of the Zhaozhou Bridge in mechanics was confirmed from the perspective of topology optimization.

The above conclusions are just the topology of stress constraints. The topological structure of the graph is obtained by considering the effect of the buckling constraint only (Figure 4.76) or buckling constraint and the stress constraint simultaneously (Figure 4.77). As can be seen from the figure, the large arch is not a strict circle, there is only one small arch on the shoulders of the bridge, the topology becomes more robust, mainly in order to meet the needs of stability. The topology structure with buckling constraint is the same as the two kinds of constraints, which is mainly because the topology of the buckling constraints had met the needs of stress constraints.

Figure 4.75: Bridge shape abstracted from the topology configuration.

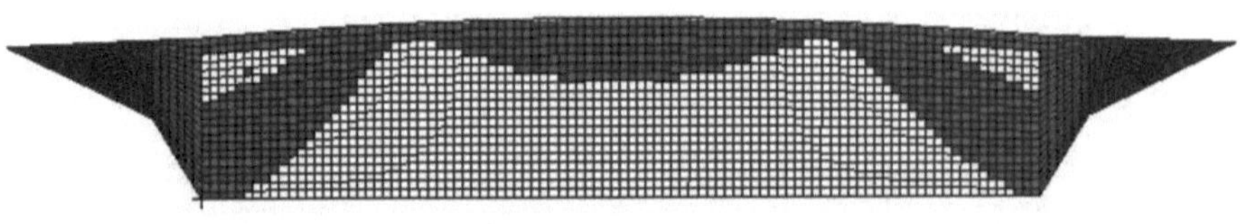

Figure 4.76: Bridge shape obtained by topology optimization with buckling constraint.

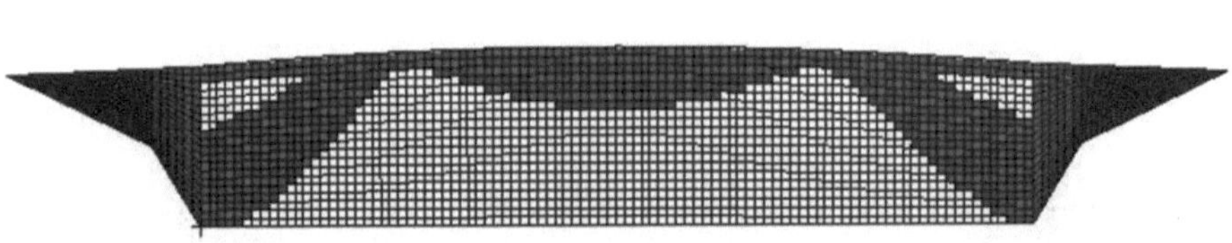

Figure 4.77: Bridge shape obtained by topology optimization with buckling and stress constraints.

4.9.3 Engineering Application 2: Research on Topology Optimization of the Drawbridge

The drawbridge is an important form in the modern bridge construction. The Chongqing Chaotianmen Yangtze River Bridge is a form of bridge shown in Figure 4.78. The bridge structure topology optimization under the distribution of force was generated to verify the structure of a bridge.

The base structure is a plane elastic body with size 16,000 × 6,000 × 6.0 mm. The elastic modulus of the structure material is $E = 210$ GPa, the Poisson's ratio is $y = 0.3$, and the material density is $\rho = 7.8 \times 10^{-6}$ kg/mm^3. A quadrilateral element with four nodes is used and the total number of elements is 2,511. A distributed force acts on the elements of the fifth layer, the elements are reserved for the nondesign region, and the two points on the lower boundary imposed a fixed support were imposed a fixed support. The finite element model is shown in Figure 4.79.

Figure 4.78: Picture of a drawbridge (source: http://image.so.com/v?q).

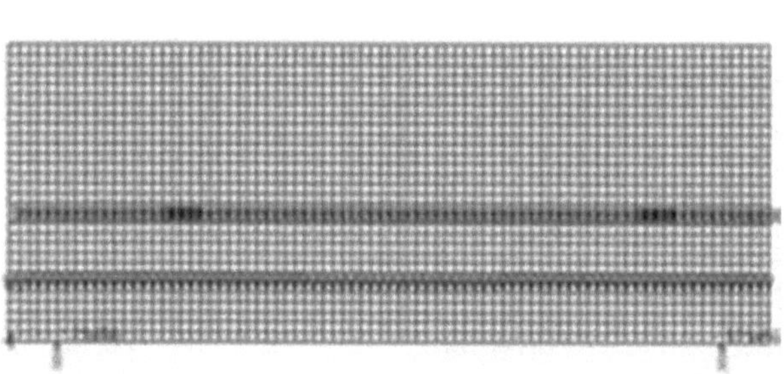

Figure 4.79: Finite element model of a drawbridge.

As can be seen from the stress constraint under the effect of topology optimization graphics Figure 4.80, the bridge topology shape is very similar to the original bridge, the outside structure is the arc frame, and the middle of the bridge is the sling between the frame and the bridge floor. Thus, the bridge shape conforms to mechanical properties.

Considering only buckling constraints, it can be seen from the topology graph Figure 4.81 that the bridge also has the external frame, just central sling disappeared. The bottom of the support strengthened, and the stability of the structure was enhanced. When the buckling constraint and the stress constraint acted at the same time, the central part of sling became thick, and the topological structure was different from the actual bridge shape (Figure 4.82).

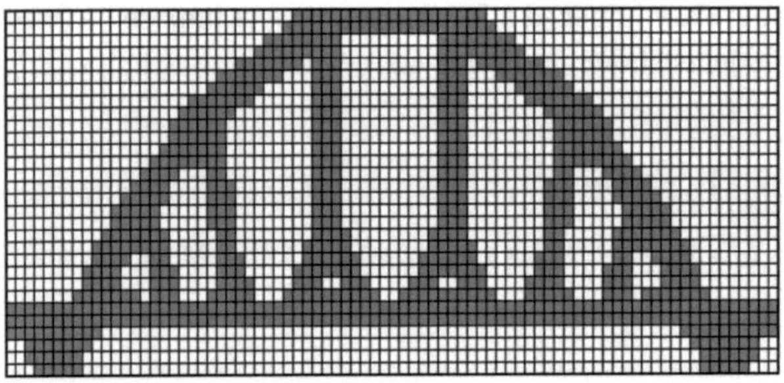

Figure 4.80: Bridge shape obtained by topology optimization with stress constraint.

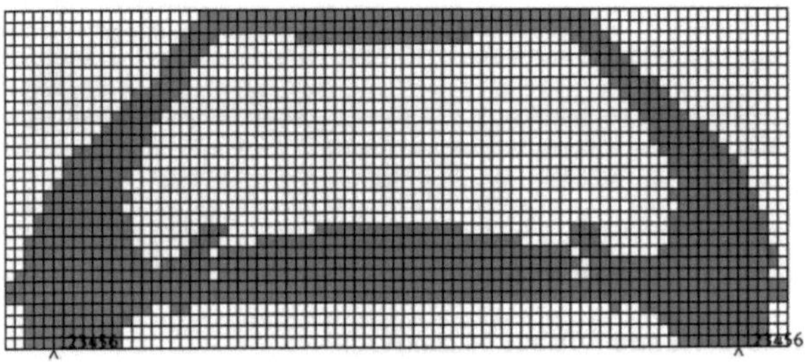

Figure 4.81: Bridge shape obtained by topology optimization with buckling constraint.

Figure 4.82: Bridge shape obtained by topology optimization with buckling and stress constraints.

4.9.4 Numerical Example of a Three-dimensional Solid Structure

Example 1

As illustrated in Figure 4.83, the base structure is a 3D elastic body with size 300 × 100 × 10 mm. The elastic modulus of the structure material is $E = 70 \times 10^3$ MPa, the Poisson's ratio is $y = 0.30$, and the material density is $\rho = 2.7 \times 10^{-6}$ kg/mm^3. A concentrated force $P = 9{,}000$ N along the up-to-down (-y) direction is located at three midpoints of the upper surface of the structure. The right- and left-hand underpart corner points of the structure are clasped. A solid element with eight nodes is used and the total number of elements is 60 × 20 × 2. The quality of the base structure is 0.81 kg.

The maximum von Mises stress of the structure is 382 MPa, the buckling factor is 3.185, and the critical force is 28,665 N. The upper limit of the stress is 150 MPa. The upper limit of the buckling critical force is 15,000 N. After 45 iterations, the model converges. After optimization, the quality of the structure is 0.19 kg (Table 4.11).

It can be seen that from Figure 4.84 to Figure 4.88 the topology is relatively thin and uniform, and the structure can meet the requirements of stress constraints, but the stability will be poor. In the topological structure of buckling constraints, the topology has better stability. When considering the stress constraints and buckling constraints, it can be seen from the graph that the topology structure is changed with respect to the single constraint, and is not the simple sum of the two topologies. At the intersection of the horizontal direction and the vertical branch, the topological structure becomes rough, the stability of the whole structure is strengthened, and the quality of the topological structure is also increased.

Figure 4.83: Base structure.

Table 4.11: Comparison of topological results with different constraints.

Constraint type	Initial quality (kg)	Optimized quality (kg)	Iteration number
Stress	0.81	0.095	41
Buckling	0.81	0.185	36
Stress + Buckling	0.81	0.19	45

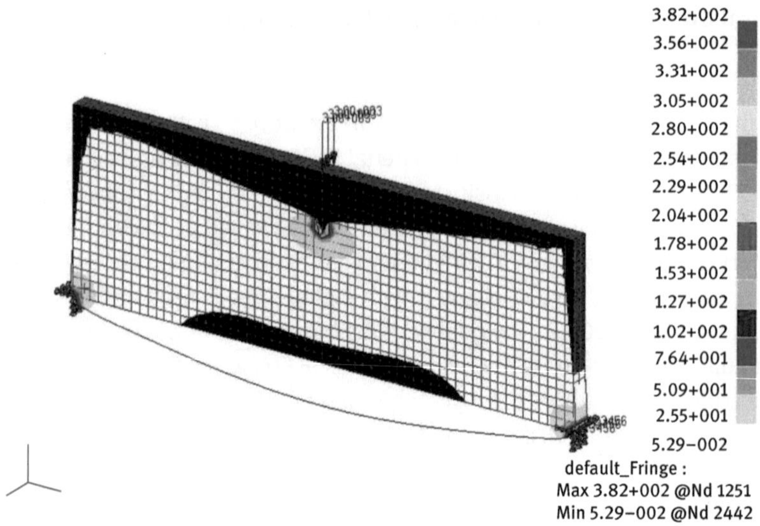

Figure 4.84: Stress fringe result of a base structure.

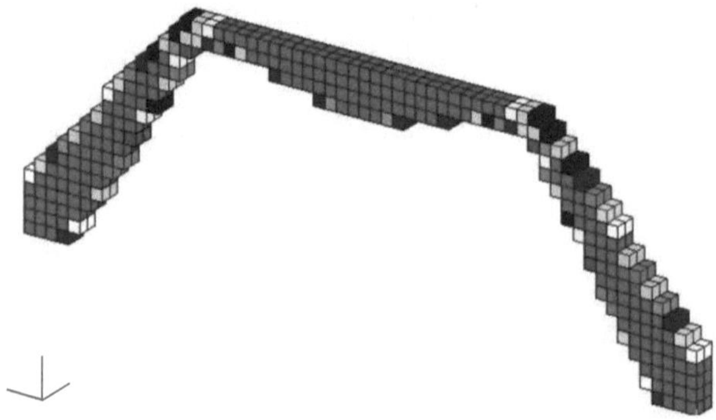

Figure 4.85: Optimal topological configuration with buckling constraints.

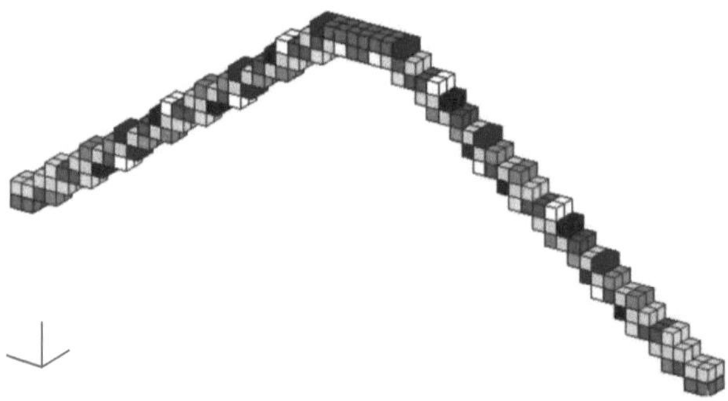

Figure 4.86: Optimal topological configuration with stress constraints.

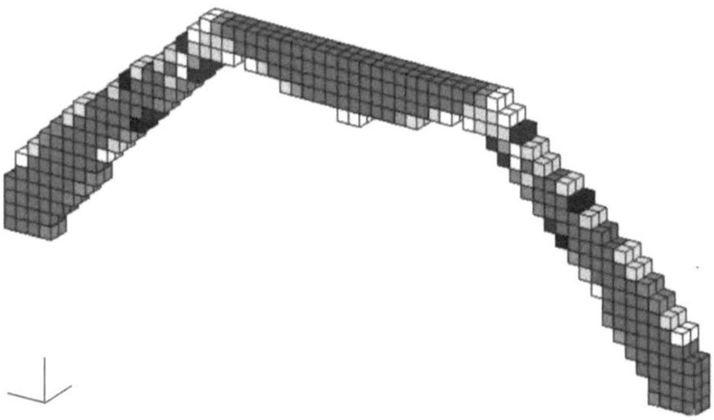

Figure 4.87: Optimal topological configuration with buckling and stress constraints.

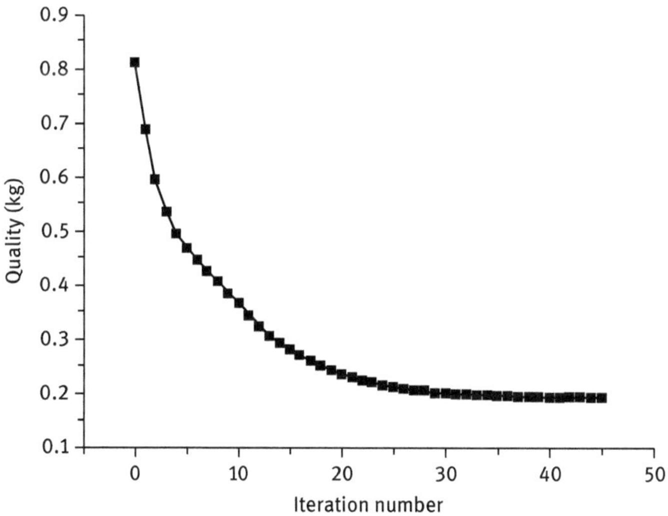

Figure 4.88: Iterative history of quality.

Example 2

As illustrated in Figure 4.89, the base structure is a 3D elastic body with size 120×45×8 mm. The elastic modulus of the structure material is $E = 70 \times 10^3$ MPa, the Poisson's ratio is $y = 0.30$, and the material density is $\rho = 2.7 \times 10^{-6}$ kg/mm^3. A concentrated force $P = 9{,}000$ N along the up-to-down ($-y$) direction is located at five midpoints of the upper boundary of the structure. The four corner points of the structure are clasped. A solid element with eight nodes is used and the total number of elements is 40 × 15 × 4. The quality of the base structure is 0.117 kg.

The maximum von Mises stress of the structure is 468 MPa, the buckling factor is 40.41, and the critical force is 202,050 N. The upper limit of the stress is 120 MPa. The

Figure 4.89: Base structure.

Table 4.12: Comparison of topological results with different constraints.

Constraint type	Initial quality (kg)	Optimized quality (kg)	Iteration number
Stress	0.117	0.025	35
Buckling	0.117	0.046	35
Stress + Buckling	0.117	0.0464	35

upper limit of the buckling critical force is 90,000 N. Iteration convergence after 35 times. After optimization, the quality of the structure is 0.0464 kg (Table 4.12).

The topology of stress constraints is slender and uniform, and the stability of the structure is poor (Figure 4.90 to Figure 4.91). The buckling constraint topology has good stability (Figure 4.92). When considering the stress and buckling constraints, the basic topology was the same as the one with buckling constraints (Figure 4.93 to Figure 4.94). This is mainly because four corners of the structure are fixed, and the structure with buckling constraint can satisfy the stress constraint.

Example 3

As illustrated in Figure 4.95, the base structure is a 3D elastic body with size 300 mm × 100 mm × 10 mm. The elastic modulus of the structure material is E = 68,890 MPa, the Poisson's ratio is y = 0.30, and the material density is $\rho = 2.7 \times 10^{-6}$ kg/mm^3. A concentrated force P = 9,000 N along the up-to-down (-y) direction is located at five midpoints of the upper boundary of the structure. The two corners of the lower edge are the fixed support. A solid element with eight nodes is used and the total number of elements is 60 × 20 × 2. The quality of the base structure is 0.81 kg.

Case 1: A concentrated force P = 27,000 N is located on nine nodes on the upper boundary, and each node force is 3,000 N, respectively.

The maximum von Mises stress of the structure is 473 MPa, the buckling factor is 1.204454, and the critical force is 32,521.5 N. The upper limit of the stress is 150 MPa. The upper limit of the buckling critical force is 22,000 N (Figure 4.96 to Figure 4.100). Iteration convergence after 44 times. After optimization, the quality of the structure is 0.215 2 kg (Table 4.13).

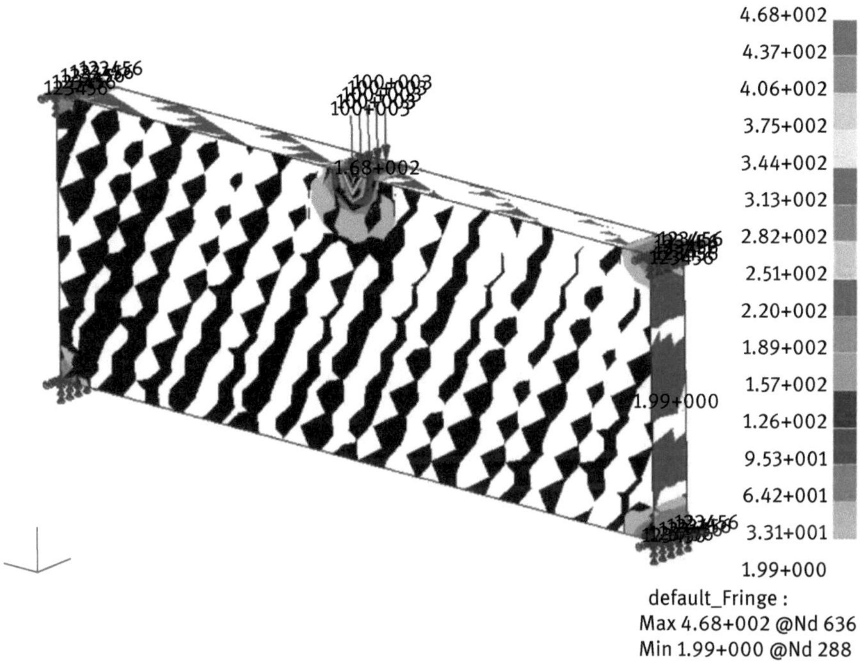

Figure 4.90: Stress fringe result of a base structure.

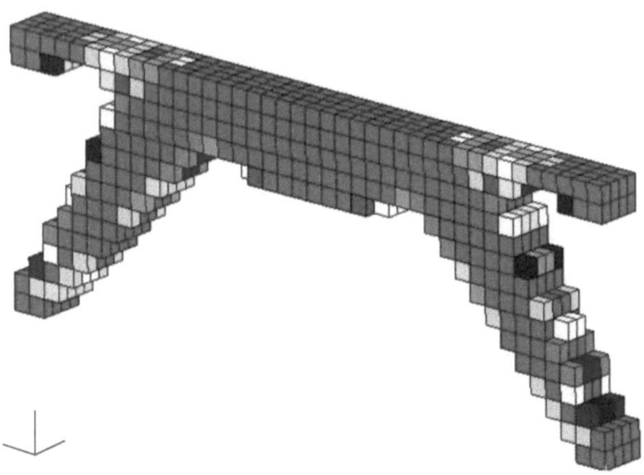

Figure 4.91: Optimal topological configuration with buckling constraints.

Case 2: A concentrated force P = 27,000 N is located on nine nodes on the upper boundary—three middle node forces are 5,000 N, and left and right three node forces are 2,000 N (Figure 4.101). Central force is much higher than that on both sides of the applied force.

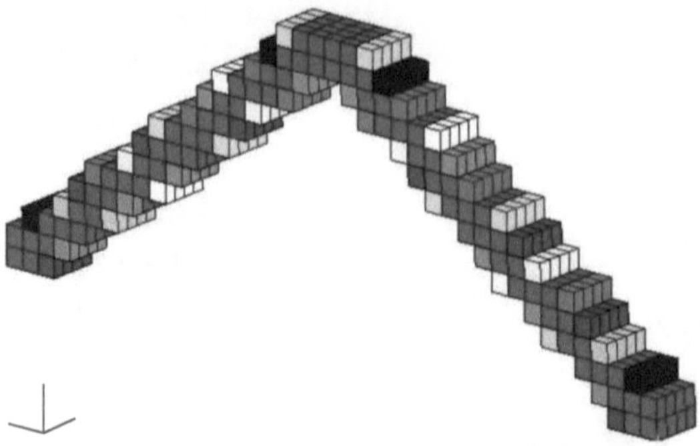

Figure 4.92: Optimal topological configuration with stress constraints.

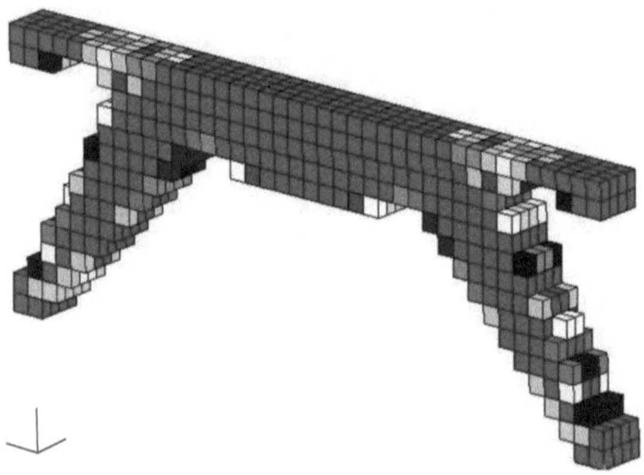

Figure 4.93: Optimal topological configuration with buckling and stress constraints.

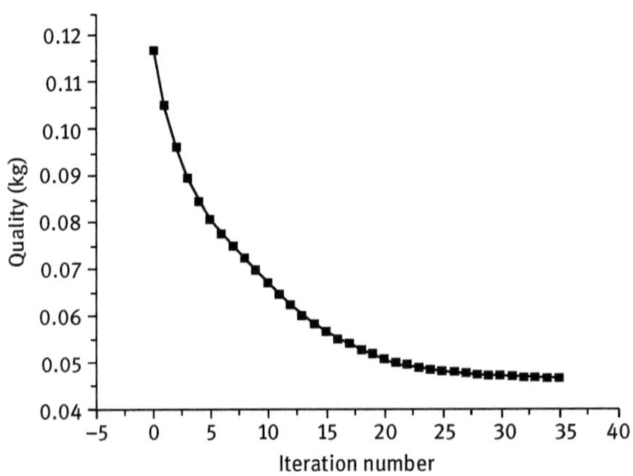

Figure 4.94: Iterative history of quality.

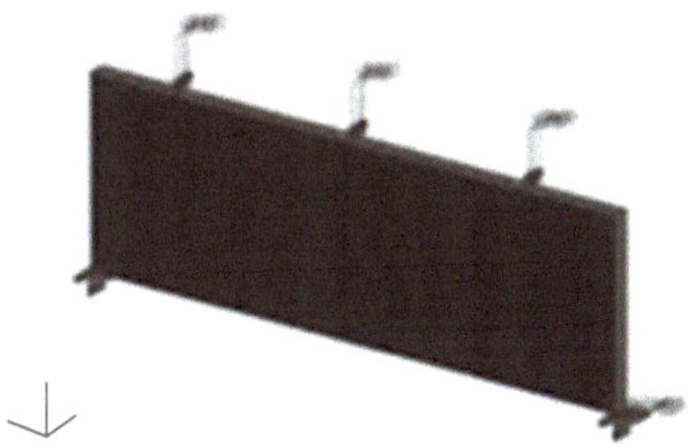

Figure 4.95: Base structure.

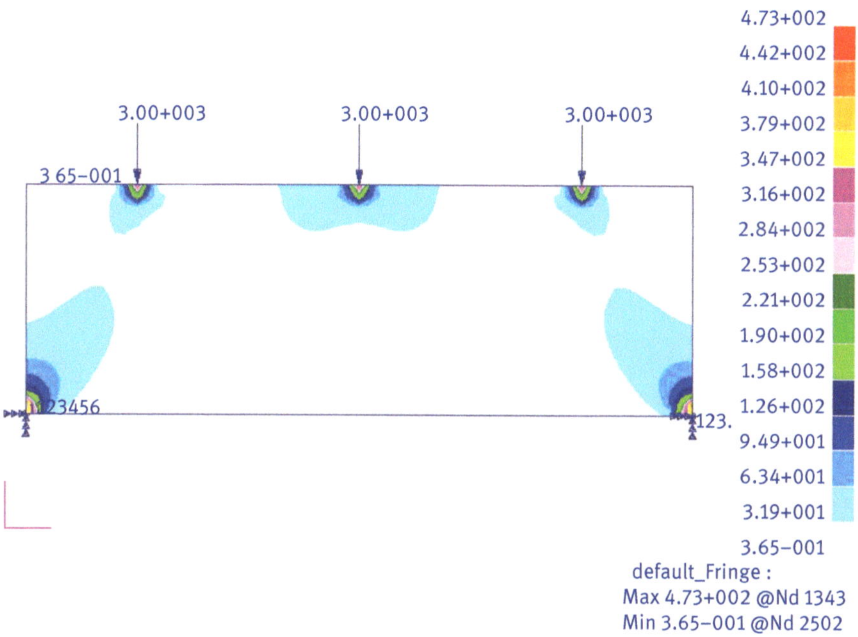

Figure 4.96: Stress fringe result of a base structure.

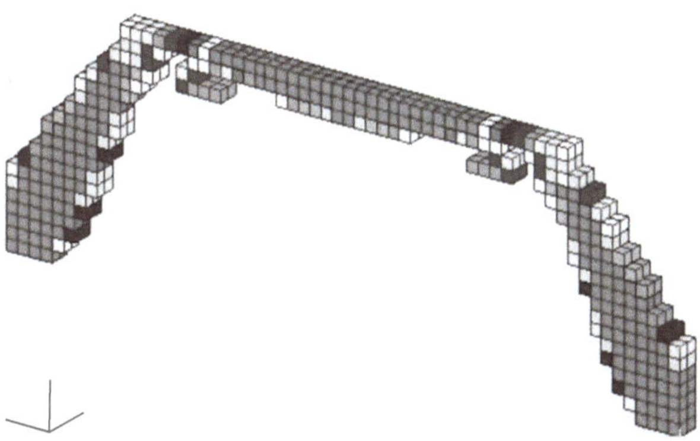

Figure 4.97: Optimal topological configuration with buckling constraints.

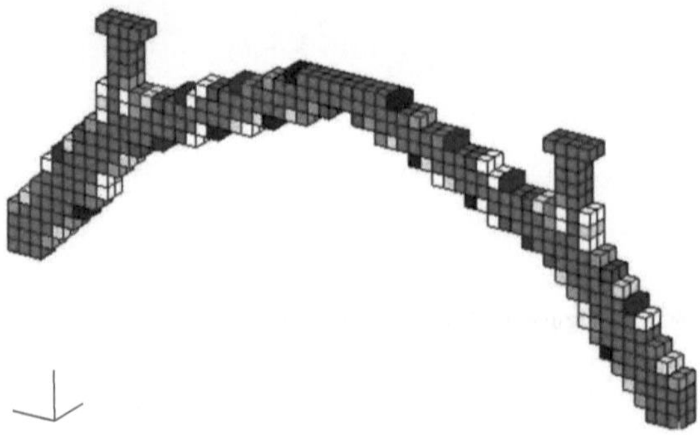

Figure 4.98: Optimal topological configuration with stress constraints.

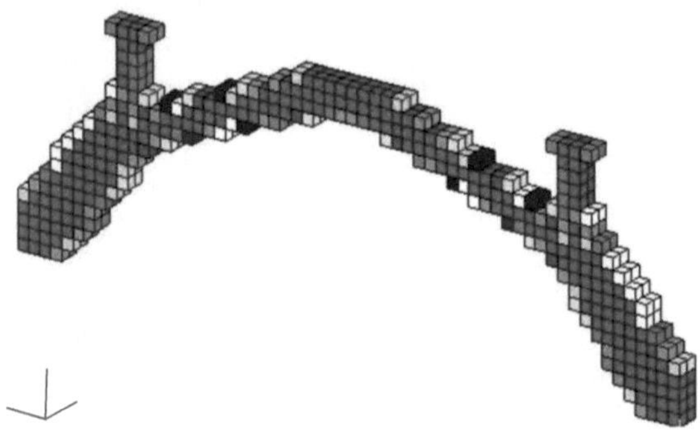

Figure 4.99: Optimal topological configuration with buckling and stress constraints.

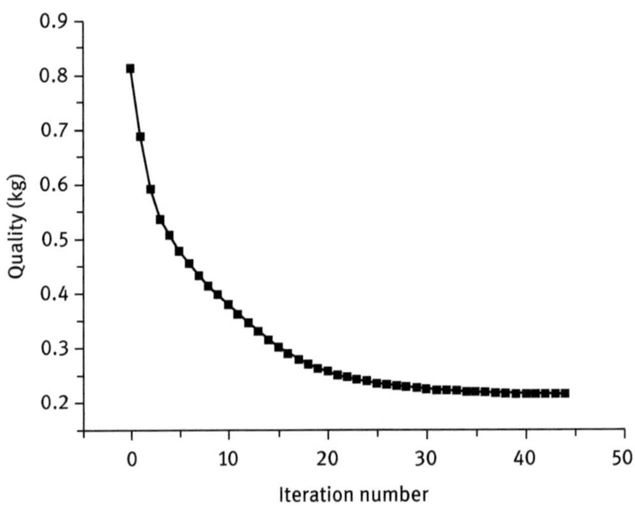

Figure 4.100: Iterative history of quality.

Table 4.13: Comparison of topological results with different constraints.

Constraint type	Initial quality (kg)	Optimized quality (kg)	Iteration number
Stress	0.81	0.21	41
Buckling	0.81	0.195	33
Stress + Buckling	0.81	0.2152	44

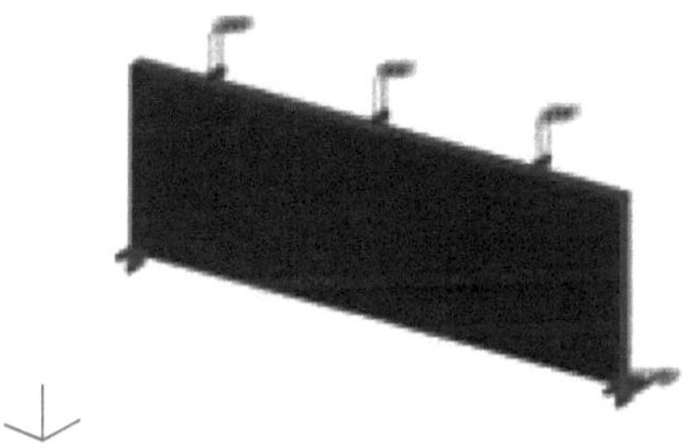

Figure 4.101: Base structure.

Table 4.14: Comparison of topological results with different constraints.

Constraint type	Initial quality (kg)	Optimized quality (kg)	Iteration number
Stress	0.81	0.277	48
Buckling	0.81	0.235	35
Stress + Buckling	0.81	0.2782	33

The maximum von Mises stress of the structure is 635 MPa, the buckling factor is 1.1569, and the critical force is 312,36.3 N. The upper limit of the stress is 150 MPa. The upper limit of the buckling critical force is 22,000 N. Iteration convergence after 33 times. After optimization, the quality of the structure is 0.2782 kg (Table 4.14).

In contrast to the two kinds of force, case 1: due to the equal force, the topology of arch is larger and the thickness of the middle beam is relatively uniform; case 2: the forces influence that the topological structure was larger. The arch of the topological structure becomes small and the middle crossbeam becomes bold, so the stability is enhanced (Figure 4.102 to Figure 4.106).

Under the single condition, the topology structure with the buckling constraint is thicker than that with the stress constraint alone, and the stability is enhanced.

Example 4

As illustrated in Figure 4.107, the base structure is a 3D elastic body with thickness 500 mm. The elastic modulus of the structure material is E = 28.6 GPa, the Poisson's ratio is y = 0.15, and the material density is ρ = 2,400 kg/m^3. A distribution force P =

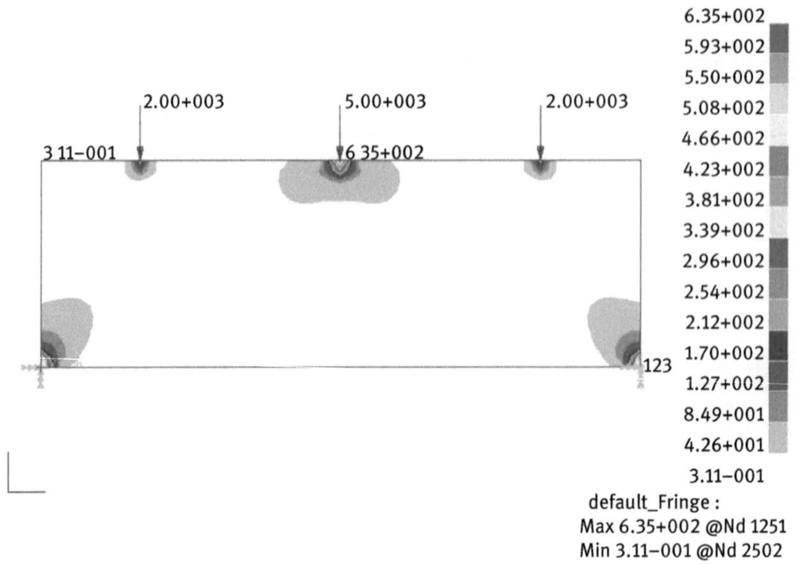

Figure 4.102: Stress fringe result of a base structure.

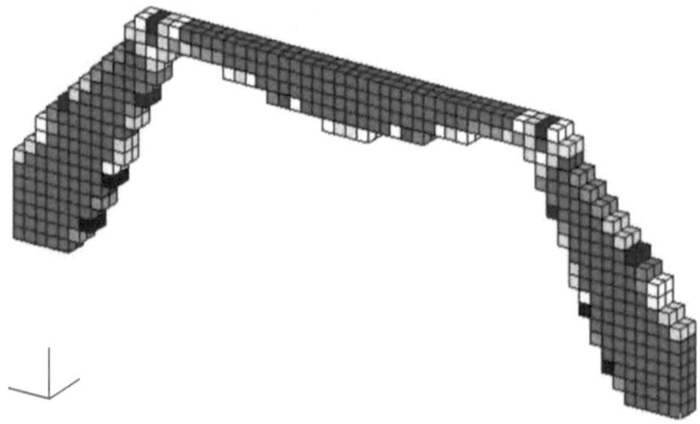

Figure 4.103: Optimal topological configuration with buckling constraints.

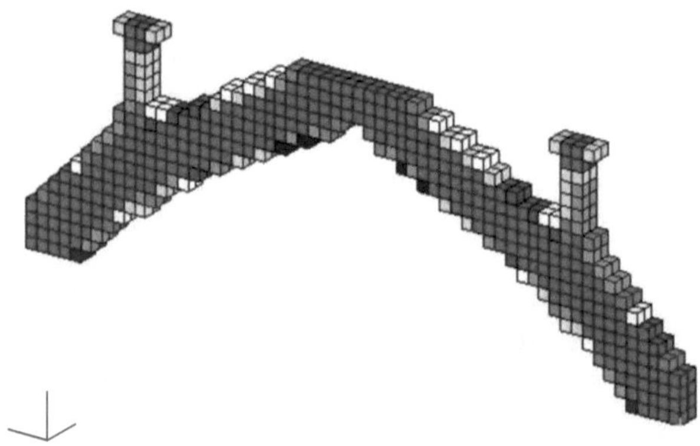

Figure 4.104: Optimal topological configuration with stress constraints.

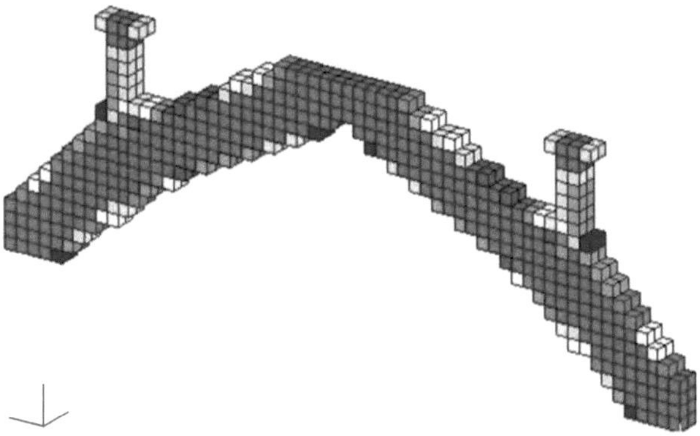

Figure 4.105: Optimal topological configuration with buckling and stress constraints.

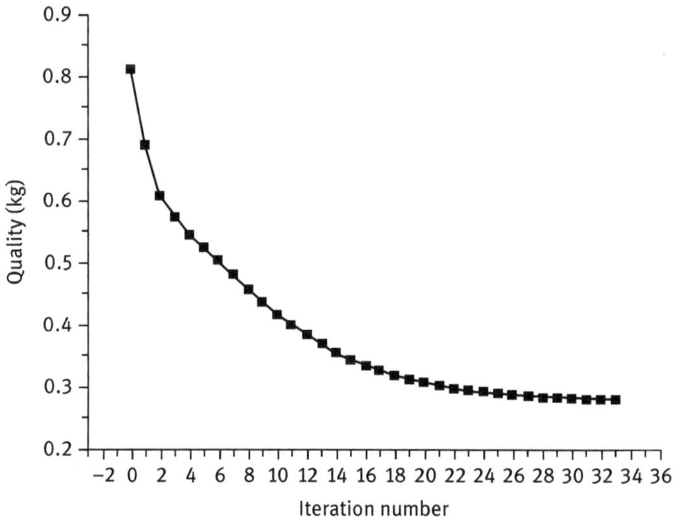

Figure 4.106: Iterative history of quality.

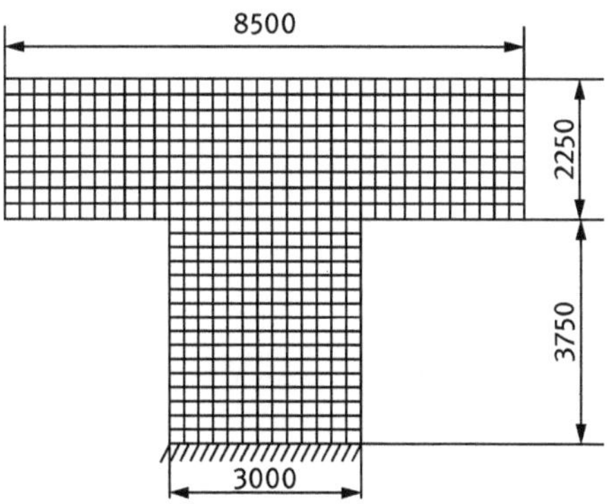

Figure 4.107: Base structure.

Table 4.15: Comparison of topological results with different constraints.

Constraint type	Initial quality (kg)	Optimized quality (kg)	Iteration number
Stress	36450	15287	44
Buckling	36450	18371	39
Stress + Buckling	36450	19970	40

1,000,000 N along the up-to-down (-y) direction is located on the upper plane of the structure. The total amount of the bearing force is 1.05×10^8 N. The bottom plane has all fixed support. A solid element with eight nodes is used and the number of total elements is 1,674. The quality of the base structure is 36,450 kg.

The maximum von Mises stress of the structure is 200 MPa, the buckling factor is $\varsigma_1 = 5.857891 \times 10^5$, and the critical force is $\lambda_1 = 6.15 \times 10^{13}$N. The upper limit of the stress is 400 MPa. The upper limit of the buckling critical force is $\bar{\lambda}_1 = 3.8 \times 10^{13}$ N. Iteration convergence after 40 times. After optimization, the quality of the structure is 19,970 kg (Table 4.15).

The middle of the bottom of the stress constraint topological structure is interrupt, but not the structure of the buckling constraint. By increasing the material, the requirements of buckling constraints can be met by the topology structure (Figure 4.108 to Figure 4.112).

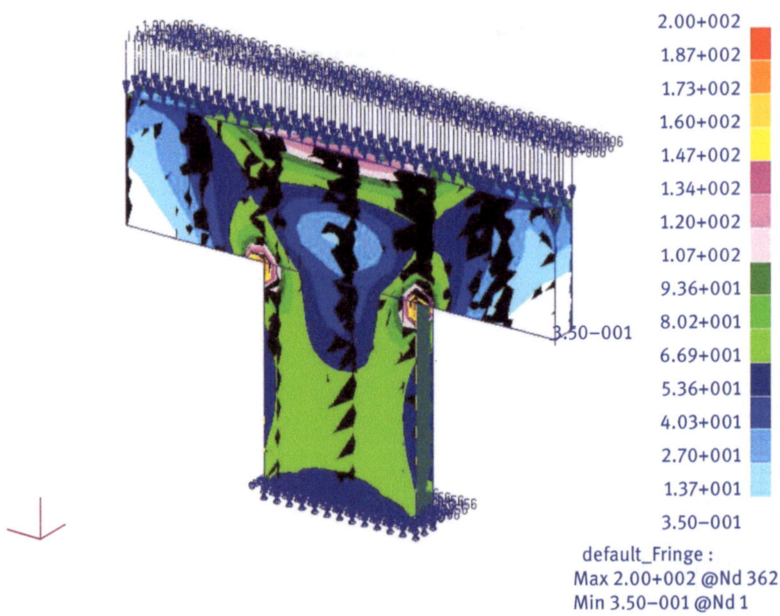

Figure 4.108: Stress fringe result of a base structure.

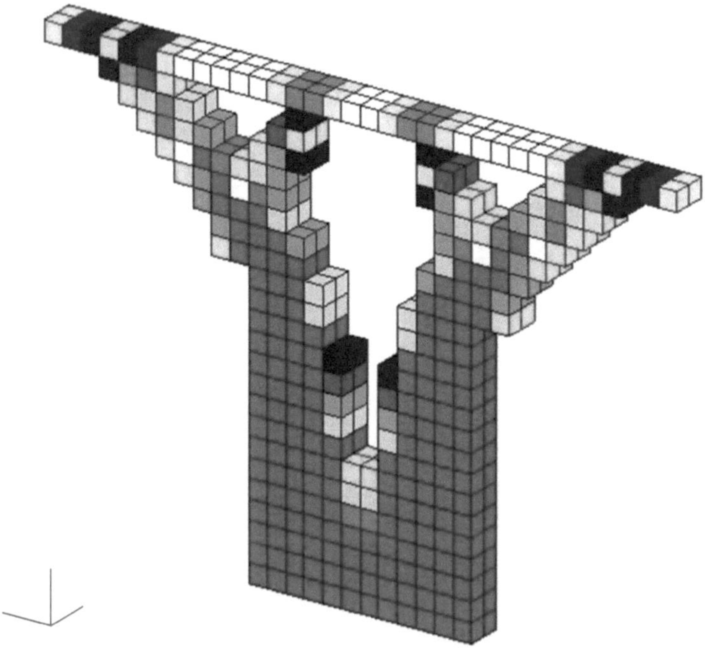

Figure 4.109: Optimal topological configuration with buckling constraints.

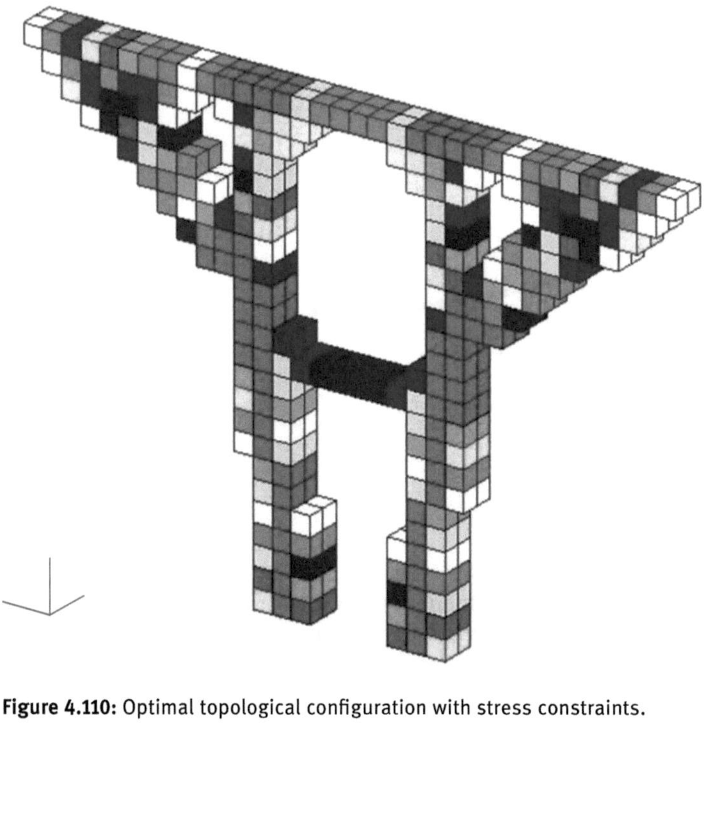

Figure 4.110: Optimal topological configuration with stress constraints.

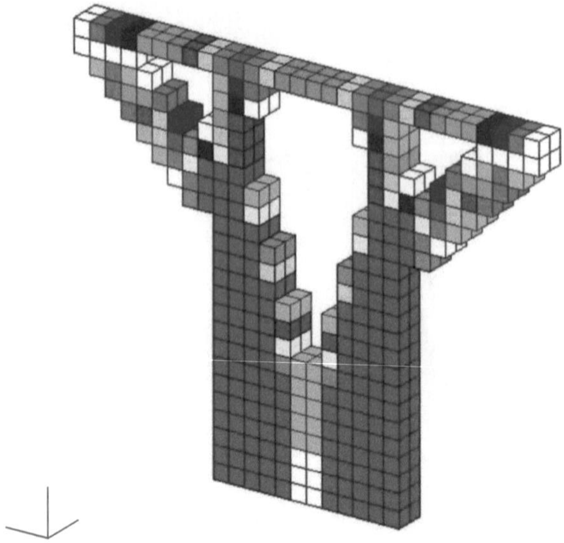

Figure 4.111: Topological configuration with buckling and stress constraints.

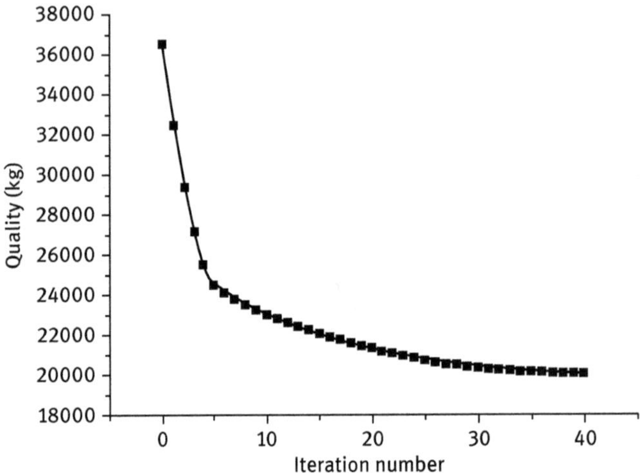

Figure 4.112: Iterative history of quality.

4.10 Summary

In this chapter, according to the ICM method, the topology optimization model for the continuum structure was constructed. The model had the minimized quality as the objective function subjected to the buckling constraints and stress constraints. The continuous independent topological variables were used in this problem. Based on the Taylor expansion and the filtering function, the objective function was approximately expressed as a second-order expression. Based on the Rayleigh quotient, the Taylor expansion and the filtering function, the buckling constraints were approximately expressed as an explicit function. Using the globalization method of stress constraints and the von Mises yield criterion in mechanics of materials, the local stress constraints were transformed into the whole structure strain energy constraint. Thus, the analysis quantity of the sensitivity was decreased. The number of variables was reduced and the model's scale was diminished. Numerical examples show that this method can solve the topology optimization problem of continuum structures with the buckling and stress constraints efficiently and give more reasonable structural topologies.

The concept of "stress constraint ratio" was defined. The "quality ratio" and the "stress constraint ratio" function curves of the topology optimization structure were established, and the "stress–quality constraint ratio" curve was fitted by the least squares method. The "quality–buckling constraint ratio" is compared with the "quality–stress constraint ratio," which provides a functional relationship for the selection of buckling constraints and stress constraints.

By using the safety factor of the material, the correction factor of the structural strain energy constraint is determined, the correction coefficient has practical engineering significance, and the allowable strain energy is more reasonable. So the structure satisfies maximum the local stress constraints, and the optimization model of the continuum structure is improved by using the structural strain energy, instead of the stress constraint.

In order to solve the optimization model, the objective function and the buckling force constraint were made an approximate treatment. The local stress constraints are transformed into the whole structure strain energy constraint by the fourth strength theory. The reasonable topological shape and the clear topology structure were obtained by solving the dual programming.

Finally, the method is applied to two-dimensional and three-dimensional continuum structure topology optimization problems with single-load and multiloadoperating conditions, and successfully verifies the optimal topology of the Zhaozhou Bridge and the drawbridge.

The results of several examples show that the method is effective and feasible. The optimization algorithm is implemented on the second development platform of MSC.Patran/Nastran software.

Numerical examples show that the global nature of the strain energy constraints, instead of the local stress constraints, can be better to get the load transfer path. From the application of several engineering examples, it is concluded that the stress global method is an effective method to solve the problem of local stress constraints.

5 Topological Optimization for Continuous Structure with Static Constraint

5.1 Introduction

If the minimum quality of the structure is the goal function, the form of the constraint can be varied in the topology optimization of the continuum structure. The types of constraints can be divided into global and local, according to the nature of constraints. The global constraints include displacement constraints, frequency constraints, global stability constraints, etc. The local constraints include stress constraints, local stability constraints, etc. If classification is done according to the structure load, the constraints can be divided into static constraints and dynamic constraints. The static constraints include stress constraints, displacement constraints, and buckling constraints, etc. The dynamic constraints include frequency constraints, etc.

At present, the research focuses on the common function of one or two kinds of constraints. The structures of one or two kinds of constraints are designed, which cannot meet the requirements of the actual working conditions. In view of the above reasons, this chapter makes the topology optimization of continuum structures with total static constraints based on the ICM method.

The continuum structure topology optimization model with total static constraints considers the displacement constraints, the stress constraints and the stability constraints. The displacement constraint is explicitly typed by the unit virtual load method. Using the idea of global change, the local stress constraint is transformed into the global strain energy constraint problem. Using the Rayleigh quotient, the Taylor expansion and the filter function, the stability constraint will be bound to approximate explicit methods, to avoid the sensitivity of the calculation. Finally, the optimization model is transformed into dual programming, and the number of design variables is reduced, the solving scale of the model is reduced and the computational efficiency is improved by using sequential two quadratic programming.

5.2 Model Establishment and Solution

5.2.1 Continuum Topology Optimization Model

According to the ICM method, based on the idea of independent topological variables and continuous mapping transformation, a continuum topology optimization model which minimizes the quality of the structure is established:

https://doi.org/10.1515/9783110462272-005

$$\begin{cases} \text{find} & \boldsymbol{t} \in (t_1, \cdots, t_N)^{\mathrm{T}} \\ \min & W = \sum_{i=1}^{N} f_w(t_i) w_i^0 \\ \text{s.t.} & \lambda_{jl}(f_k(t_i),\ f_g(t_i)) \le \bar{\lambda}_{jl} \\ & \sum_{i=1}^{N} u_{irjl}(f_k(t_i)) \le \bar{u}_{rjl} \\ & Q_{jl}(f_k(t_i)) \le \bar{Q}_{jl} \\ & 0 \le t_i \le 1 \qquad (i = 1,\ldots, N; r = 1,\ldots, R; j = 1,\ldots, J; l = 1,\ldots, L) \end{cases} \tag{5.1}$$

where $\boldsymbol{t}$ is topology design variables; W is the total quality of the structure; w_i and N are the quality of i-th element and the total number of elements, respectively; $\bar{\lambda}_j$ is the upper limit of the critical load; J is the total number of the buckling mode; Q_{jl} is the l-th load case strain energy of the j-th buckling mode and $\bar{Q}_{jl}$ is the l-th load case strain energy of the upper limit j-th buckling mode; u_{irjl} is the i-th element's r-th displacement j-th buckling mode l-th load case's displacement and $\bar{u}_{rjl}$ is the upper limit of r-th displacement j-th buckling mode l-th load case's displacement; J is the buckling mode to calculate; N is the total number of units; and L is the condition number.

5.2.2 Explicit Approximate Function of the Buckling Constraint

The buckling constraint approximate explicit is the same as provided in the second chapter; here it is not repeated. Uniform expression formula of buckling constraints can be obtained:

$$\sum_{i=1}^{N} d_{ijl}\, x_i \le h_{jl} \tag{5.2}$$

If we define $A_{ijl}^{(k)} = \left(U_{ijl}^{(k)} + \lambda_{jl} V_{ijl}^{(k)}\right) / V_{\Sigma jl}$, the left-hand side of in equation (5.2) can be rewritten as

$$\sum_{i=1}^{N} d_{ijl} x = \sum_{i=1}^{N} x_i \left(U_{ijl}^{(k)} + \lambda_{jl} V_{ijl}^{(k)}\right) / V_{\Sigma jl} x_i^{(k)} = \sum_{i=1}^{N} A_{ijl}^{(k)} x_i / x_i^{(k)} = \sum_{i=1}^{N} \left(t_i^{(k)}\right)^b A_{ijl}^{(k)} x_i \tag{5.3}$$

If we define $d_{ijl} = \left(t_i^{(k)}\right)^b A_{ijl}^{(k)}$, the right-hand side of the in equation can be rewritten as

$$h_{jl} = \bar{\lambda}_{jl} - \lambda_{jl}\left(\boldsymbol{x}^{(k)}\right) + \sum_{i=1}^{N} \left(U_{ijl}^{(k)} + \lambda_{jl} V_{ijl}^{(k)}\right) / V_{\Sigma jl} = \bar{\lambda}_{jl} - \lambda_{jl}\left(\boldsymbol{x}^{(k)}\right) + \sum_{i=1}^{N} A_{ijl}^{(k)} \tag{5.4}$$

Here, $U_{ijl} = \boldsymbol{u}_{jl}^{\mathrm{T}}\boldsymbol{k}_i\boldsymbol{u}_{jl}/2$ is the modal strain energy of element i in the l mode of the j mode, $V_{ijl} = \boldsymbol{u}_{jl}^{\mathrm{T}}\boldsymbol{g}_i\boldsymbol{u}_{jl}/2$ is the modal geometrical strain energy of element i in the l mode of the j mode, and $V_{ijl} = \boldsymbol{u}_{jl}^{\mathrm{T}}\boldsymbol{g}_i\boldsymbol{u}_{jl}/2$ is the total geometric strain energy of the structure under the l work condition of mode j.

5.2.3 Explicit Approximate Function of the Displacement Constraint

The displacement constraint explicit approximate function is the same as provided in the third chapter; here it is not repeated. Displacement constraints can be expressed as

$$u_r = \sum_{i=1}^{N} \int (\sigma_i^{\mathrm{V}})^{\mathrm{T}} (\boldsymbol{\varepsilon}_i^{\mathrm{R}})\ dv = \sum_{i=1}^{N} (\boldsymbol{F}_i^{\mathrm{R}})^{\mathrm{T}} \boldsymbol{u}_i^{\mathrm{V}} \tag{5.5}$$

According to the theory of the ICM method, the filter function is introduced into the displacement constraint:

$$u_r = \sum_{i=1}^{N} \left(\boldsymbol{F}_i^{\mathrm{R}}\right)^{\mathrm{T}} \boldsymbol{u}_i^{\mathrm{V}} = \sum_{i=1}^{N} \frac{\left(t_i{}^{(K)}\right)^b}{t_i^b} \left(\boldsymbol{F}_i^{\mathrm{R}}\right)^{\mathrm{T}} \boldsymbol{u}_i^{\mathrm{V}} \tag{5.6}$$

Also base $x_i = \frac{1}{t_i^b}$

$$u_r = \sum_{i=1}^{N} \left(\boldsymbol{F}_i^{\mathrm{R}}\right)^{\mathrm{T}} \boldsymbol{u}_i^{\mathrm{V}} = \sum_{i=1}^{N} \frac{x_i}{x_i^{(K)}} \left(\boldsymbol{F}_i^{\mathrm{R}}\right)^{\mathrm{T}} \boldsymbol{u}_i^{\mathrm{V}} \tag{5.7}$$

Let $\left(\boldsymbol{F}_i^{\mathrm{R}}\right)^{\mathrm{T}} \boldsymbol{u}_i^{\mathrm{V}}/x_i^{(k)} = c_{ir}$; the displacement constraint was expressed as an explicit approximate function:

$$u_r = \sum_{i=1}^{N} \frac{x_i}{x_i{}^{(k)}} \left(\boldsymbol{F}_i{}^{\mathrm{R}}\right)^{\mathrm{T}} \boldsymbol{u}_i{}^{\mathrm{V}} = \sum_{i=1}^{N} c_{ir} x_i \le \bar{u}_r \tag{5.8}$$

5.2.4 Explicit Approximate Function of the Stress Constraint

The strain energy constraint's explicit approximate function is the same as provided in the fourth chapter; here, it is not repeated.

The strain energy constraint can be expressed as

$$\sum_{i=1}^{N} \frac{\left(t_i^b\right)^{(k)}}{t_i^b} e_i^{(k)} \le \sum_{i=1}^{N} \gamma(1+\mu)(\bar{\sigma}^2) V_i/(3E) \tag{5.9}$$

The left- and right-hand sides of the in equation are rewritten as $\sum_{i=1}^{N}(1+\mu)(\bar{\sigma}^2)V_i/(3E)$ $=\sum_{i}^{N}\overline{e_i}=q,\ (t_i^b)^{(k)}e_i^{(k)}=s_i$. The strain energy constraints can be written as follows:

$$\sum_{i=1}^{N}\frac{s_i}{t_i^b}\le\gamma q \tag{5.10}$$

The topology variables are transformed into design variables as follows:

$$\sum_{i=1}^{N}s_i x_i\le\gamma q \tag{5.11}$$

5.2.5 Processing of Objective Function

The objective function process is the same as provided in the third chapter; here, it is not repeated. The object function is transformed into an explicit expression by second-order Taylor expansion:

$$W=\sum_{i=1}^{N}\{[w_i^0\beta(\beta+1)/2x_i^{(k)\beta+2}]x_i^2+[-w_i^0\beta(\beta+2)/x_i^{(k)\beta+1}]x_i\} \tag{5.12}$$

where $x_i^{(k)}$ is the initial design variable of the i-th element at the k-th iteration.

We define $A_i=w_i^0\beta(\beta+1)/2x_i^{(k)\beta+2}$ and $B_i=-w_i^0\beta(\beta+2)/x_i^{(k)\beta+1}$. The objective function can be written as $W=\sum_{i=1}^{N}(A_ix_i^2+B_ix_i)$.

The whole topology optimization model is changed into

$$\begin{cases}\text{find} & \boldsymbol{x}\in E^N \\ \min & W=\sum_{i=1}^{N}\left(A_ix_i^2+B_ix_i\right) \\ \text{s.t.} & \sum_{i=1}^{N}d_{ijl}x_i\le h_{jl} \qquad (j=1,\ldots,J;\ l=1,\ldots,L) \\ & \sum_{i=1}^{N}c_{irjl}x_i\le\bar{u}_{rjl} \qquad (r=1,\ldots,R) \\ & \sum_{i=1}^{N}s_{ijl}x_i\le\gamma q_{jl} \\ & 1\le x_i\le\bar{x}_i \qquad (i=1,\ldots,N)\end{cases} \tag{5.13}$$

5.2.6 Solution of the Topological Optimization Model

According to the relation between the topology and the design variables, the topology optimization model is transformed into the model with the design variables. In order to reduce the number of variables and the model's scale, the optimization model is translated into a dual programming:

$$\begin{cases} \text{find} & \chi \\ \max & \boldsymbol{\Phi}(\chi) \\ \text{s.t.} & \chi \geq 0 \end{cases} \tag{5.14}$$

If $\chi = \begin{bmatrix} \mu \\ \eta \\ \zeta \end{bmatrix}$, $\boldsymbol{\Phi}(\mu, \eta, \zeta) = \min\limits_{1 \leq x_i \leq \bar{x}_i} L(\boldsymbol{x}, \mu, \eta, \zeta)$,

$$\begin{aligned} L(\boldsymbol{x}, \mu, \eta, \zeta) = \sum_{i=1}^{N} (A_i x_i^2 + B_i x_i) \quad + \sum_{j=1}^{J} \sum_{l=1}^{L} \mu_{jl} \left(\sum_{i=1}^{N} d_{ijl} x_i - h_{jl} \right) \\ + \sum_{j=1}^{J} \sum_{l=1}^{L} \eta_{jl} \left(\sum_{i=1}^{N} s_{ijl} x_i - \gamma q_{jl} \right) + \sum_{j=1}^{J} \sum_{l=1}^{L} \zeta_{jl} \left(\sum_{i=1}^{N} c_{irjl} x_i - \bar{u}_{rjl} \right) \end{aligned} \tag{5.15}$$

According to the Kutta condition, the derivative of an augmented Lagrangian functionwith respect to the design variable can be written as follows:

$$\partial L(\boldsymbol{x}, \mu, \eta, \zeta)/\partial x_i = 2A_i x_i^* + B_i + \sum_{j=1}^{J} \sum_{l=1}^{L} \mu_{jl} d_{ijl} + \sum_{j=1}^{J} \sum_{l=1}^{L} \eta_{jl} s_{ijl} + \sum_{j=1}^{J} \sum_{l=1}^{L} \zeta_{jl} c_{irjl}$$
$$\begin{cases} \leq 0 & (x_i^* = \bar{x}_i) \\ = 0 & (1 < x_i^* < \bar{x}_i) \\ \geq 0 & (x_i^* = 1) \end{cases} \tag{5.16}$$

Let $I_a = \{i | 1 < x_i^* < \bar{x}_i\}$ be the active variable set, and there is a set of identities:

$$2A_i x_i^* + B_i + \sum_{j=1}^{J} \sum_{l=1}^{L} \mu_{jl} d_{ijl} + \sum_{j=1}^{J} \sum_{l=1}^{L} \eta_{jl} s_{ijl} + \sum_{j=1}^{J} \sum_{l=1}^{L} \zeta_{jl} c_{irjl} \equiv 0 \tag{5.17}$$

For the identities of both sides of μ_{ml}, μ_{jm}, η_{nl}, η_{jn}, $\zeta_{\tau l}$, $\zeta_{j\tau}$, we get the following derivations:

$$\begin{aligned} &\partial x_i^* / \partial \mu_{ml} = -d_{iml}/2A_i, \ \partial x_i^* / \partial \mu_{jm} = -d_{ijm}/2A_i \\ &20\, \partial x_i^* / \partial \eta_{nl} = -s_{inl}/2A_i, \ \partial x_i^* / \partial \eta_{jn} = -s_{ijn}/2A_i, \\ &\partial x_i^* / \partial \zeta_{\tau l} = -c_{ir\tau l}/2A_i, \ \partial x_i^* / \partial \zeta_{j\tau} = -c_{irj\tau}/2A_i \end{aligned} \tag{5.18}$$

Because it is difficult to obtain the explicit expression of $\boldsymbol{\Phi}(\chi)$, it will be transformed into a two-order Taylor expansion approximation.

According to the dual theory, the design variable's one- and two-order sensitivity calculation of formula (5.16), the objective function is as follows:

$$\begin{aligned} &\partial\boldsymbol{\Phi}/\partial\mu_{jl} = \sum_{i=1}^{N} d_{ijl}x_i^* - h_{jl},\ \partial\boldsymbol{\Phi}/\partial\eta_{jl} = \sum_{i=1}^{N} s_{ijl}x_i^* - \gamma q_{jl} \\ &\partial\boldsymbol{\Phi}/\partial\zeta_{jl} = \sum_{i=1}^{N} c_{irjl}x_i^* - \bar{u}_{rjl} \end{aligned} \tag{5.19}$$

$$\partial^2\boldsymbol{\Phi}/(\partial\mu_{jl}\partial\mu_{ml}) = \sum_{i=1}^{N} d_{ijl}\partial x_i^*/\partial\mu_{ml},\ \partial^2\boldsymbol{\Phi}/(\partial\eta_{jl}\partial\eta_{nl}) = \sum_{i=1}^{N} s_{ijl}\partial x_i^*/\partial\eta_{nl}$$

$$\begin{aligned} &\partial^2\boldsymbol{\Phi}/(\partial\mu_{jl}\partial\mu_{jm}) = \sum_{i=1}^{N} d_{ijl}\partial x_i^*/\partial\mu_{jm},\ \partial^2\boldsymbol{\Phi}/(\partial\eta_{jl}\partial\eta_{jn}) = \sum_{i=1}^{N} s_{ijl}\partial x_i^*/\partial\eta_{jn} \\ &\partial^2\boldsymbol{\Phi}/(\partial\zeta_{jl}\partial\zeta_{\tau l}) = \sum_{i=1}^{N} c_{irjl}\partial x_i^*/\partial\zeta_{\tau l},\ \partial^2\boldsymbol{\Phi}/(\partial\zeta_{jl}\partial\zeta_{j\tau}) = \sum_{i=1}^{N} c_{irjl}\partial x_i^*/\partial\zeta_{j\tau} \end{aligned} \tag{5.20}$$

Taking formula (5.18) into formula (5.20), we obtain

$$\begin{aligned} &\partial^2\boldsymbol{\Phi}/(\partial\mu_{jl}\partial\mu_{ml}) = -\sum_{i=1}^{N} d_{ijl}d_{iml}/2A_i,\ \partial^2\boldsymbol{\Phi}/(\partial\eta_{jl}\partial\eta_{nl}) = -\sum_{i=1}^{N} s_{ijl}s_{inl}/2A_i \\ &\partial^2\boldsymbol{\Phi}/(\partial\mu_{jl}\partial\mu_{ml}) = -\sum_{i=1}^{N} d_{ijl}d_{iml}/2A_i,\ \partial^2\boldsymbol{\Phi}/(\partial\eta_{jl}\partial\eta_{nl}) = -\sum_{i=1}^{N} s_{ijl}s_{inl}/2A_i \\ &\partial^2\boldsymbol{\Phi}/(\partial\zeta_{jl}\partial\zeta_{\tau l}) = -\sum_{i=1}^{N} c_{irjl}c_{i\tau l}/2A_i,\ \partial^2\boldsymbol{\Phi}/(\partial\zeta_{jl}\partial\zeta_{j\tau}) = -\sum_{i=1}^{N} c_{irjl}c_{ij\tau}/2A_i \end{aligned} \tag{5.21}$$

$$\begin{aligned} &\partial^2\boldsymbol{\Phi}/(\partial\mu_{jl}\partial\eta_{jl}) = \partial^2\boldsymbol{\Phi}/(\partial\eta_{jl}\partial\mu_{jl}) = -\sum_{i=1}^{N} d_{ijl}s_{ijl}/2A_i \\ &\partial^2\boldsymbol{\Phi}/(\partial\mu_{jl}\partial\zeta_{jl}) = \partial^2\boldsymbol{\Phi}/(\partial\zeta_{jl}\partial\mu_{jl}) = -\sum_{i=1}^{N} d_{ijl}c_{irjl}/2A_i \\ &\partial^2\boldsymbol{\Phi}/(\partial\eta_{jl}\partial\zeta_{jl}) = \partial^2\boldsymbol{\Phi}/(\partial\zeta_{jl}\partial\eta_{jl}) = -\sum_{i=1}^{N} s_{ijl}c_{irjl}/2A_i \end{aligned} \tag{5.22}$$

We can obtain the second-order programming model:

$$\begin{cases} \text{find} & \chi \\ \min & -\boldsymbol{\Phi}(\chi) = \dfrac{1}{2}\chi^{\mathrm{T}}\mathbf{D}\chi + (\mathbf{H}^{\mathrm{T}} - \chi^{0\mathrm{T}}\mathbf{D})\chi \\ \text{s.t.} & \chi \geq 0 \end{cases} \tag{5.23}$$

where $\boldsymbol{D} = -\nabla^2\boldsymbol{\Phi}$ and $\boldsymbol{H} = -\nabla\boldsymbol{\Phi}$. The second-order programming model was solved, and we can obtain $\boldsymbol{\chi}$. The active variable set was updated. Accordingly, this process repeats until

$$\left\|\boldsymbol{x}^{(k+1)} - \boldsymbol{x}^{(k)}\right\| / \left\|\boldsymbol{x}^{(k)}\right\| \le \varepsilon$$

The optimum values of design variables $\boldsymbol{x}^*$ are solved; the optimum values of topological variables $\boldsymbol{t}^*$ are accordingly obtained. With the new topological variables, the structure is modified. Then the structural analysis and topological optimization are renewable and begin until the following convergence condition is satisfied:

$$\Delta W = \left|(W^{(k+1)} - W^{(k)})/W^{(k+1)}\right| \le \varepsilon \tag{5.24}$$

where $W^{(k)}$ and $W^{(k+1)}$ are the structural quality of the previous iteration and the current iteration, respectively, and ε is the convergence precision. In the present study, $\varepsilon = 0.001$ is employed.

5.3 Upper Limit for Constraint

Static total constraints include buckling constraints, displacement constraints and stress constraints. The appropriate selection of the upper bound of the three kinds of constraints has great significance. It not only directly affects the structure of the optimal topology and the quality of the optimal topology, but also directly affects the efficiency of the topology design. The curve of quality ratio and constraint ratio of the three constraints, the function relationship and the selection rules of the upper bound are used in front of the content; here, it is not repeated. The three kinds of constraints of the curve of quality ratio and constraint ratio are given in Figure 5.1.

5.4 Optimization Algorithms

The optimization algorithms of the static total constraints are based on the second development platform of MSC.Patran/Nastran software, and the use of the development tool PCL language to design the optimization algorithm program and the user input interface. The specific steps of optimization are as follows:

(1) Finite element model: The geometric model of the continuum structure is established on the MSC.Patran platform, which is divided into finite element mesh, and defines the material properties, boundary conditions, and the properties of the element. Due to the structural topology optimization

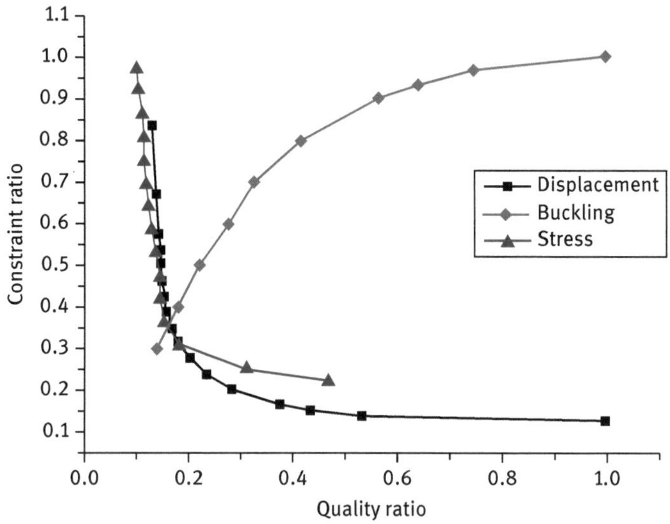

Figure 5.1: Curve of quality ratio and constraint ratio.

by buckling constraints, the properties of planar elements are set to shell elements.

(2) Basic optimization parameter input: It includes the convergence accuracy, the radius of the filter, the external force, etc.
(3) Stress constrained optimization parameter input: It includes the name of constraint, stress value, etc.
(4) Buckling constraint optimal parameter input: It Includes the sum of the total number of buckling modes, the name of the modal constraint, the numerical value of the buckling constraint and the number of corresponding buckling modes.
(5) Displacement constraint optimization parameter input: It including the name of constraint, the displacement of the control node, displacement constraint value, etc.
(6) Establish virtual load condition.
(7) Submit MSC.Nastran solution: The linear static finite element analysis of the structure and the finite element analysis of the buckling mode are carried out. The linear static finite element analysis solver is SOL101, and the buckling mode finite element analysis solver is SOL105.
(8) The corresponding finite element analysis results are calculated and extracted.
(9) The effective stress constraint, the displacement constraint and the buckling constraint are selected.
(10) Generate input files required for the dual quadratic programming.
(11) Call dual quadratic programming to solve the optimization model.

(12) According to the iterative convergence criterion to determine whether to meet the convergence conditions, if not satisfied, go back to step (7); if satisfied, then stop the iteration and record the data, determine the optimal topology and inversion topology variables, get the optimal topology graph, and end the optimization calculation.

The block diagram of the optimization program is shown in Figure 5.2.

5.5 Numerical Simulation and Case Study

5.5.1 Numerical Example of Two-dimensional Plane Continuum

Example 1

As illustrated in Figure 5.3, the base structure is a plane elastic body with size 80.0 × 50.0 × 2.0 mm. The elastic modulus of the structure material is E = 68.89 × 10^3 MPa, the Poisson's ratio is y = 0.3, and the material density is ρ = 2.7 × 10^{-6} kg/mm^3. The right- and left-hand underpart corner points of the structure are clasped. A concentrated force P = 7,000 N along the up-to-down (-y) direction is located on the top of structure. A quadrilateral element with four nodes is used and the total number of elements is 52 × 26. The quality of the base structure is 0.73 kg.

The displacement of the key node is 0.348 mm. The buckling factor is 0.017. The initial critical load is 119 N. The allowable stress of the structure is 380 MPa. In this example, the upper limit maximal value of the critical load is 75 N and the upper limit maximal value of the displacement is 1.2 mm. The upper limit allowable stress of the structure is 150 MPa. The convergence precision is 0.001, after 49 times of iteration. The quality of the optimized topology is 0.2 kg (Figure 5.4 to Figure 5.8). Table 5.1 lists the comparison of the results of topology optimization with different constraints.

Example 2

As illustrated in Figure 5.9, the base structure is a plane elastic body with size 32.0 × 10.0 × 2.0 mm. The elastic modulus of the structure material is E = 215 MPa, the Poisson's ratio is y = 0.33, and the material density is ρ = 2.7×10^{-6} kg/mm^3. The right- and left-hand underpart corner points of the structure are clasped. A concentrated force P = 900 N along the up-to-down (-y) direction is located on the top of structure. A quadrilateral element with four nodes is used and the total number of elements is 72 × 24. The quality of the base structure is 1.728 kg. The vertical displacement of A, B and C are −0.0127 mm, −0.01565 mm and −0.0127 mm, respectively, and B is the middle point. The maximum von Mises stress of the structure is 1,130 MPa, the buckling factor is −14.165, and the critical force is 12,748 N.

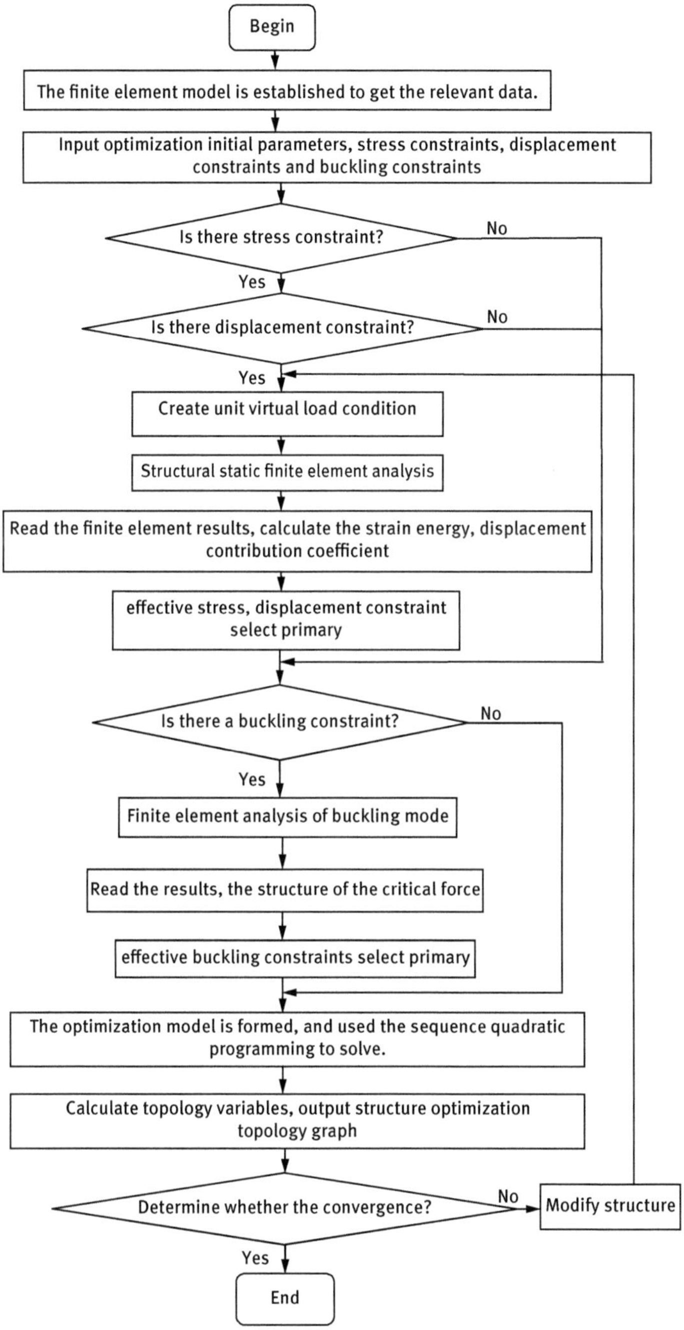

Figure 5.2: Flow chart for the algorithm of topology optimization of continuum structure subjected to total static constraints.

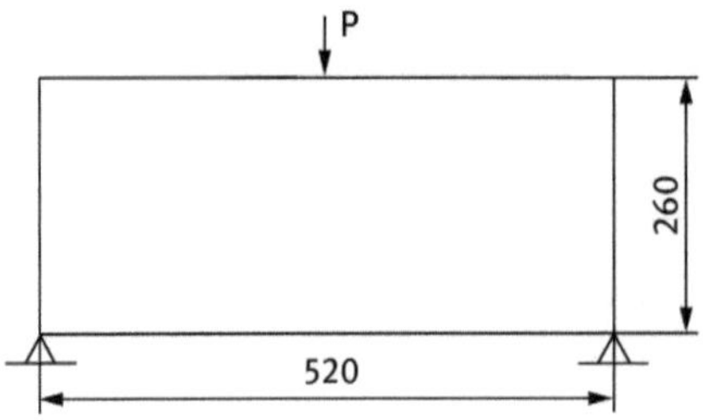

Figure 5.3: Base structure.

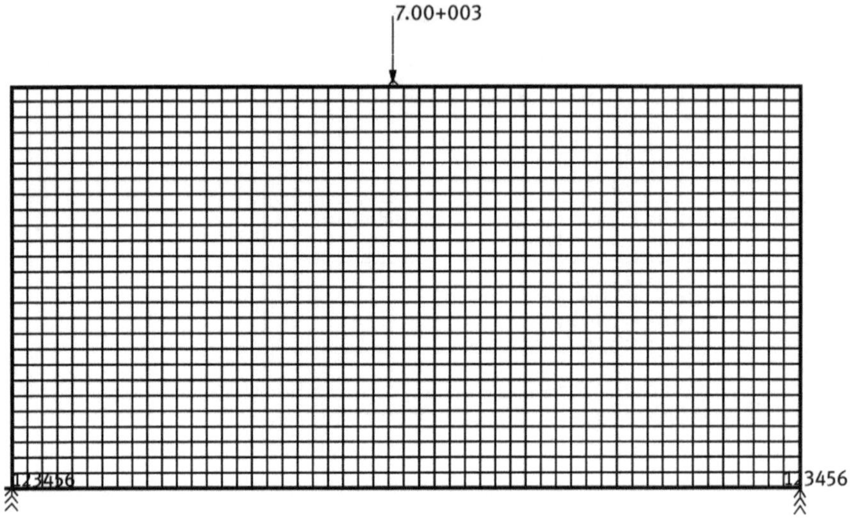

Figure 5.4: Finite element model.

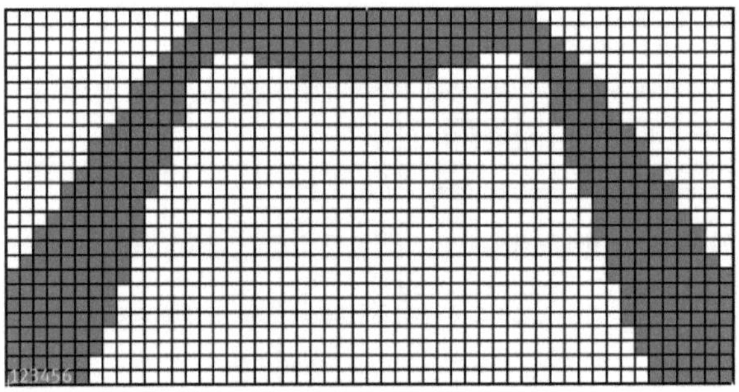

Figure 5.5: Optimal topological configuration with buckling constraint.

Case 1: The upper limit of the vertical downward displacement at the B point is 0.08 mm. The upper limit of the vertical downward displacement at the A,C point is 0.06 mm. The stress upper limit is 1,200 MPa, without considering the influence of buckling (Figure 5.10 to Figure 5.14).

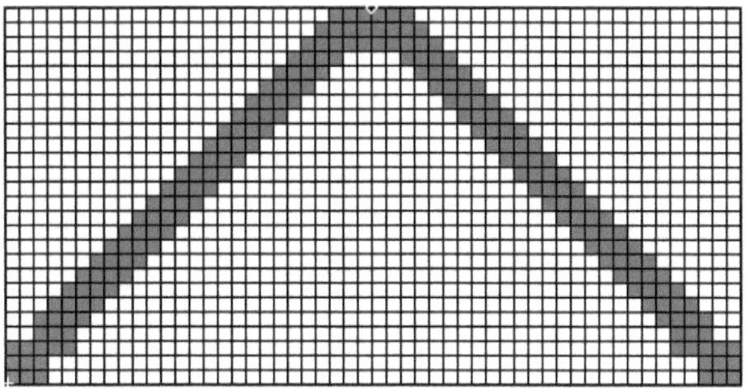

Figure 5.6: Optimal topological configuration with displacement constraint.

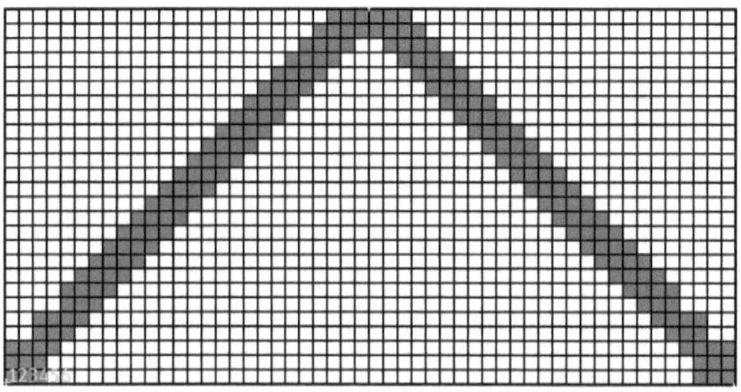

Figure 5.7: Optimal topological configuration with stress constraint.

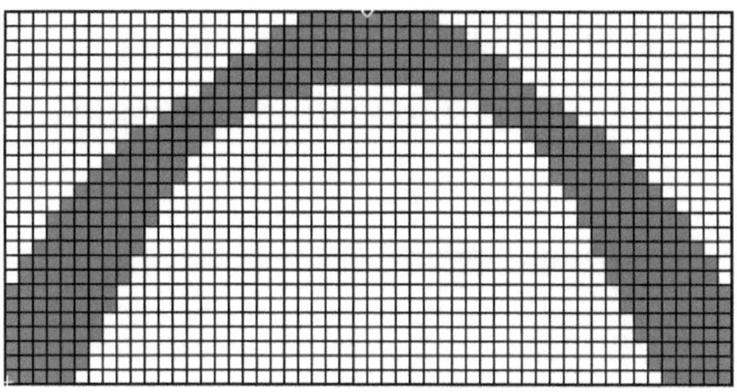

Figure 5.8: Optimal topological configuration with total static constraints.

Table 5.1: Comparison of topological results with different constraints.

Constraint type	Initial quality (kg)	Optimized quality (kg)	Iteration number
Stress	0.73	0.11	37
Displacement	0.73	0.11	37
Buckling	0.73	0.22	44
Stress + displacement + Buckling	0.73	0.21	49

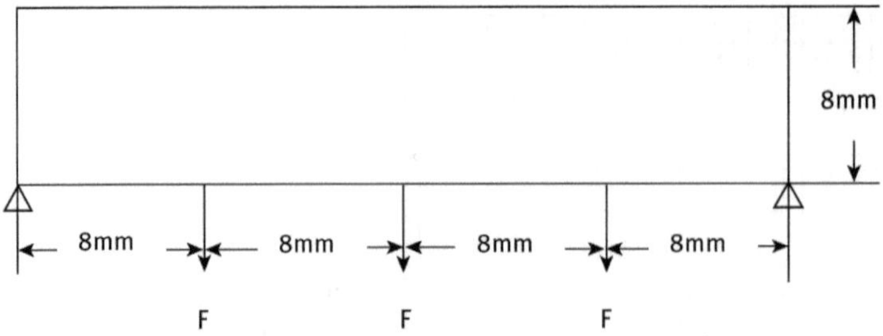

Figure 5.9: Base structure.

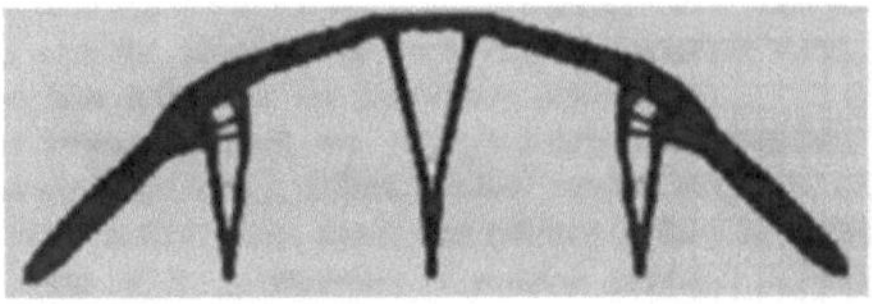

Figure 5.10: Optimal topology configuration in [64].

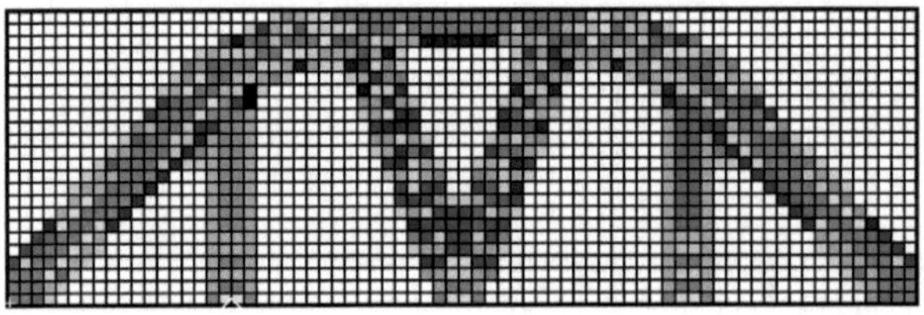

Figure 5.11: Topological configuration with displacement and stress constraints before inversion for case 1.

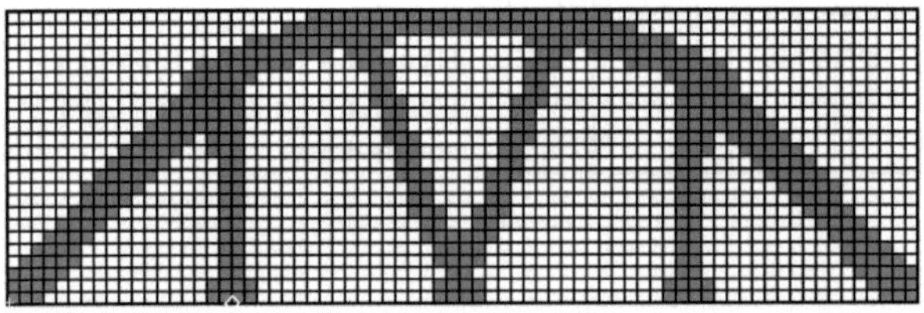

Figure 5.12: Topological configuration with displacement and stress constraints after inversion for case 1.

Case 2: The upper limit of the vertical downward displacement at the B point is 0.05 mm. The upper limit of the vertical downward displacement at the A,C point is 0.06 mm. The upper limit of the buckling critical force is 5,000 N, and the upper limit of the stress is 2,000 MPa (Figure 5.15 to Figure 5.20).

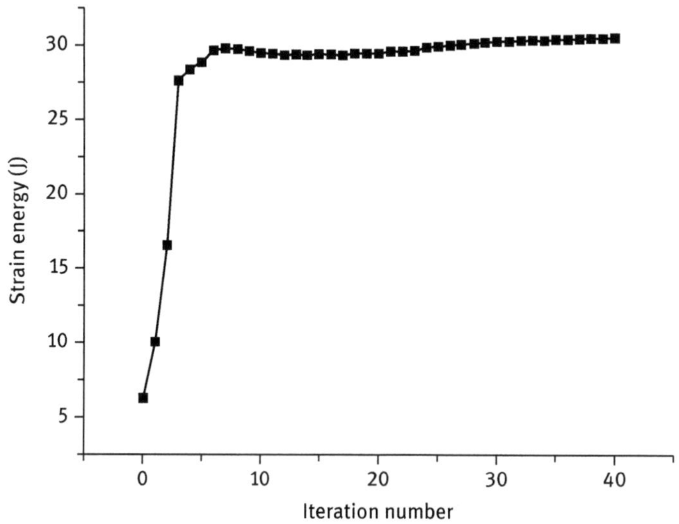

Figure 5.13: Iterative history of strain energy for case 1.

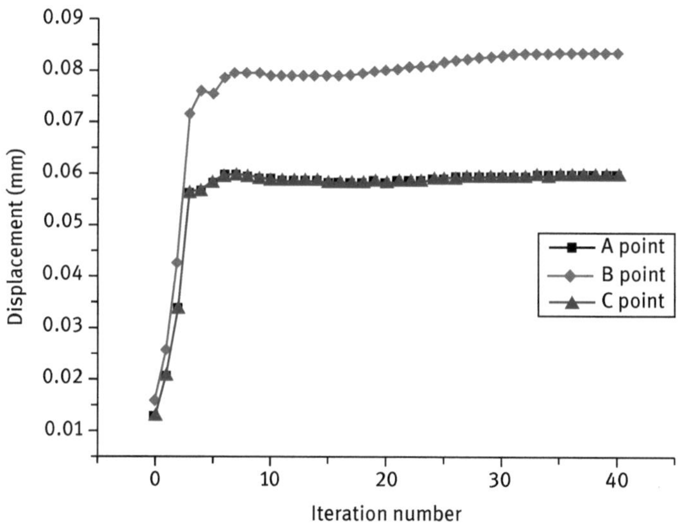

Figure 5.14: Iterative history of displacement for case 1.

Case 3: The upper limit of the vertical downward displacement at the B point is 0.08 mm. The upper limit of the vertical downward displacement at the A,C point is 0.06 mm. The upper limit of the buckling critical force is 5,000 N, and the upper limit of the stress is 2,000 MPa (Figure 5.21 to Figure 5.26).

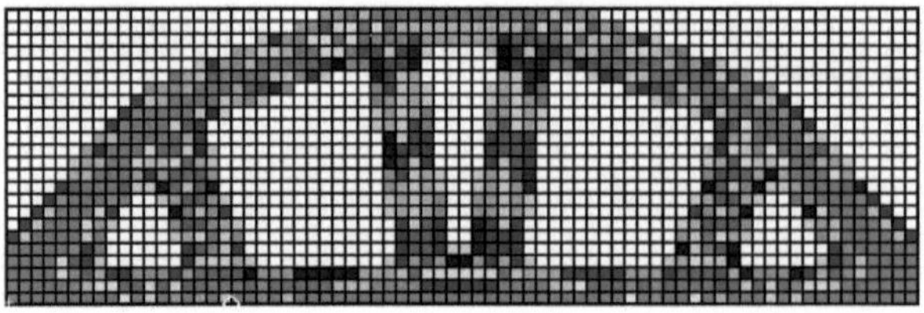

Figure 5.15: Topological configuration with total static constraints before inversion for case 2.

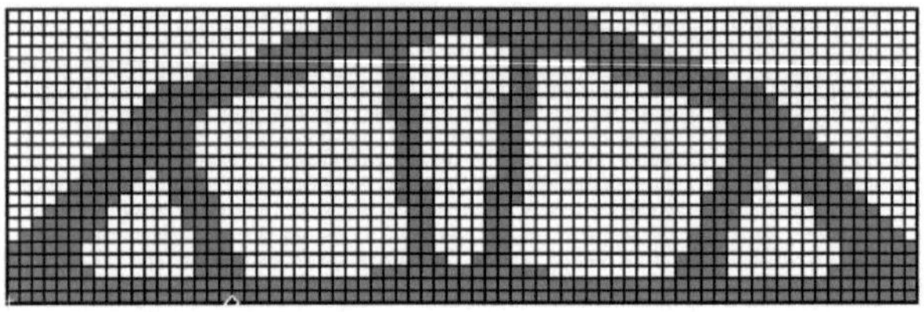

Figure 5.16: Topological configuration with total static constraints after inversion for case 2.

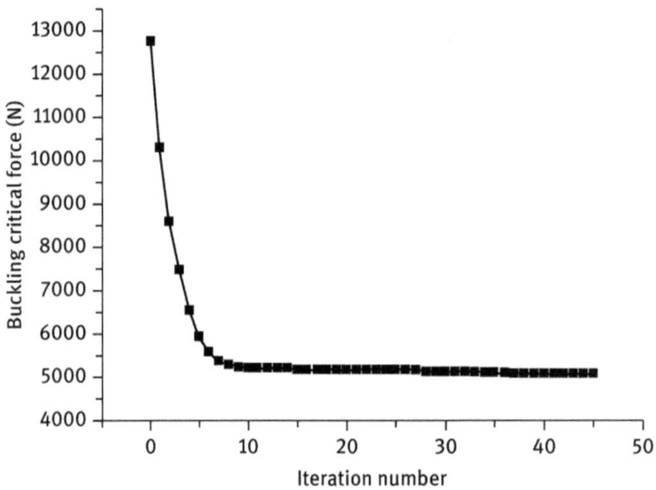

Figure 5.17: Iterative history of critical buckling critical load for case 2.

Compared with the above cases, the whole topological structure is relatively long and slender, and the results are similar to those obtained from the literature [65] when considering only the displacement and stress constraints.When the buckling constraint is added, the B point takes smaller displacement constraints, and the topology changes greatly. The fixed support increases the topology structure. The support of

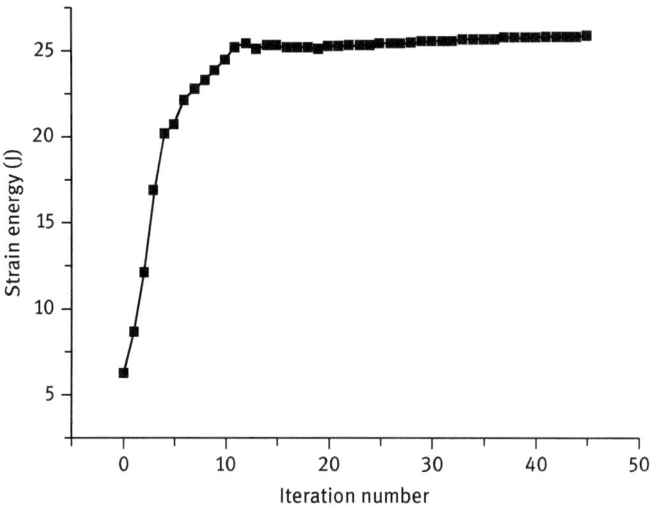

Figure 5.18: Iterative history of strain energy for case 2.

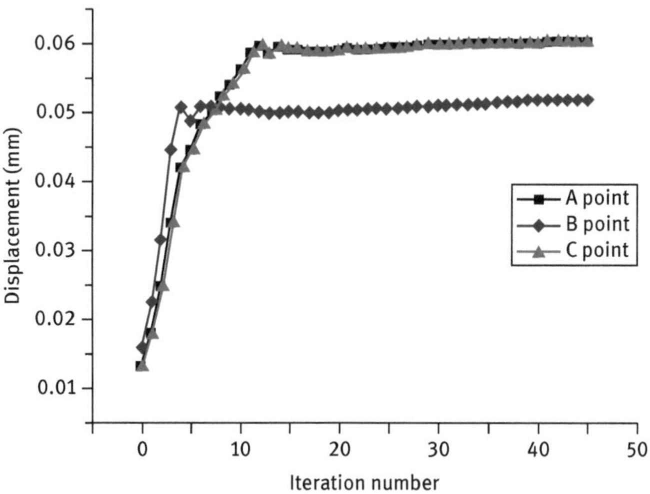

Figure 5.19: Iterative history of displacement for case 2.

the middle part is divided into two branches, which enhance the stability of the structure. Taking a larger displacement constraint to point B, the optimization structure is actually weakened, and the central support structure is weakened. Because the buckling constraint is added, there is still a topological connection between the two fixed supports.

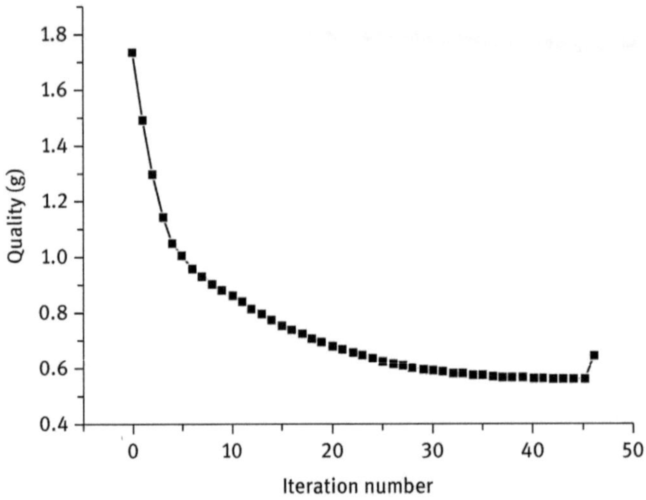

Figure 5.20: Iterative history of quality for case 2.

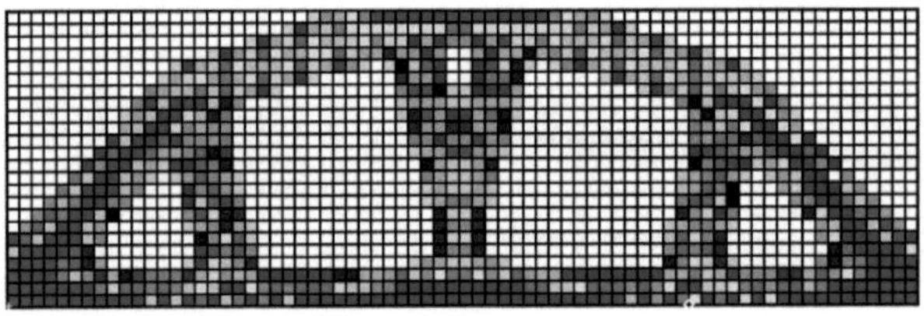

Figure 5.21: Topological configuration with total static constraints before inversion for case 3.

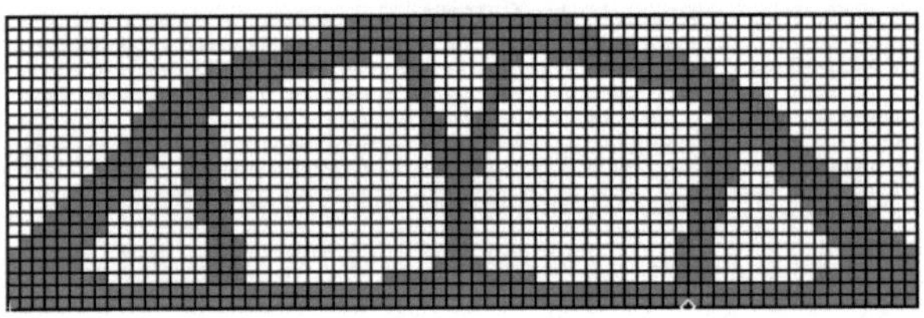

Figure 5.22: Topological configuration with total static constraints after inversion for case 3.

Example 3

(1) The model of force in the hole

As illustrated in Figure 5.27, the base structure is a plane elastic body with size 800 × 400 × 2.0 mm. The elastic modulus of the structure material is E = 68.89 GPa, the Poisson's ratio is y = 0.30, and the material density is $\rho = 2.7{\times}10^{-6}$ kg/mm^3. The right- and left-hand underpart corner points of the structure are clasped. A concentrated force P = 12,000 N along the up-to-down (-y) direction is located at the hole of

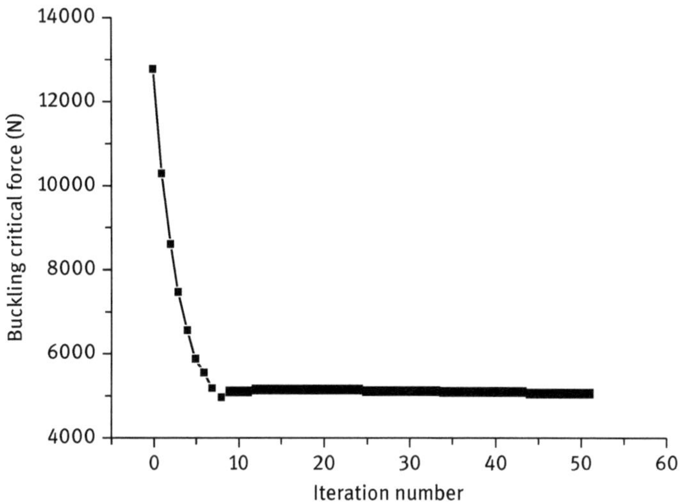

Figure 5.23: Iterative history of critical buckling critical load for case 3.

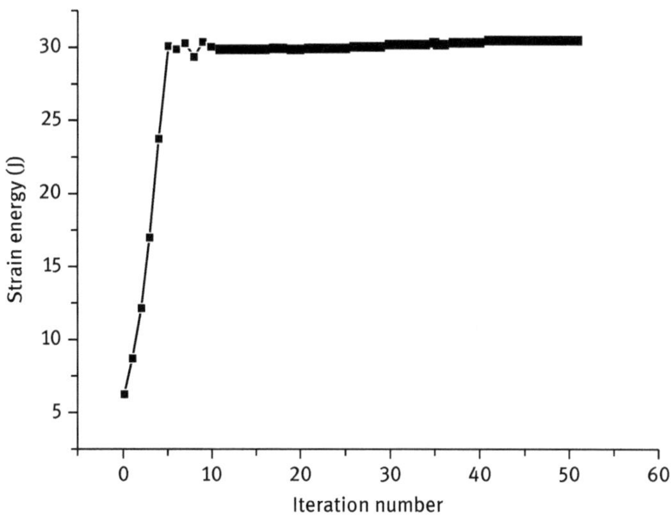

Figure 5.24: Iterative history of strain energy for case 3.

structure. The displacement of the force acting point is −1.56 mm. A quadrilateral element with four nodes is used and the total number of elements is 600. The quality of the base structure is 1.296 kg. The maximum von Mises stress of the structure is 355 MPa, the buckling factor is 0.013, and the critical force is 153 N.

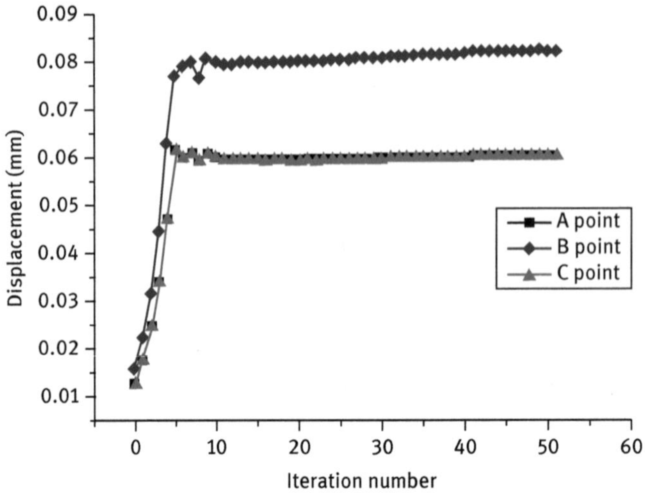

Figure 5.25: Iterative history of displacement for case 3.

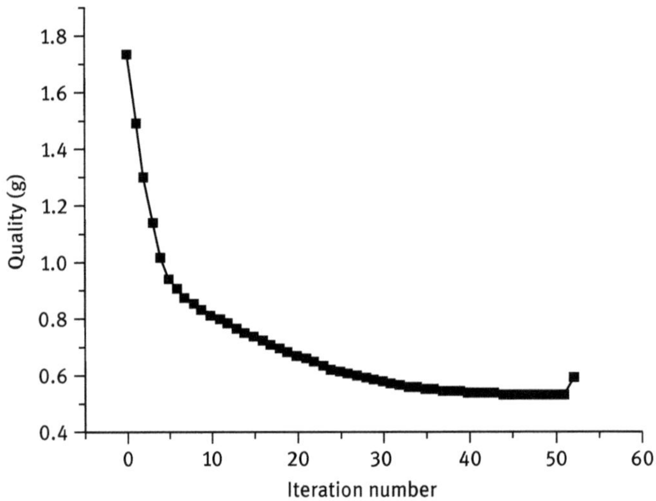

Figure 5.26: Iterative history of quality for case 3.

The upper limit of the displacement of the force action point is −2.8 mm, and the upper limit of the stress is 220 MPa. The upper limit of the buckling critical force is 80 N. After 40 iterations, the mathematical model converges.

The topological structure of each of the various constraints is compared. Considering only the buckling constraints, the whole topology only under the square hole, the topological structure is relatively short and thick. When the stress and displacement constraints are added, the topological structure increases the

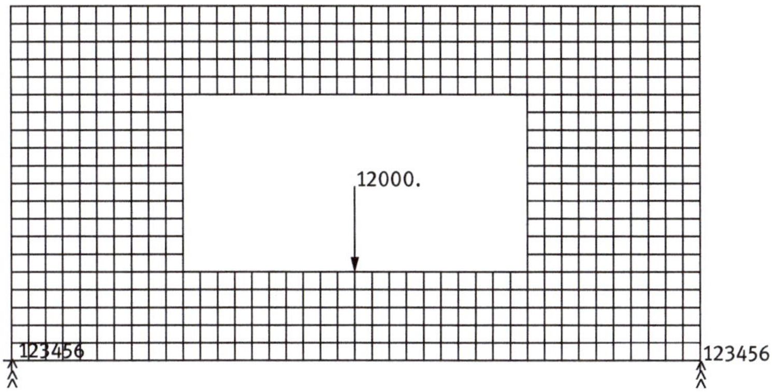

Figure 5.27: Base structure.

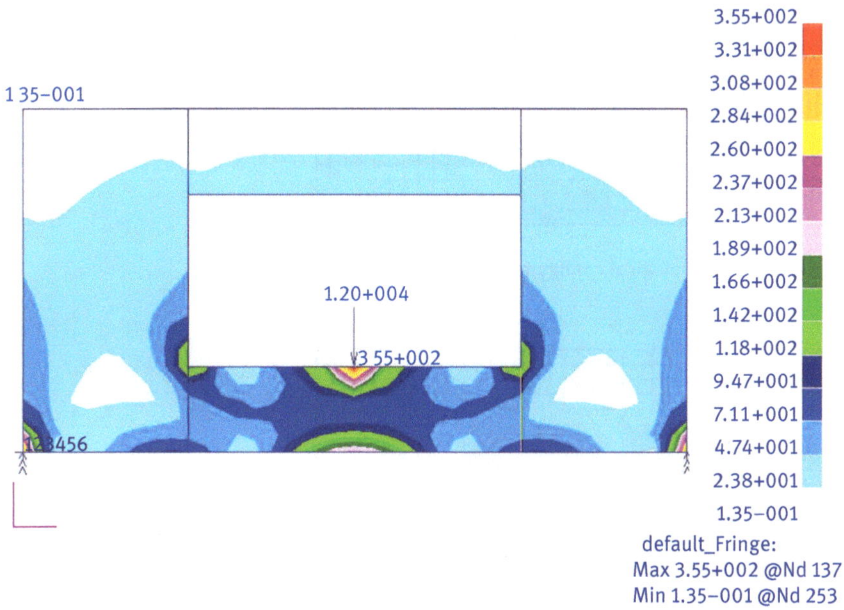

Figure 5.28: Stress fringe result of a base structure.

upper part of the square hole, and the overall size of the topological structure is larger. From this example, we can see that a low-height structure has the better stability (Figure 5.28 to Figure 5.32).

(2) The model of force out the hole

A concentrated force P = 12,000 N along the up-to-down (-y) direction is located at the midpoint of the upper boundary of the structure. The right- and left-hand underpart corner points of the structure are clasped. The displacement of the

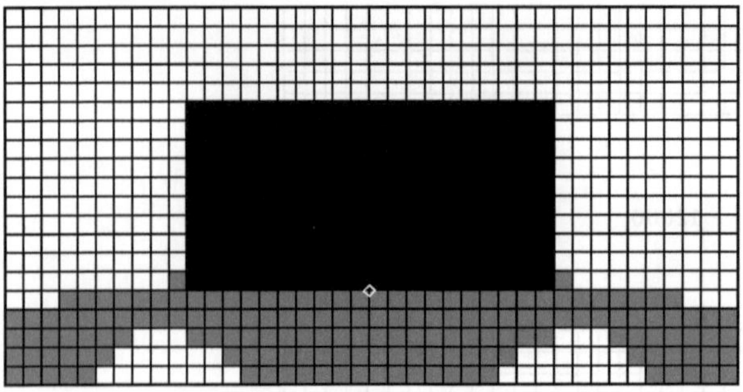

Figure 5.29: Optimal topological configuration with buckling constraint.

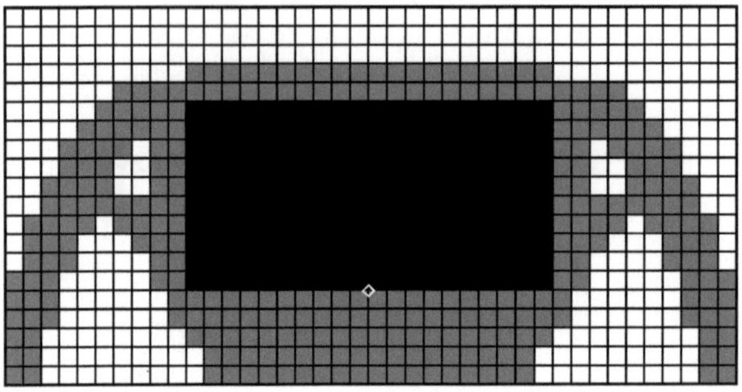

Figure 5.30: Optimal topological configuration with displacement constraint.

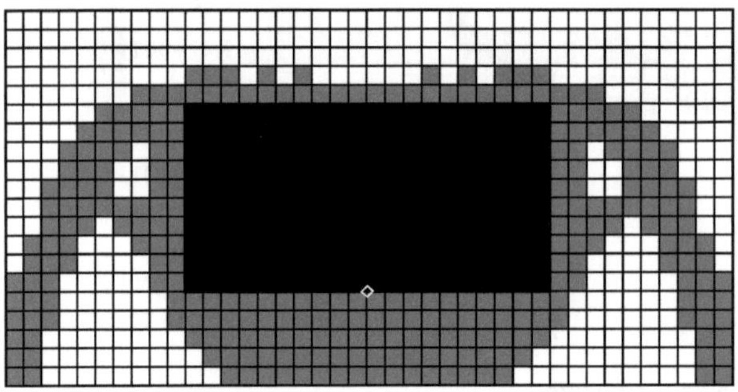

Figure 5.31: Optimal topological configuration with stress constraint.

force acting point is –1.39 mm. The maximum von Mises stress of the structure is 362 MPa, the buckling factor is 0.0060258, and the critical force is 72 N. The quality of the base structure is 1.296 kg.

The upper limit of the displacement of the force action point is –2.8 mm, and the upper limit of the stress is 200 MPa. The upper limit of the buckling critical force is 38 N. After 34 iterations, the mathematical model converges.

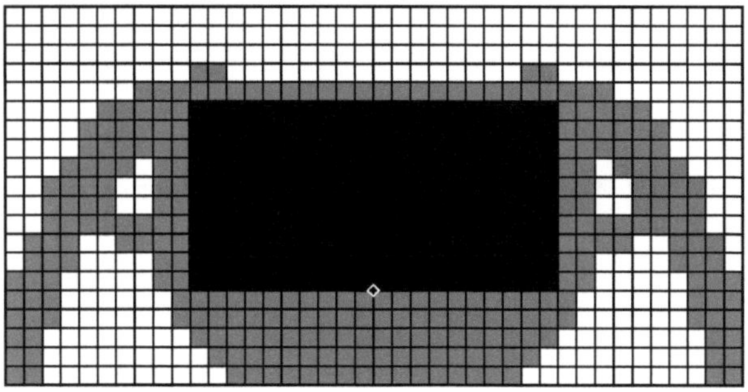

Figure 5.32: Optimal topological configuration with total static constraints.

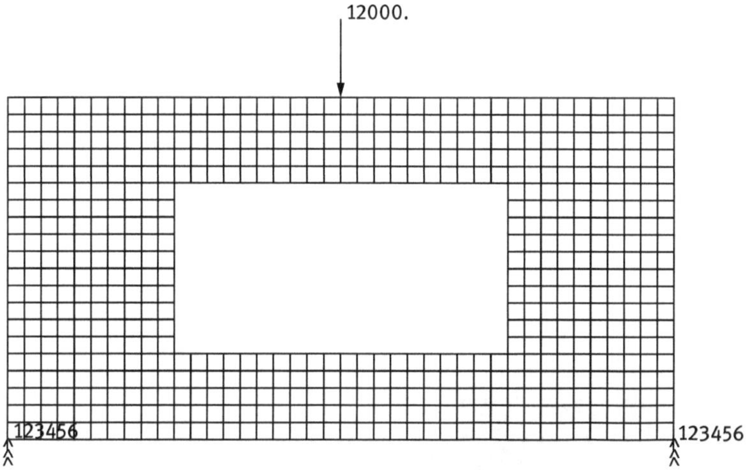

Figure 5.33: Base structure.

Comparing the above topological structure, the fixed support at the bottom of the topological structure is relatively robust, while the top is relatively thin, when considering the bucking constraint. When considering only the stress or displacement constraints, the two topology structures are almost the same. When all static constraints are applied, the topological structure only increases the strength of the fixed support at the bottom of the base of the stress constraint structure (Figure 5.33 to Figure 5.38).

4. Example of multiwork conditions

As illustrated in Figure 5.39, the base structure is a plane elastic body with size 600 × 260 × 2.0 mm. The elastic modulus of the structure material is $E = 68.89$ GPa, the Poisson's ratio is $y = 0.30$, and the material density is $\rho = 2.7\times10^{-6}$ kg/mm^3. The right- and left-hand underpart corner points of the structure are clasped. A

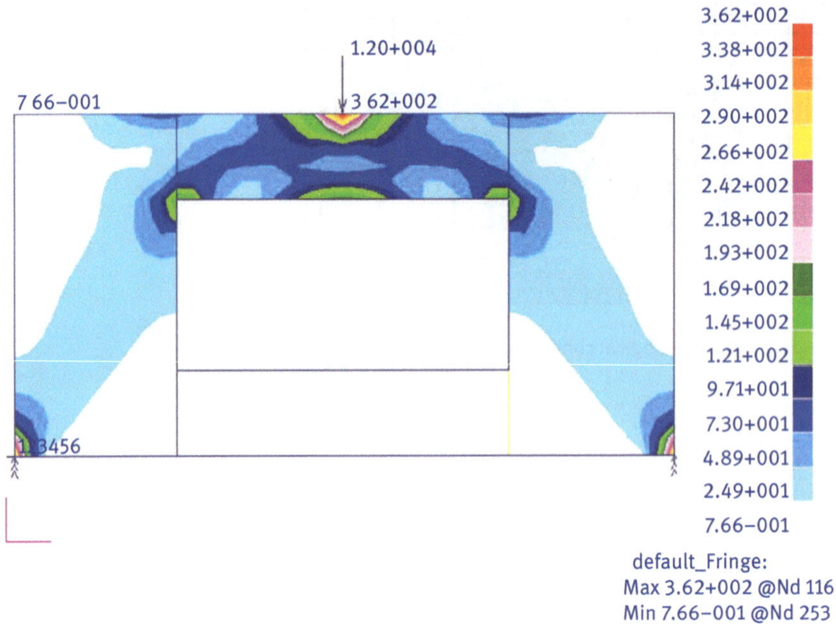

Figure 5.34: Stress fringe result of a base structure.

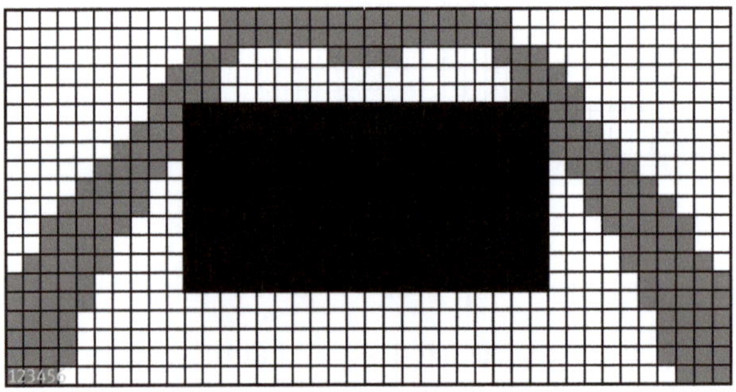

Figure 5.35: Optimal topological configuration with buckling constraint.

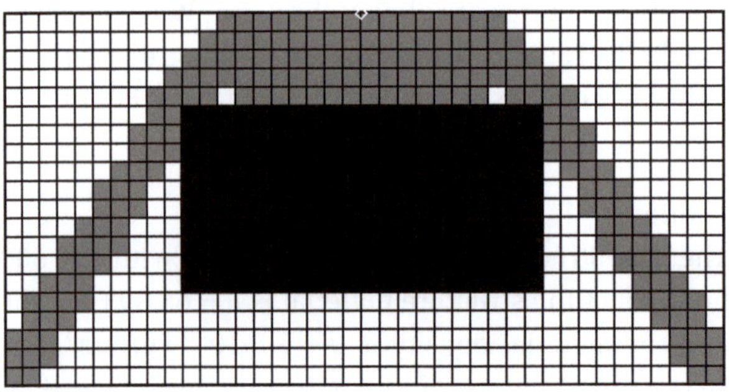

Figure 5.36: Optimal topological configuration with displacement constraint.

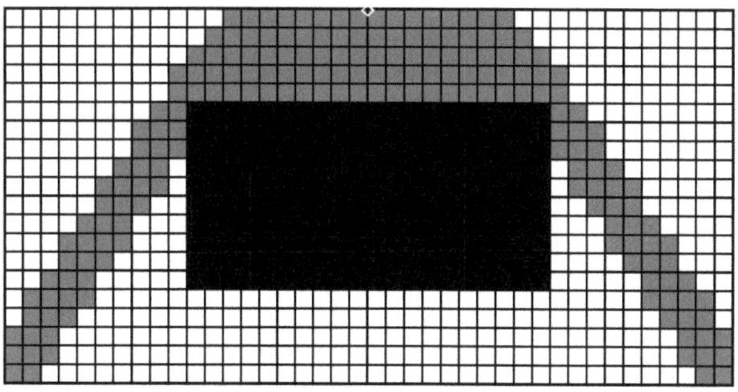

Figure 5.37: Optimal topological configuration with stress constraint.

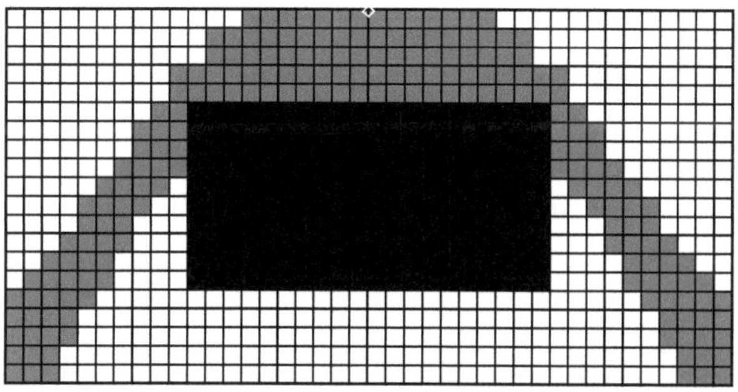

Figure 5.38: Optimal topological configuration with total static constraints.

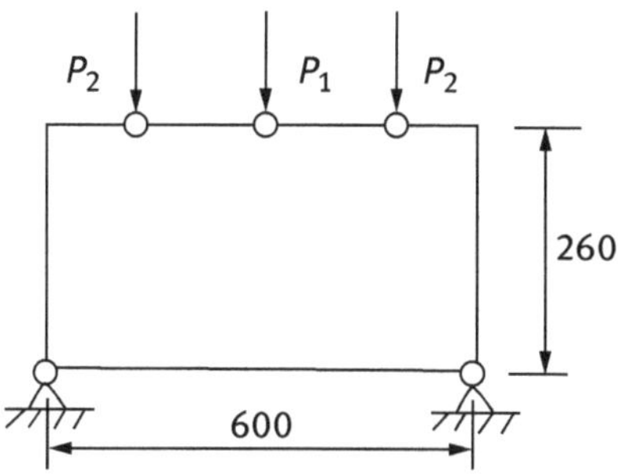

Figure 5.39: Base structure.

quadrilateral element with four nodes is used and the total number of elements is 60 × 26. Condition 1: a concentrated force P = 12,000 N along the up-to-down (-y) direction is located at the midpoint of the upper boundary of the structure. Condition 2: two concentrated forces $P2$ = 3,500 act on the upper boundary point which distance is 220 mm to the midpoint. The first-order buckling force is 112 N, the upper limit critical buckling force is 60 N, the second-order buckling critical force is 792 N, and the upper limit critical buckling force is 400 N. The key displacement point is the force action point. The upper limit of the operating point displacement is 2.0 mm under condition 1. The upper limit of the operating point displacement is 1.2 mm under condition 2. The upper limit of the stress is 150 MPa. The quality of the base structure is 0.84 kg, and the convergence accuracy is 0.001 (Table 5.2).

Table 5.2: Comparison of topological results with different constraints and load cases.

Constraint type	Initial quality (kg)	Optimized quality (kg)	Iteration number
All constraint of single-work condition	0.73	0.20832	48
All constraints of multiwork conditions	0.73	0.35513	54

The main research of this example is: the topology optimization of continuum structure with all static constraint interaction under the single condition and multi-conditions. As can be seen from the graph, the topological structure with buckling constraints is "robust." Higher order constraint plays a major role in considering the multi-order buckling constraints (Figure 5.40 to Figure 5.44).

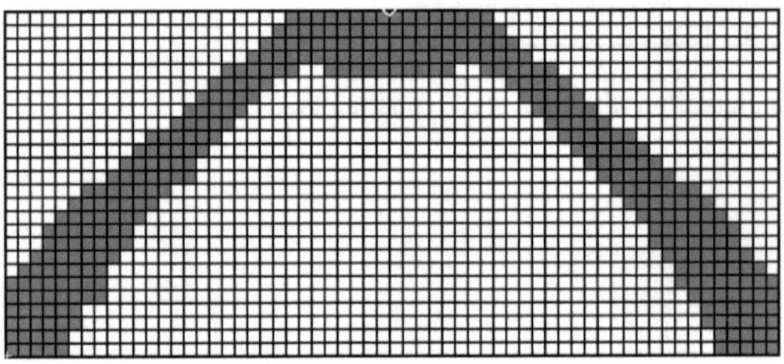

Figure 5.40: Topological configuration of load case 1 with total static constraints.

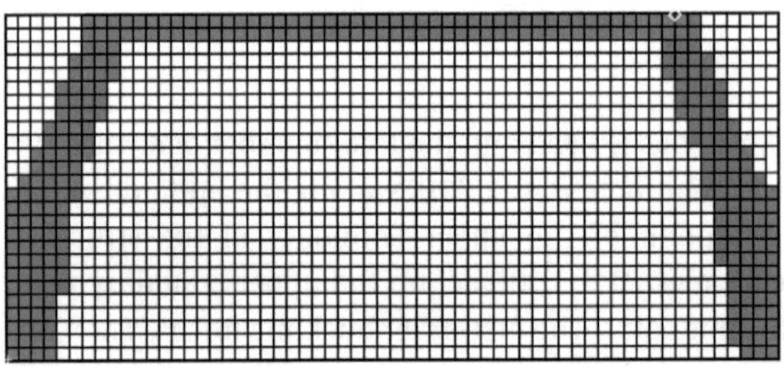

Figure 5.41: Topological configuration of load case 2 with total static constraints.

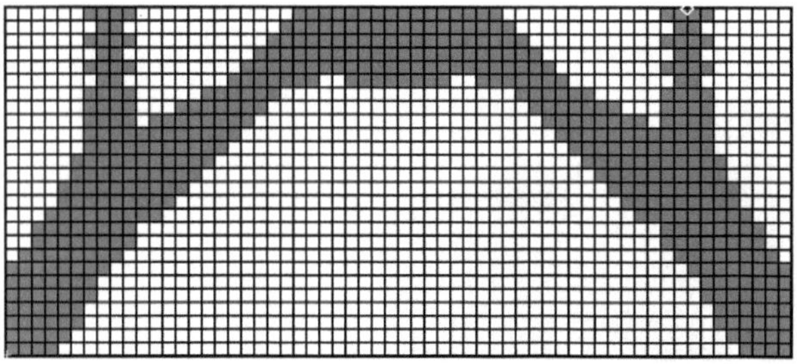

Figure 5.42: First-order mode topological configuration of multiple load cases with total static constraints.

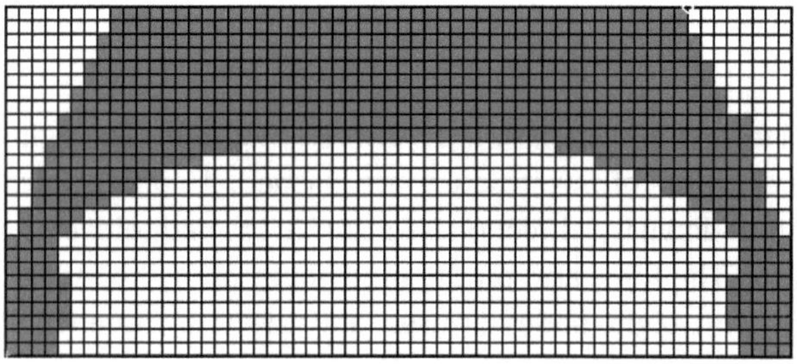

Figure 5.43: Second-order mode topological configuration with multiple load cases $\bar{\lambda}_2$ = 400 N.

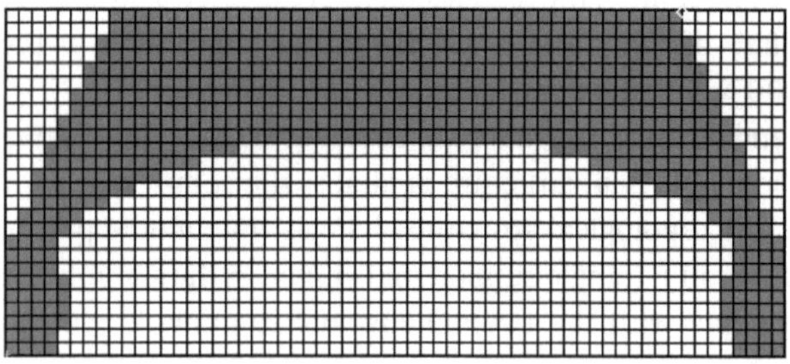

Figure 5.44: First- and second-order mode topological configuration with multiple load cases $\bar{\lambda}_1$ = 60 N, $\bar{\lambda}_2$ = 400 N.

For example, the topological structures of the first- and second-order buckling constraints are the same as the structure of the second-order buckling constraints.

5.5.2 Numerical Example of a Three-dimensional Solid Structure

Example 1

As illustrated in Figure 5.45, the base structure is a 3D elastic body with size 300 × 100 × 10 mm. The elastic modulus of the structure material is $E = 70 \times 10^3$ MPa, the

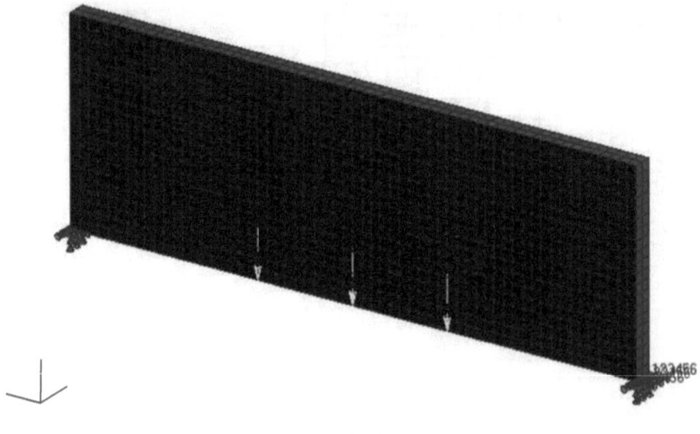

Figure 5.45: Base structure.

Poisson's ratio is $y = 0.30$, and the material density is $\rho = 2.7 \times 10^{-6}$ kg/mm^3. The right- and left-hand underpart corner points of the structure are clasped. A concentrated force $P = 27{,}000$ N along the up-to-down (-y) direction is located at the midpoint of the lower boundary of the structure. The force acts on the nine nodes of the boundary, respectively. A solid element with eight nodes is used and the total number of elements is 60 × 20 × 2. The quality of the base structure is 0.81 kg.

The displacement of the force acting point is −0.291 mm. The maximum von Mises stress of the structure is 462 MPa, the buckling factor is 2.2857, and the critical force is 61,714 N. The upper limit of the displacement of the force action point is −1.6 mm, and the upper limit of the stress is 150 MPa. The upper limit of the buckling critical force is 35,000 N. Iteration convergence after 37 times.

When the displacement and stress constraints are considered, the whole structure is relatively thin, and the topological structure is almost the same. When the buckling constraint is added, the topological structure changes: the middle part of the support points changes more robust single support and enhances the stability. By comparing the numerical example with the two-dimensional example 2, we can found that the two ends of the bottom are disconnected, and in the two-dimensional example 2 are continuous, which is caused by the different positions of the force (5.46 to Figure 5.50).

Example 2

As illustrating in Figure 5.51, the base structure is a 3D elastic body with size 120 × 45 × 8 mm. The elastic modulus of the structure material is $E = 70 \times 10^3$ MPa, the Poisson's ratio is $y = 0.30$, and the material density is $\rho = 2.7 \times 10^{-6}$ kg/mm^3. The right- and left-hand underpart corner points of the structure are clasped. A distribution force $P = 14{,}350$ N along the up-to-down (-y) direction is located at all points of the upper boundary of the structure. The force acts on the nine nodes of

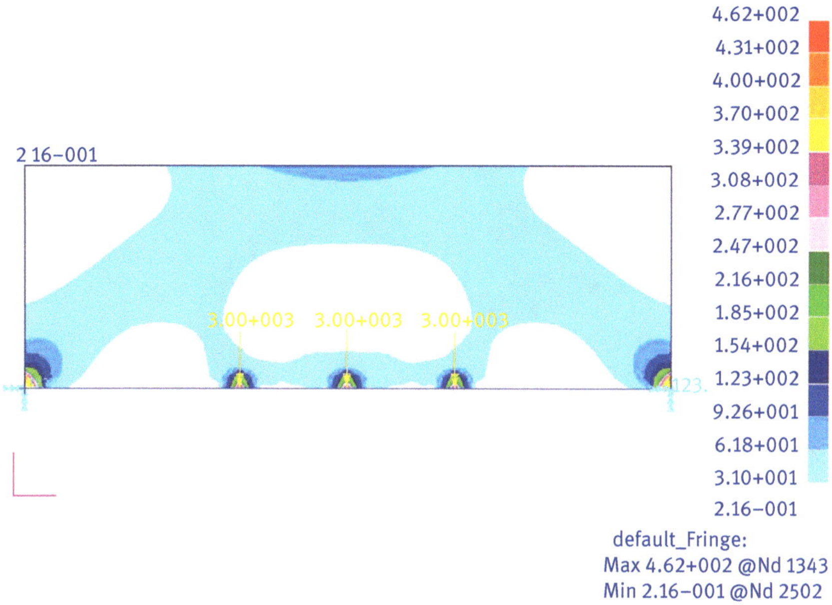

Figure 5.46: Stress fringe result of a base structure.

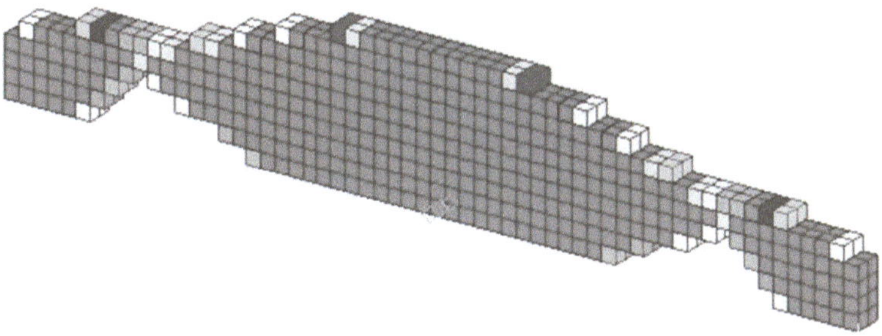

Figure 5.47: Optimal topological configuration with buckling constraint.

the boundary, respectively. A solid element with eight nodes is used and the total number of elements is 40 × 15 × 4. The quality of the base structure is 0.117 kg (Figure 5.52 to Figure 5.60).

The displacement of the force acting point is –0.102 mm. The maximum von Mises stress of the structure is 516 MPa, the buckling factor is 2.7123, and the critical force is 38,922 N. The upper limit of the displacement of the force action point is –0.12 mm, and the upper limit of the stress is 160 MPa. The upper limit of the buckling critical force is 22,000 N. Iteration convergence after 35 times. Due to the consideration of multiple constraints, the static fully constrained topological structure is stronger than a single constraint.

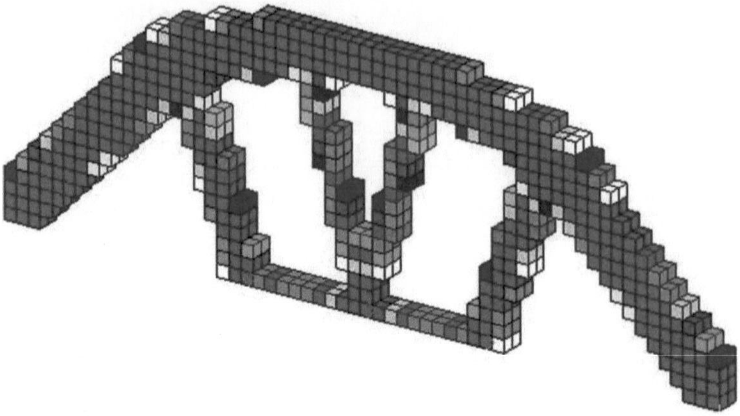

Figure 5.48: Optimal topological configuration with displacement constraint.

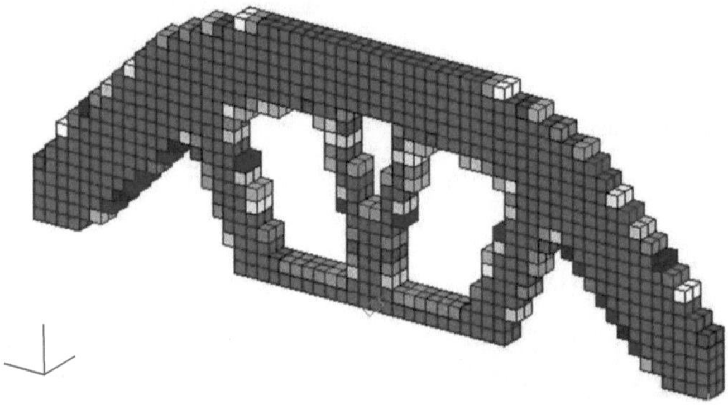

Figure 5.49: Optimal topological configuration with stress constraint.

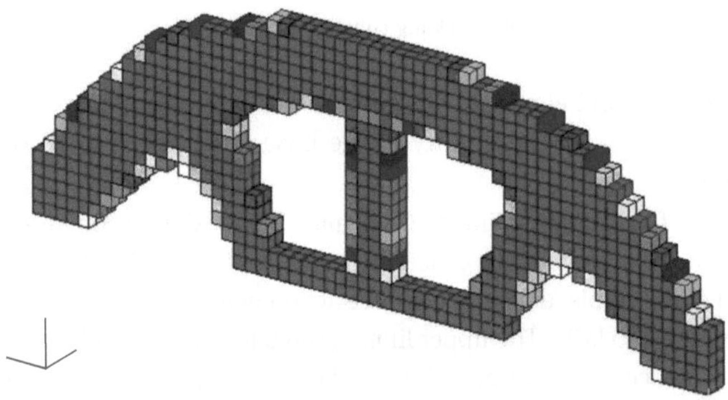

Figure 5.50: Optimal topological configuration with total static constraints.

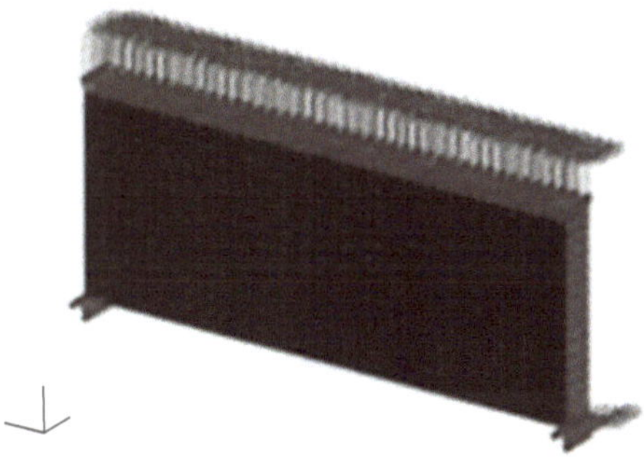

Figure 5.51: Base structure.

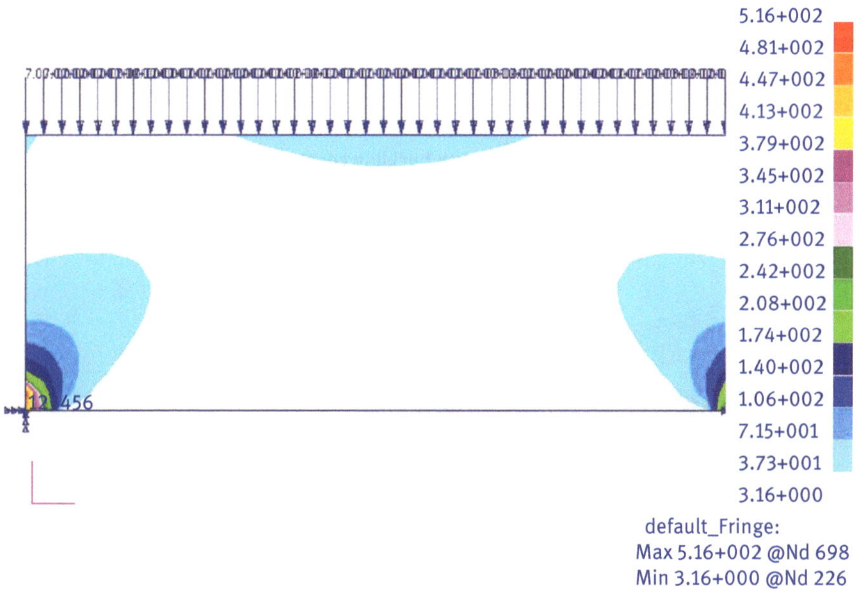

Figure 5.52: Stress fringe result of a base structure.

5.6 Summary

In this chapter, according to the ICM method, the topology optimization model for the continuum structure was constructed. The model had the minimized quality as the objective function subjected to the buckling constraints, displacement constraints and stress constraints. The continuous independent topological variables were used in this problem. Based on the Taylor expansion and the filtering function, the objective function was approximately expressed as a second-order expression. Based on the Rayleigh quotient, the Taylor expansion and the filtering function, the buckling constraints were approximately expressed as an explicit function. Based

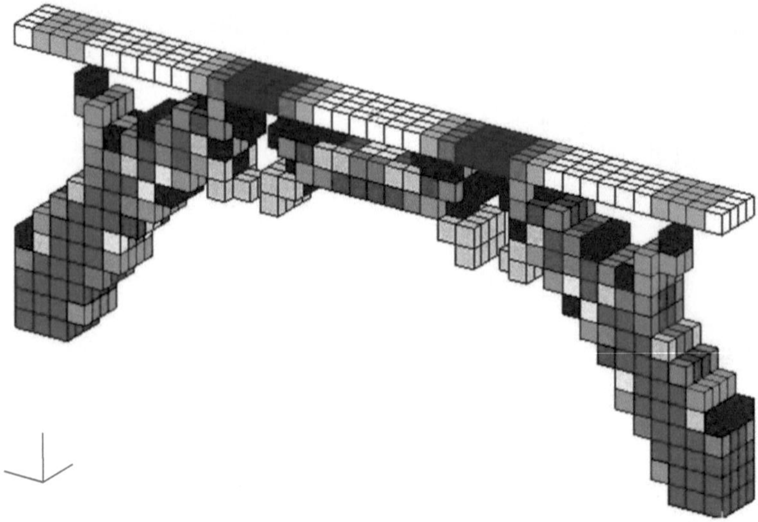

Figure 5.53: Optimal topological configuration with buckling constraint.

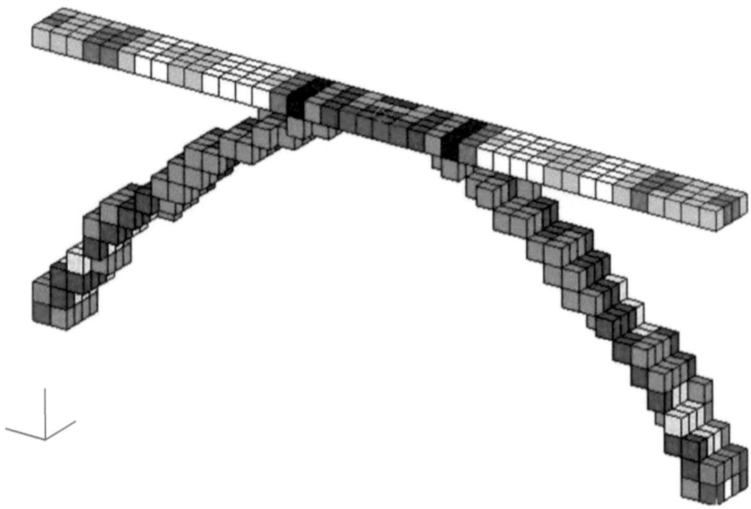

Figure 5.54: Optimal topological configuration with displacement constraint.

on the filter function, the displacement constraints are expressed approximately by Mohr theorem. Using the globalization method of stress constraints and the von Mises yield criterion in mechanics of materials, the local stress constraints were transformed into the whole structure strain energy constraint. Thus, the analysis quantity of the sensitivity was decreased. The optimization model was translated into

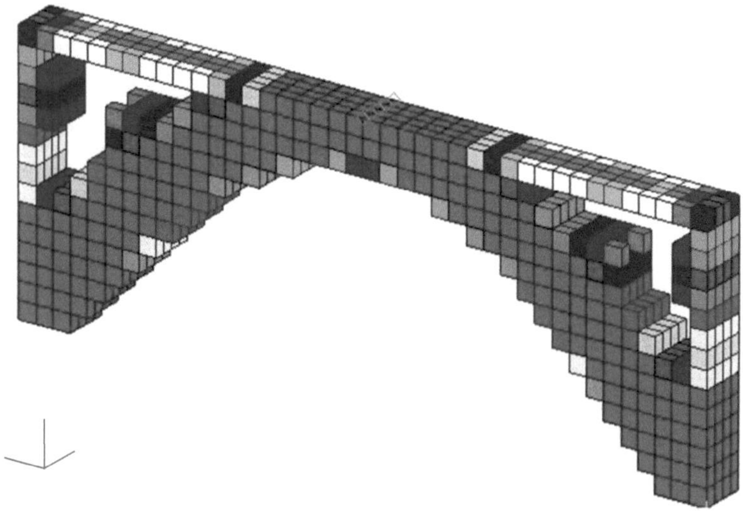

Figure 5.55: Optimal topological configuration with stress constraint.

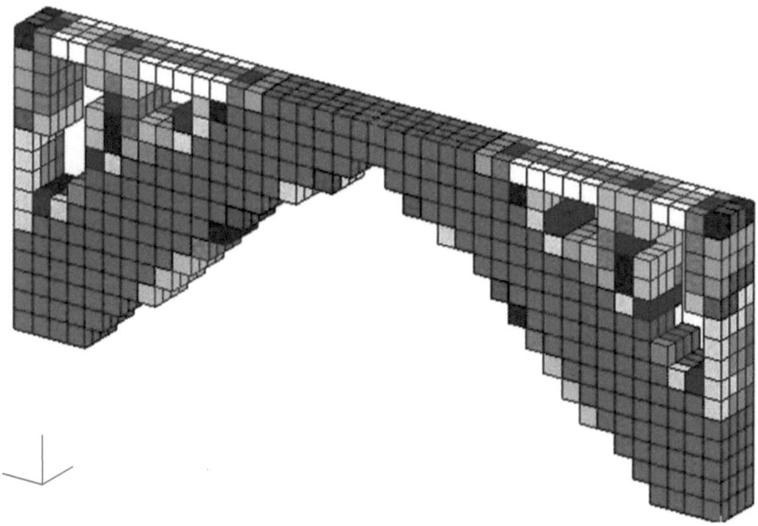

Figure 5.56: Optimal topological configuration with total static constraints.

a dual programming and solved by the sequence second-order programming. The number of variables was reduced and the model's scale was diminished. Numerical examples show that this method can solve the topology optimization problem of continuum structures with the buckling constraints, displacement constraints and stress constraints efficiently and give more reasonable topology structures.

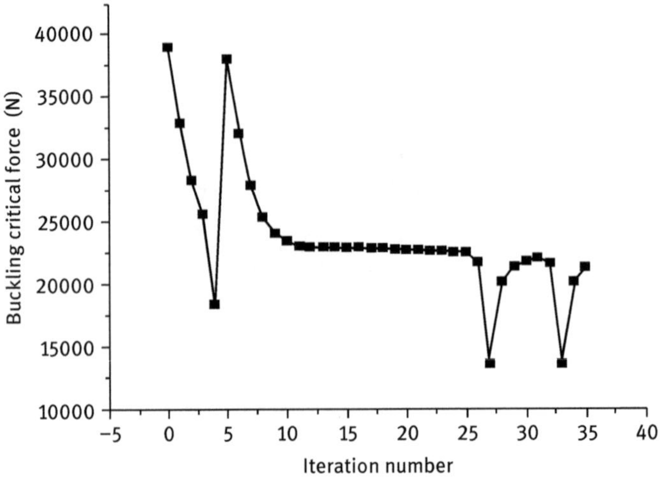

Figure 5.57: Iterative history of critical buckling critical load.

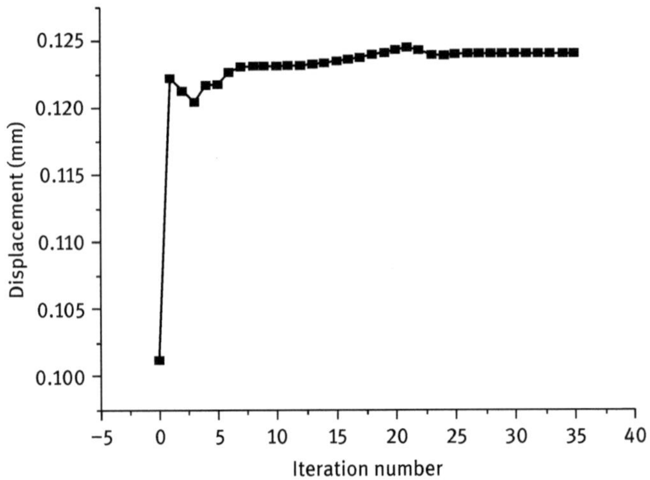

Figure 5.58: Iterative history of displacement.

The function curves of the "quality ratio" and the whole constraint ratio of the topology optimization structure are established, which provide a set of functional relations for the selection of the constraint values in all static constraint topology optimization. The topology optimization model is solved by the dual programming method, and the reasonable topology shape and clear topology structure are obtained.

This method is applied to two-dimensional and three-dimensional continuum structure topology optimization problems with single- and multiple-loading conditions,

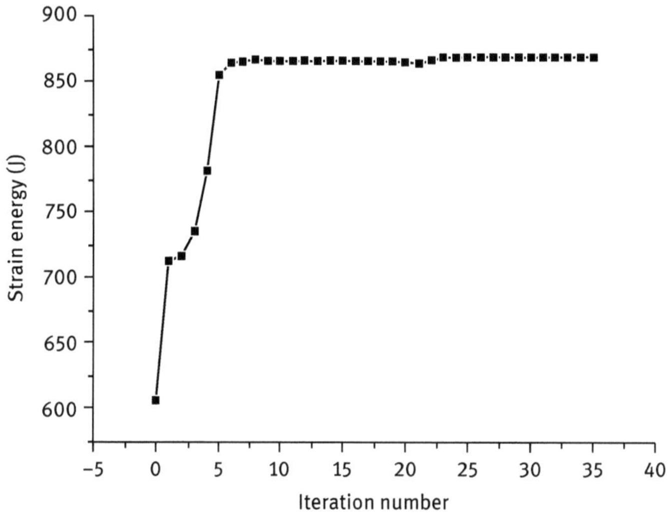

Figure 5.59: Iterative history of strain energy.

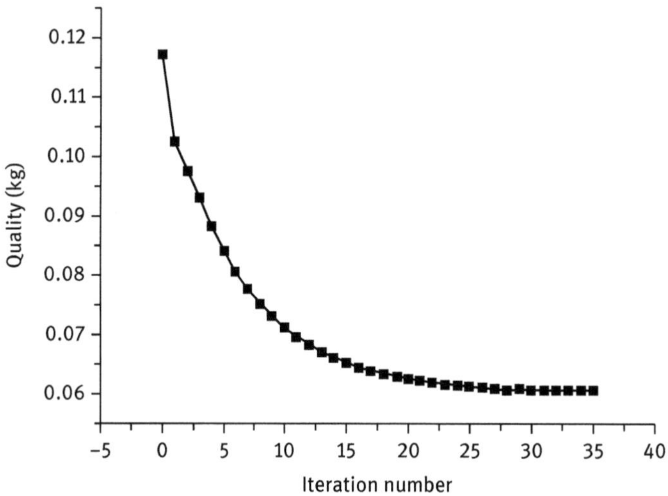

Figure 5.60: Iterative history of quality.

the topological structure with all constraints and case with single constraint are compared, and it is suitable for all the static constraint structures' optimal design. The results of several examples show that the method is effective and feasible. The optimization algorithm is implemented on the second development platform of MSC. Patran/ Nastran software.

6 Software and User Instructions

6.1 Introduction

There are two main development and design methods of finite element analysis and computer simulation software. One is a comprehensive development and design method, including the development of the core algorithm, the design of a preprocessing system, the design of a postprocessing system, etc. The advantage of this method is that the creators have all of the software right. The disadvantage is that the workload is more, covering a wide range of knowledge and a long period of the development time. Another method is developing the use of a user interface based on the existing commercial software; the software developers will add the main algorithm, using some mature software modules to achieve the desired functionality for developers. The advantage of this method is that it can meet the functional requirements of the software in a short time, the development time is shorter, and the development cost of software is also low, whereas the disadvantage is that the software cannot fully own the copyright.

All the topology optimization procedures are based on MSC.Patran/Nastran development language PCL, which is the finite element analysis software. Options menu, the user input parameters from the interface and the results of the output interface are provided in software. Users need to enter the corresponding topology optimization parameters. It can be two-dimensional or three-dimensional structure topology optimization.

The topology optimization program can solve the following problems:

(1) Two-dimensional and three-dimensional structures of the continuum under buckling constraints in a single-working condition, multimodal and multi-working condition topology optimization problem
(2) Buckling constraint, displacement constraint interaction, single-working condition, multiworking condition topology optimization problem
(3) Buckling constraints, stress constraints, the continuum of two-dimensional and three-dimensional structures of a single-working condition, multicondition topology optimization problem
(4) Two-dimensional and three-dimensional structures of a continuum under the full constraint in a single-working condition, multicondition topology optimization problem.

In order to make users to master the software as soon as possible, the program has a detailed documentation and application examples.

https://doi.org/10.1515/9783110462272-006

6.2 Compiling and Linking of a User Program

In the root directory of the MSC. Patran installation, there is an initial file init.pcl. The optimization module menu can be added to the main interface by modifying this file. In the init.pcl file, add the path of the user optimize program; the user functions are on the path.

Add a program to use the following statement:

```
file_add_path(999, “user_path”).
```

Add a Patran path, for example, as follows:

```
STRING topology_home[128]=“d:\userpath\disp_stress_buck”
file_add_path(999, topology_home)
```

The init.pcl file contains two commands:

```
!! INPUT p3prolog.pcl NOERROR
!! INPUT p3epilog.pcl NOERROR
```

Define global variables in the p3prolog.pcl file. Add the functions of the compiler in the P3epilog.pcl file. While placing the two p3prolog.pcl and p3epilog.pcl files, as well as all the functions in the path of the Patran, the program will automatically compile all the user function links, when the Patran start.

Writing subroutine and subfunction is the core of the developing work. There are a large number of internal functions in the PCL language, and the user can directly call them. At the same time in order to meet the needs of users, you can also define functions to complete the special function: these functions are called the *custom function*.

The basic structure of the custom function in the PCL language is as follows:

```
FUNCTION function_name(Parameter variable)
defined variable
Procedural subject
END FUNCTION
```

6.3 Setup for Analysis and Output

First, the geometric model is established in MSC.Patran, and the finite element mesh is divided. Due to the topology optimization under buckling constraint in this program, the type of the element is shell element. After defining the material property and adding boundary conditions, the topology optimization type is selected in the topology optimization menu, and the optimized data is input in the optimization parameter interface to solve the model. When we do finite element analysis using Nastran, the buckling analysis solver SOL105 is added to submit analysis. At the same time, the relevant results are needed to use the finite element analysis, so the type of output results must be the appropriate settings. When the data is output, the element strain energy and the grid point force are needed.

The corresponding PCL program is shown below:

```
jobfile.writei("CASE INPUT 0", 3)
jobfile.writec("CASE INPUT 1","ESE=ALL")
jobfile.writec("CASE INPUT 2","ELFORCE=ALL")
jobfile.writec("CASE INPUT 3","GPFORCE=ALL")
```

The file suffix for MASTER, DBALL is the default output file of finite element analysis, which stores a large amount of analysis information and occupies a lot of space. These files are not useful in topology optimization of continuum structures. Because the topology optimization model design variables are more and the memory space in the operation of the machine is limited, the program will stop. We must add the following sentences when reading the structure analysis file:

```
Jobfile.writec("FMS INPUT 0", 2)
jobfile.writec("FMS INPUT 1","ASSIGN DBALL='111.DBALL',TEMP, DELETE")
jobfile.writec("FMS INPUT 2","ASSIGN MASTER='111.MASTER',TEMP,DELETE")
```

The function of the statement is to delete the *.DBALL files and *.MASTER files generated by each iteration of the optimization, so that you can save memory and hard disk space.

6.4 *Topological Optimization* Menu and Inputs

The main window of the topology optimization software is shown in Figure 6.1. In the main window of MSC.Patran/Nastran, respectively, add the menu and dialog box. The software mainly includes three parts, considering the buckling of the continuum structure topology optimization, optimization results output, and the help document. Three parts correspond to the three menus of the main window: the *topology optimization, Results* and *Optimization's help*.

Click the *topology optimization* option in the main menu, and the drop-down menu will pop up. This menu contains four submenu items: *Buckling optimization, Displacement and buckling optimization, Stress and buckling optimization,* and *Stress displacement and buckling optimization* (Figure 6.2). The *Buckling optimization* menu item is used to solve the problem of topology optimization of continuum structures with buckling constraints; the *Displacement and buckling optimization* menu item is used to solve the problem of topology optimization of continuum structures with the interaction of displacement constraints and buckling constraints; the *Stress and buckling optimization* menu item is used to solve the problem of topology optimization of continuum structures with stress and buckling constraints; and *Stress displacement and buckling optimization* menu item solves the problem of topology optimization of continuum structures, which are subjected to displacement, stress and buckling constraints. The following four submenu items are explained next.

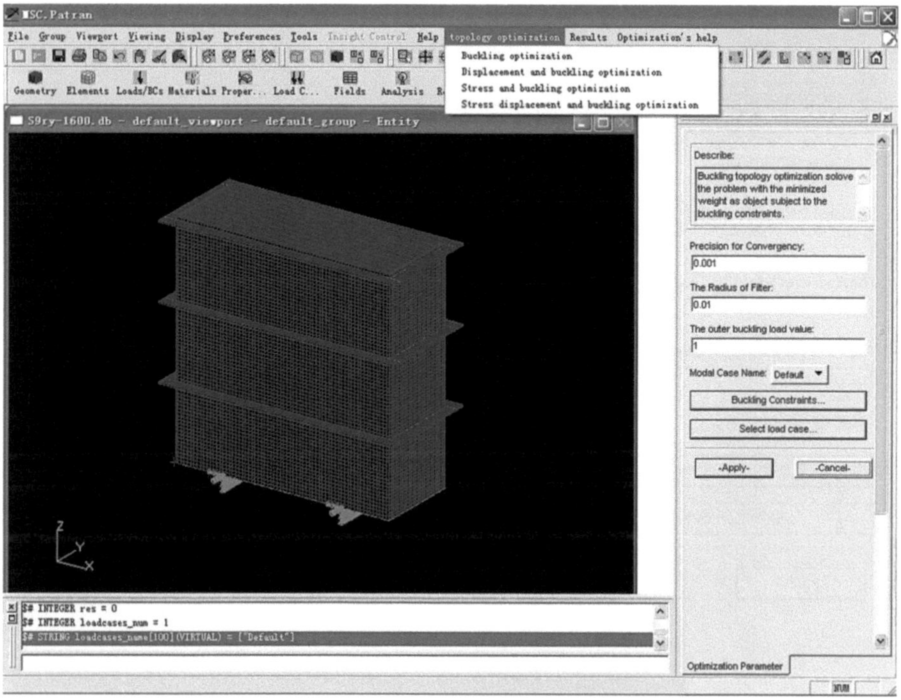

Figure 6.1: The main window of the program.

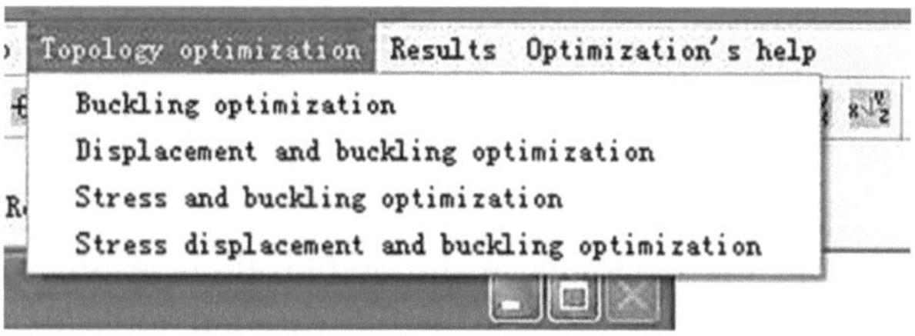

Figure 6.2: *Topology optimization* menu.

6.4.1 *Buckling Optimization* Menu Item

Click the *Buckling optimization* menu item, and then the data entry window will appear as shown in Figure 6.3. These data are used to solve the problem of topology optimization of continuum structures with buckling constraints. The input parameters are explained one by one.

Precision for Convergency is the convergence accuracy of the optimize target, which is usually taken to be 0.001. *The Radius of Filter* is the filter radius of the structural unit, which is usually taken 1.5 times the unit length. The data is used

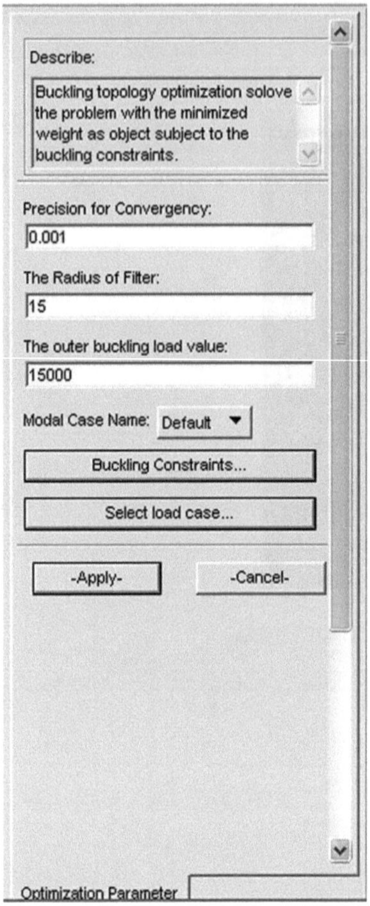

Figure 6.3: Window of buckling topology optimization.

to eliminate the grid problem and the dependence the optimization results to the grid. *The outer buckling load value* is the size of the external force that causes the buckling of the structure. Because the forces causing buckling are not necessarily external forces,it need to be judged by designers. Therefore, automatic extraction is not needed here, and manual input is needed.

Button *Apply* function is: clicking *Apply* program will optimize the structure based on user input parameters. The function of the *Cancel* button is to reject the previous input and close the window.

The setting interface of buckling constraints is given in Figure 6.4.

Number of the Calculated Buckling Modality denotes the number of modes needed for the buckling analysis. The structure cannot work after reaching the first-order buckling; usually only one-order buckling can be solved. This program can also solve the high-order buckling mode optimization. *Constraint Name* is the name of the constraint.

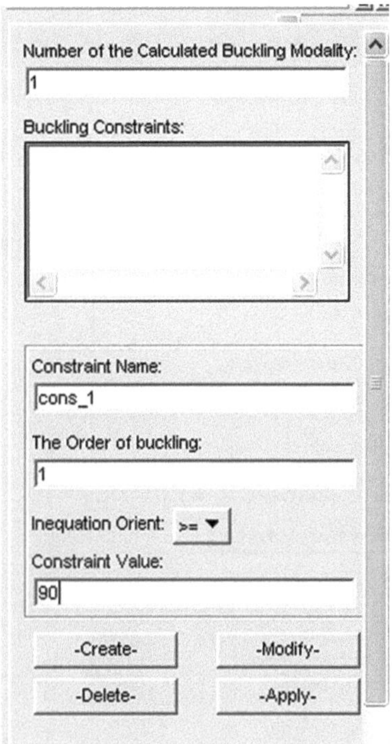

Figure 6.4: Window of buckling constraint parameter.

The Order of buckling is the order of buckling constraints, also known as the number of modes. The values corresponding to buckling mode constraints need to be set if the multiple order buckling mode to be solved.

Inequation Orient is the direction of constraint inequality.

Constraint Value is the value of the critical force of buckling constraints.

Click the *Apply* button to create the constraint.

Select load case is the selection of working conditions button, selecting the condition in the dialog box, and then click the *Apply* button to optimize.

6.4.2 *Displacement and Buckling Optimization* Menu Item

Click *Displacement and buckling optimization* menu, and the data entry window will pop up as shown in Figure 6.5. These data are used to solve the problem of topology optimization of continuum structures, which are subjected to displacement constraints and buckling constraints. The parameters in the data entry window are the same as those in the window of the buckling constraints, but the displacement constraint is added to the set button.

Displacement Constraints is the displacement constraint button.

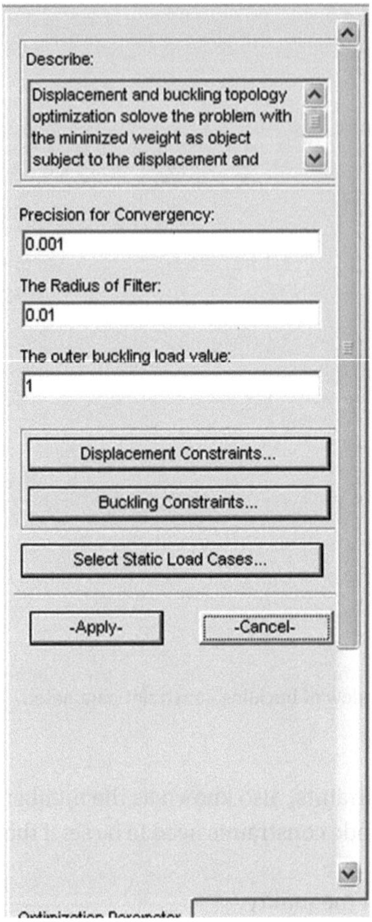

Figure 6.5: Window of buckling and displacement constraints.

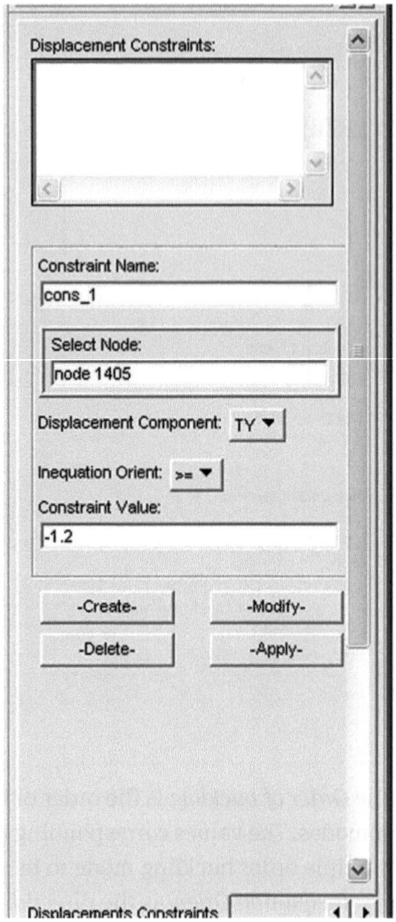

Figure 6.6: Window of a displacement constraint parameter.

Click the button and the displacement constraint parameter settings interface will pop up as shown in Figure 6.6.

In Figure 6.6, *Constraint Name* is the name of the constraint; *Select Node* is the node number needed to control the displacement. *Displacement Component* is the component of the displacement constraint. The other parameters are the same as provided in the *Optimization Buckling* menu item.

In the *Constraint Name:* text box, the displacement constraint name is input, the system will give a default value, such as Cons_1, and the user can also define the constraint name. Using the mouse to define the node in the structure in the left work space, the *Select Node:* node selection box will display the number of selected nodes. In the *Displacement Component:* drop-down menu, the direction of displacement constraints is defined, such as TY. Then the value of the displacement constraint is entered in the text box of the *Constraint limit value:* Click the *Create* button to create *Cons_1* constraint name, the constraint node is the 525th node, the constraint direction is the Y axis, and the displacement constraint value is –0.5 mm. Use the same method to create other displacement constraints.

If the user finds the error defined in the process of the displacement constraint value, direction or other information, first select the constraint in the *Constraints: Displacement* list box, and then click *Modify* to modify the constraint information.

If the user finds the error defined in the process of the displacement constraint value, direction or other information, first select the constraint in the *Constraints: Displacement* list box and then click *Delete* to delete the constraint information.

When the displacement constraint is created, click the *Apply* button to close the displacement constraint definition window. With the previous introduction, the buckling constraint parameters are set, and then click OK.

Then click the *Apply* button in the parent window, and the optimize program begins. If you want to cancel the job, click *Cancel.*

6.4.3 *Stress and Buckling Optimization* Menu Item

Click the *Stress and buckling optimization* menu item, and *buckling and stress constraints of the structure topology optimization* window will pop up, as shown in Figure 6.7. The meaning of the parameters in the window is the same as provided in the *Buckling optimization* menu item. The target convergence accuracy, the maximum side length and the stress of the unit are input in the window. Then click the *Stress Constraints* button, enter the stress constraint definition window (the stress input unit is MPa) and click confirm (Figure 6.8). After selecting the work condition, click the *Apply* button to run the topology optimization program.

6.4.4 *Stress Displacement and Buckling Optimization* Menu Item

Click the *Stress displacement and buckling optimization* menu item, and the *static full constraint structure topology optimization* window will pop up, as shown in Figure 6.9. The meaning of the parameters in the window is the same as provided in the *Buckling optimization* menu item. The target convergence accuracy, stress, maximum side length of the unit, and external force are input in this window, click the *Stress constraints* button, entering the stress constraint definition window, input stress, unit MPa, click *Apply* button, click *Displacement constraint* button to set the displacement constraints, specific parameters as above; click *Buckling Constraints* to set the parameters of the buckling constraints, specific parameters as above. At the

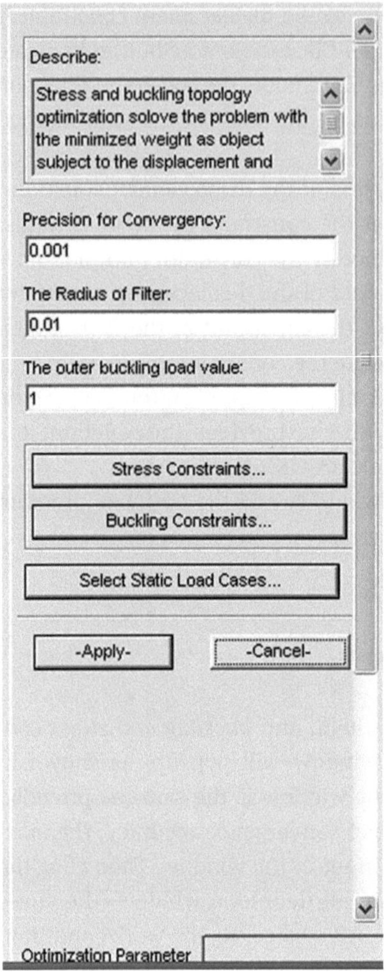

Figure 6.7: Window of buckling and stress constraints.

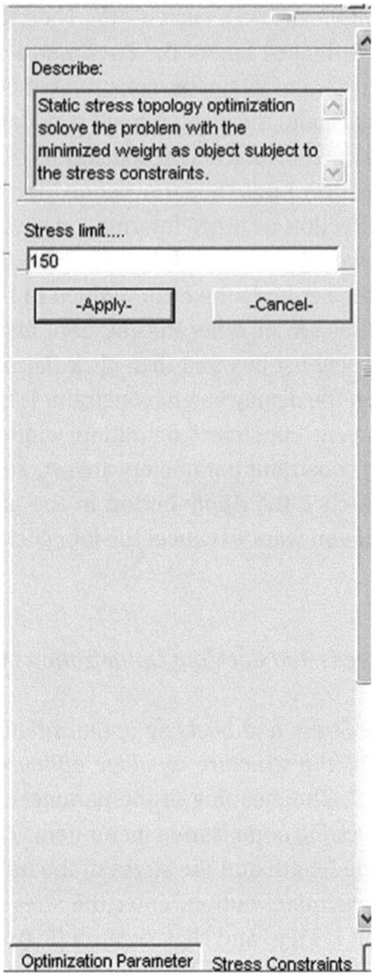

Figure 6.8: Window of a stress constraint parameter.

end of the selection process, click the *Apply* button to run the topology optimization program.

6.5 Output and Results Menu

After the end of the structure topology optimization, the user clicks the *Results* menu and the drop-down menu will pop up, as shown in Figure 6.11.

The drop-down menu contains two menu items: *Text Result* and *Graphic Result*.

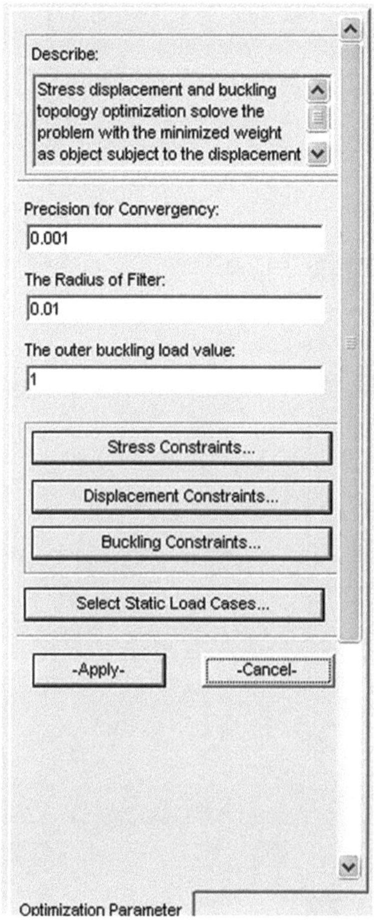

Figure 6.9: Window of all static constraints.

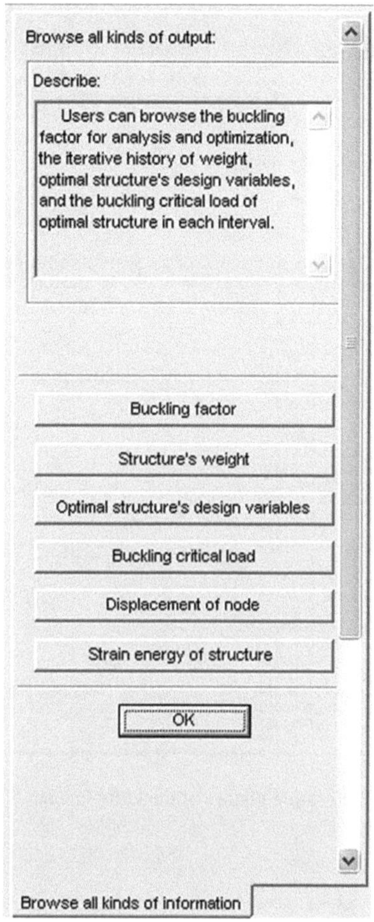

Figure 6.10: Output window of *Text Result*.

6.5.1 *Text Result* Menu Item

Click the *Text Result* menu item, and then the text output dialog box will pop up, as shown in Figure 6.10.

1. Buckling factor iterative history

Click the *Buckling factor* button, and the iteration history window of the *buckling factor* will pop up, as shown in Figure 6.12.

2. Structural quality iterative history

Click the *Structure's quality* button, and the *structure quality iteration history* window will pop up, as shown in Figure 6.13.

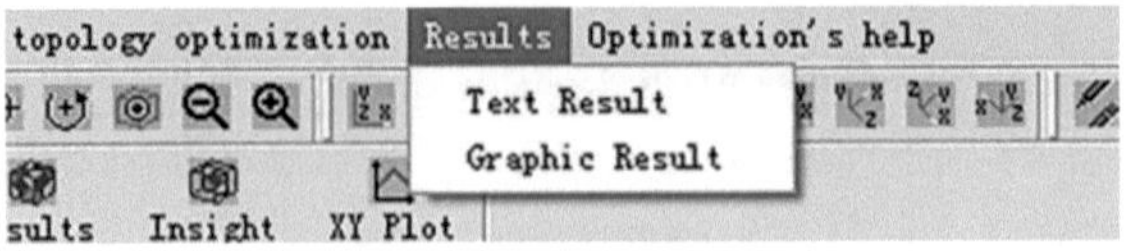

Figure 6.11: *Results* menu.

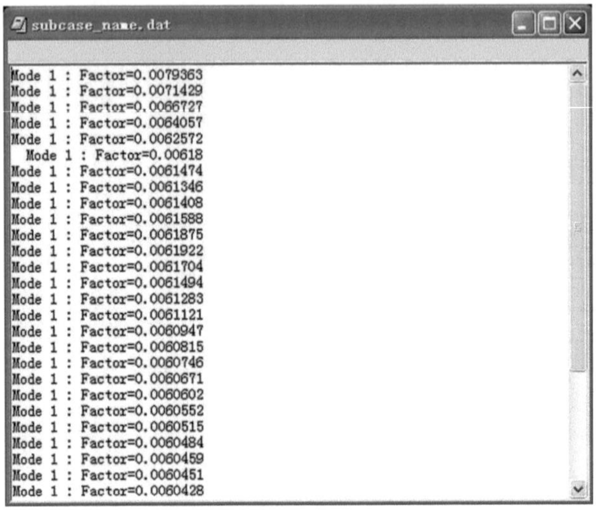

Figure 6.12: Iterative history of buckling factor.

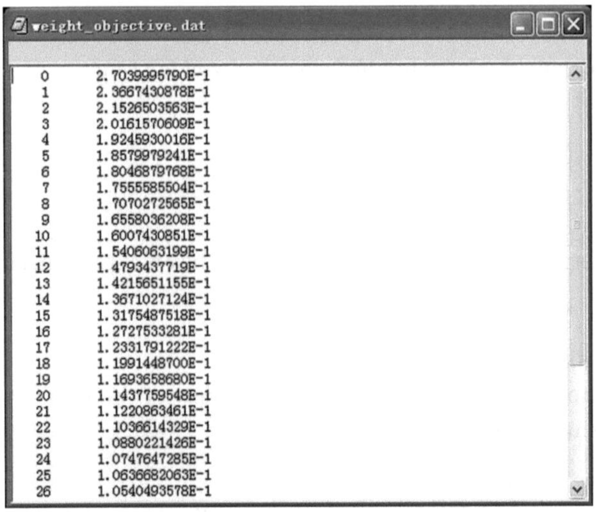

Figure 6.13: Iterative history of quality.

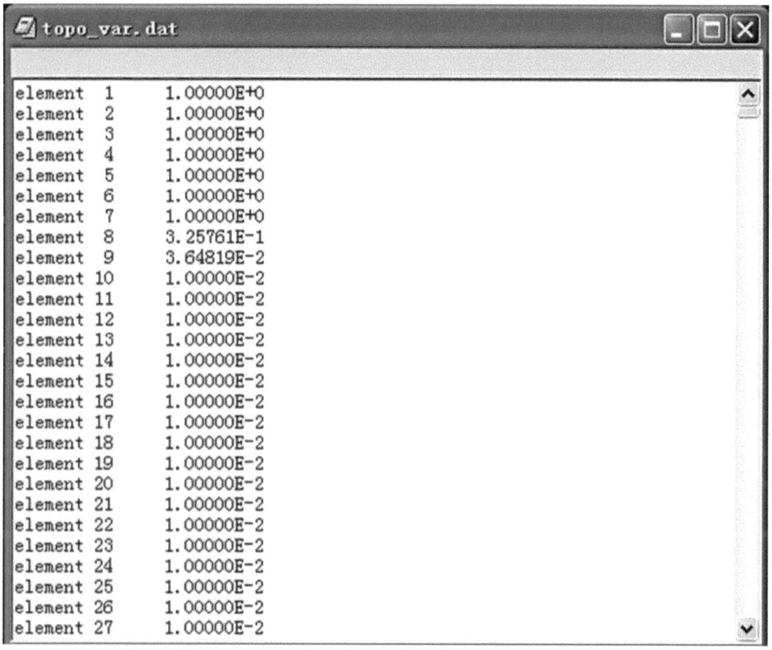

element	1	1.00000E+0
element	2	1.00000E+0
element	3	1.00000E+0
element	4	1.00000E+0
element	5	1.00000E+0
element	6	1.00000E+0
element	7	1.00000E+0
element	8	3.25761E-1
element	9	3.64819E-2
element	10	1.00000E-2
element	11	1.00000E-2
element	12	1.00000E-2
element	13	1.00000E-2
element	14	1.00000E-2
element	15	1.00000E-2
element	16	1.00000E-2
element	17	1.00000E-2
element	18	1.00000E-2
element	19	1.00000E-2
element	20	1.00000E-2
element	21	1.00000E-2
element	22	1.00000E-2
element	23	1.00000E-2
element	24	1.00000E-2
element	25	1.00000E-2
element	26	1.00000E-2
element	27	1.00000E-2

Figure 6.14: Design variables of an optimal structure.

3. Iterative history of design variables

Click the *Optimal structure's design variables*, and each unit of the topology design variables window will pop up, as shown in Figure 6.14.

4. Iterative history of buckling critical force

Click the *Buckling critical load* button, and the critical buckling force iteration history window will pop up, as shown in Figure 6.15.

5. Iterative history of control point displacement

Click the *Displacement of node* button, and the iteration history of the control point displacement window will pop up, as shown in Figure 6.16.

6. Iteration history of the strain energy

Click the *Strain energy of structure* button, and the strain energy iteration history of structure window will pop up, as shown in Figure 6.17.

6.5.2 Graphic Result Output

Click the *Graphic Result* menu item in the *Results* menu, and the output window of the graphic results will pop up (Figure 6.18).

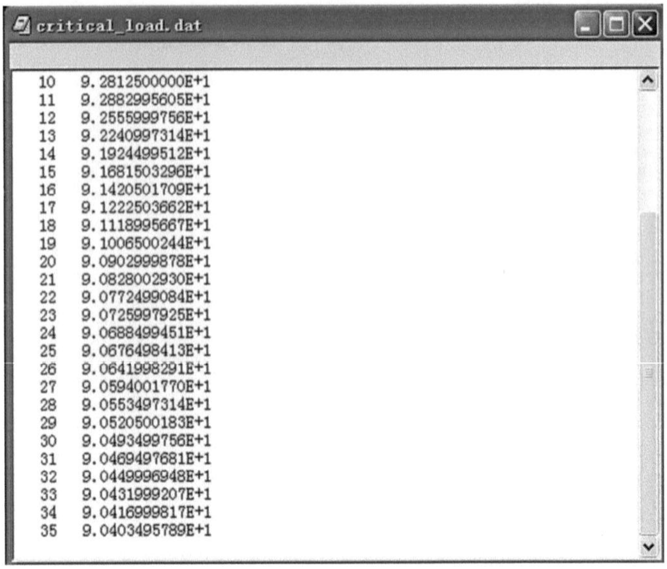
critical_load.dat

10	9.2812500000E+1
11	9.2882995605E+1
12	9.2555999756E+1
13	9.2240997314E+1
14	9.1924499512E+1
15	9.1681503296E+1
16	9.1420501709E+1
17	9.1222503662E+1
18	9.1118995667E+1
19	9.1006500244E+1
20	9.0902999878E+1
21	9.0828002930E+1
22	9.0772499084E+1
23	9.0725997925E+1
24	9.0688499451E+1
25	9.0676498413E+1
26	9.0641998291E+1
27	9.0594001770E+1
28	9.0553497314E+1
29	9.0520500183E+1
30	9.0493499756E+1
31	9.0469497681E+1
32	9.0449996948E+1
33	9.0431999207E+1
34	9.0416999817E+1
35	9.0403495789E+1

Figure 6.15: Iterative history of critical buckling critical load.

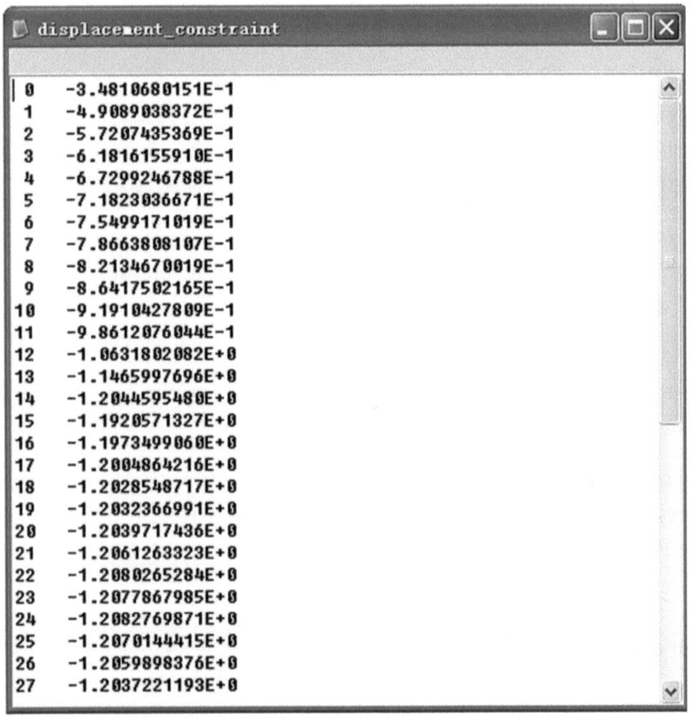
displacement_constraint

0	-3.4810680151E-1
1	-4.9089038372E-1
2	-5.7207435369E-1
3	-6.1816155910E-1
4	-6.7299246788E-1
5	-7.1823036671E-1
6	-7.5499171019E-1
7	-7.8663808107E-1
8	-8.2134670019E-1
9	-8.6417502165E-1
10	-9.1910427809E-1
11	-9.8612076044E-1
12	-1.0631802082E+0
13	-1.1465997696E+0
14	-1.2044595480E+0
15	-1.1920571327E+0
16	-1.1973499060E+0
17	-1.2004864216E+0
18	-1.2028548717E+0
19	-1.2032366991E+0
20	-1.2039717436E+0
21	-1.2061263323E+0
22	-1.2080265284E+0
23	-1.2077867985E+0
24	-1.2082769871E+0
25	-1.2070144415E+0
26	-1.2059898376E+0
27	-1.2037221193E+0

Figure 6.16: Iterative history of displacement.

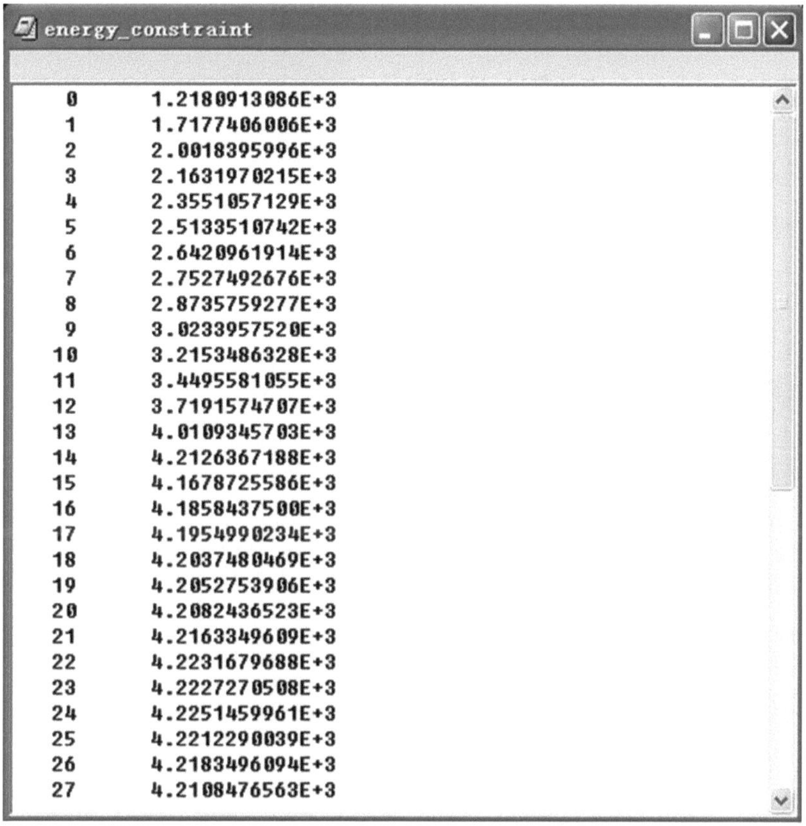

energy_constraint

0	1.2180913086E+3
1	1.7177406006E+3
2	2.0018395996E+3
3	2.1631970215E+3
4	2.3551057129E+3
5	2.5133510742E+3
6	2.6420961914E+3
7	2.7527492676E+3
8	2.8735759277E+3
9	3.0233957520E+3
10	3.2153486328E+3
11	3.4495581055E+3
12	3.7191574707E+3
13	4.0109345703E+3
14	4.2126367188E+3
15	4.1678725586E+3
16	4.1858437500E+3
17	4.1954990234E+3
18	4.2037480469E+3
19	4.2052753906E+3
20	4.2082436523E+3
21	4.2163349609E+3
22	4.2231679688E+3
23	4.2227270508E+3
24	4.2251459961E+3
25	4.2212290039E+3
26	4.2183496094E+3
27	4.2108476563E+3

Figure 6.17: Iterative history of strain energy.

1. The structural quality iteration history curve

Click the *Historical curve of* weight button, and the xy_window will pop up; the iteration history curve window of structural quality will appear, as shown Figure 6.19.

2. Iterative history of buckling critical force

Click the *Historical curve of buckling critical load* button, and the xy_window window will pop up, which is the structural buckling critical force of the iteration history curve, as shown in Figure 6.20.

3. The iteration history curve of the controlled point displacement of the

Click the *Historical curve of displacement* button, and the displacement of the iteration history curve window will pop up, as shown in Figure 6.21.

4. Iterative history curve of structural strain energy

Click the *Historical curve of strain energy* button, and the iteration history of the strain energy curve window will pop up, as shown in Figure 6.22.

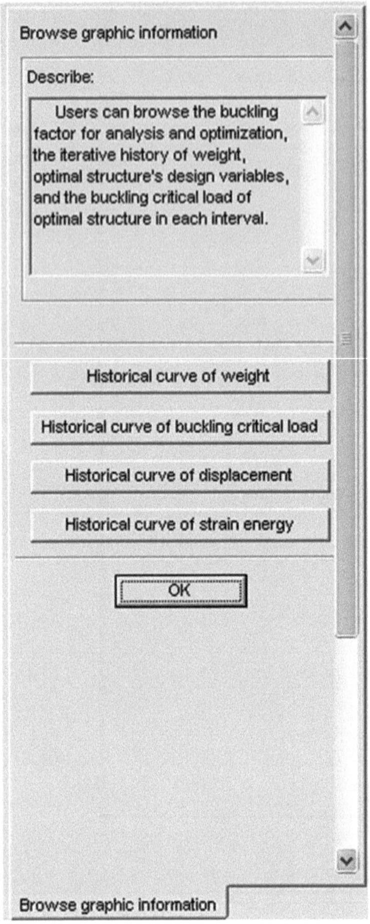

Figure 6.18: Output window of *Graphic result.*

6.5.3 Help Document: *Optimization's Help* Menu

In the Patran main window toolbar, the user clicks the *Optimization's help* menu, and then the topology optimization software random help document will pop up, as shown in Figure 6.23.

1. User instructions

Click *Directions* and the user manual help document will pop up, as shown in Figure 6.24.

2. Example of topology optimization

Click *Examples* and users can open the topology optimization example, as shown Figure 6.25.

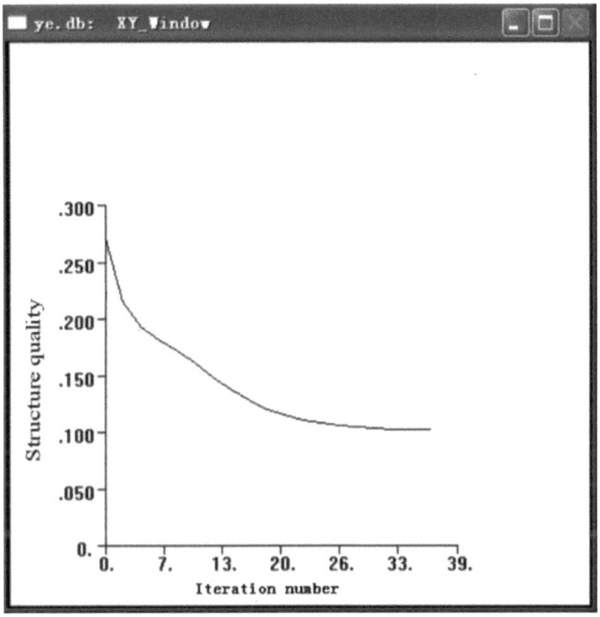

Figure 6.19: Iterative history curve of quality.

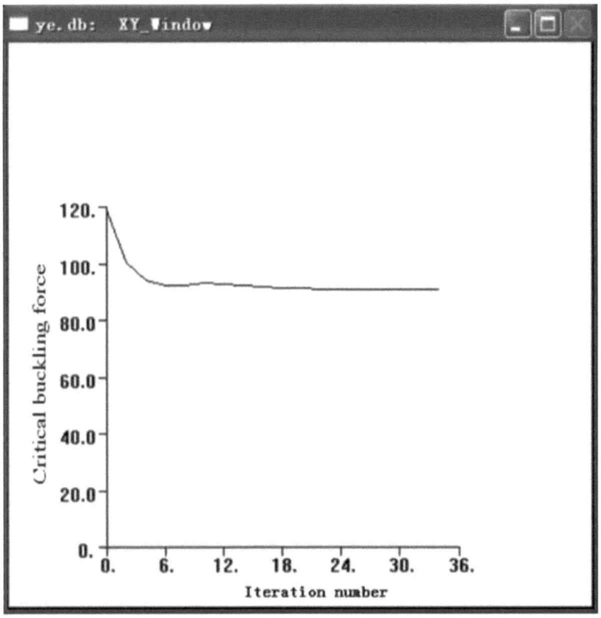

Figure 6.20: Iterative history curve of critical buckling critical load.

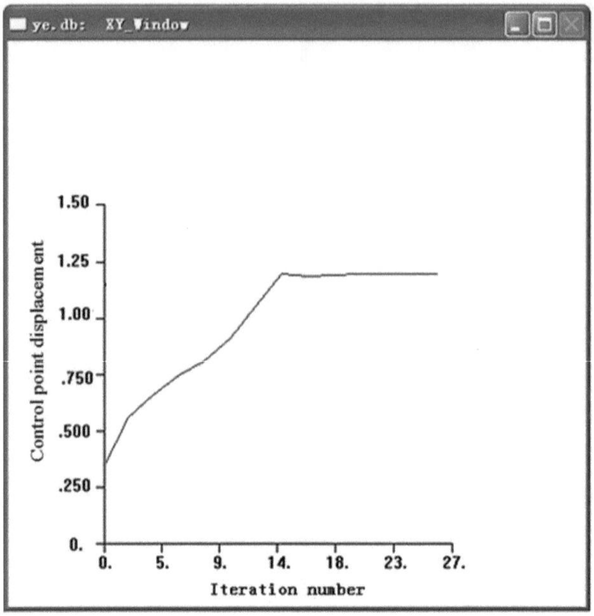

Figure 6.21: Iterative history curve of displacement.

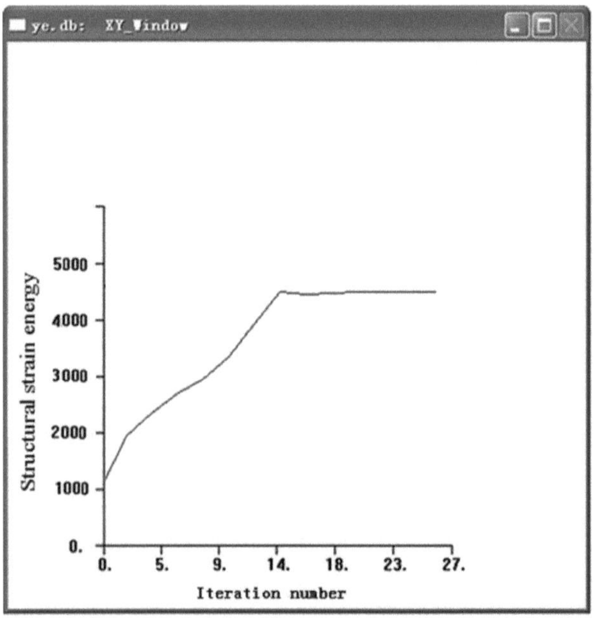

Figure 6.22: Iterative history curve of strain energy.

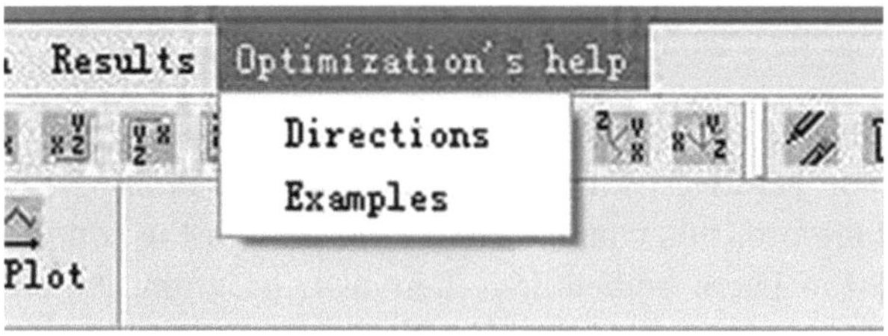

Figure 6.23: *Optimization's help* menu.

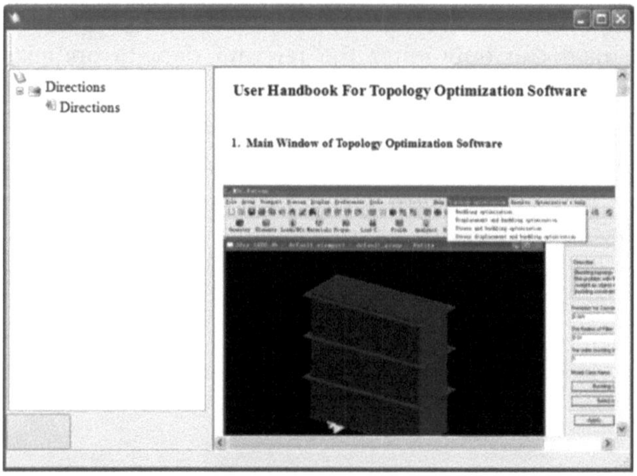

Figure 6.24: Directions to users.

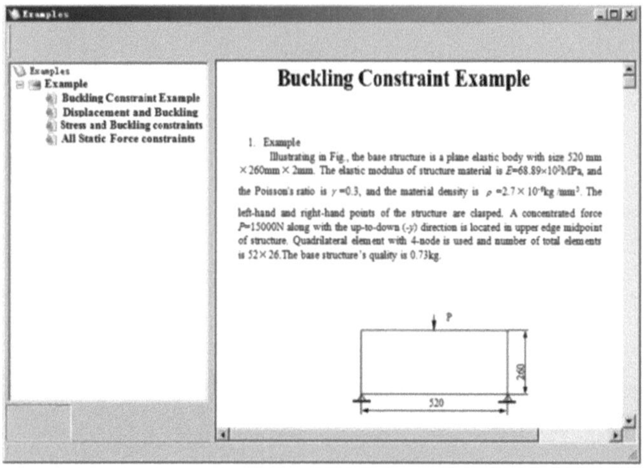

Figure 6.25: Examples of topological optimization.

6.6 Summary

Topology optimization program is based on the user development interface of the finite element analysis software Nastran MSC.Patran/, using PCL language to carry out the program. Therefore, this chapter mainly introduces the compiling and linking method of the users embedding their own program. At the same time, the result of the finite element analysis in the topology optimization program has been illustrated. All the output settings have been set up in the topology optimization program. Users do not to make any settings. In this paper, the practical significance of each constraint parameter and optimal parameter in the input interface of the topology optimization program is introduced in detail. The user can only enter the relevant data to optimize the structure.

The final text format and graph forms' output functions were added in the software. It can be exempted from the export results trouble. Finally, the user manual and the help document were joined in the software. They were facilitate to use the optimization software for the users.

7 Engineering Applications

7.1 Introduction

In the industrial production and daily life, we need to build a variety of structures such as the aerospace, space station, high buildings, large-span bridges, offshore oil drilling platforms, high-speed trains, and large hydraulic projects to meet the needs of people. The engineer should not only consider the performance of these structures, but also save the material, so that the structure can be optimized when designing these structures. Structure optimization has a variety of algorithms. Most algorithms were given by professional programmers, but these algorithms were not used conveniently by the engineer. Developing an optimal design system with friendly interface can solve the problem.

The purpose of this chapter is to develop a mixed programming engineering optimization system. Size structure optimization and topology optimization were applied in the optimization design of an electric power transformer and we achieved the better result of optimization design.

7.2 System Optimization in Industry via Mixed Programming

VC++ is now a widely used language in developing a program. This is mainly because VC++ is not only an integrated development environment, but also has a convenient and practical Microsoft basic class library MFC (Microsoft Foundation Classes), and it can easily call the Windows API function to generate friendly graphical interface and interactive interface. In addition, the OpenGL graphics library was also packaged in VC++, and users can use it to quickly develop a high-quality 3D graphics. As a result, VC++ is often used as the main language for a variety of application development.

FORTRAN language is widely popular in today's world, and is the most suitable for numerical calculation in a computer language. With the passage of time, a large number of mature and reliable FORTRAN language codes have accumulated in various fields. Therefore, it is an effective way to save time and save manpower, material and financial resources by using FORTRAN language to compile complex numerical program code and reuse of FORTRAN language code to optimize the system design.

Hybrid programming refers to mixed programming of two or more program design languages, calling each other, transferring the parameters of each other, and sharing data structure and data information, so as to form unified program entities. Mixed programming of VC and FORTRAN can make full use of friendly programming interface of VC++ object-oriented language and excellent numerical

https://doi.org/10.1515/9783110462272-007

computing power of FORTRAN language. Using these languages, the software design will be of high quality and will take less time to complete.

7.2.1 Implementation Method of Mixed Programming: VC++ and FORTRAN

There are many articles which give introduction to mixed programming methods of VC++ and FORTRAN. In sum, there are two kinds of methods in general: one method is that VC++ calls dynamic-link library (DLL) generated by FORTRAN, and another way is that VC++ calls directly the executable file generated by FORTRAN. There are many references to introduce the first method, but references giving introduction to the second method are relatively less. The advantages and disadvantages of the two methods are explained later.

The solved methods were given to VC++ directly to call the FORTRAN implementation file.

(1) VC++ call FORTRAN to generate the DLL
The subroutine SUBRUTINE and the function FUNCTION were compiled into a DLL in a FORTRAN program. The DLL was called in the VC++ to achieve preprocessing and postprocessing visualization. This method is seldom used in the preprocessing of the data, and it is relatively easy to use when the number of subprograms is less. But when the FORTRAN program has a number of subprograms, as well as multiple sub-subprograms, you need to generate a number of DLL, and the operation is relatively cumbersome. If a function has multiple parameters, a large number of parameter values and address are need to be transferred,which makes it easy to make mistakes. Programming is time-consuming and difficult to debug.

In this paper, in view of the above reasons, the method of VC++ to call FORTRAN executable file in the optimization system design will avoid the existence of the above problems.

(2) VC++ call the executable file generated by FORTRAN
Linear programming, two planning procedures, general planning procedures, as well as the duality of two planning procedures, etc., were programmed by using FORTRAN language, and the executable files were generated. Use shell execution function ShellExecute () to call the corresponding executable file in VC++. Data transfer between VC++ and FORTRAN is performed by calling text files. There are mainly two methods of data input in a program designed in FORTRAN: one method is to directly enter the initial data in the DOS window. In the second method, the initial data is written to a text file, and the FORTRAN program will automatically read the initial data from the text file. In the design of the optimization system, all the initial data was inputted from the VC++ design of the pre-processing interface and the disk

text files were generated, which were called in a FORTRAN program. The result files generated by FORTRAN were called by VC++ in postprocessing of program; a variety of optimized iterative curves were generated from the data results. There are two problems when VC++ calls the executable file generated by FORTRAN. First, If FORTRAN calculates a large amount of data, takes a long time and does not end the calculation yet, VC++ begins to call the FORTRAN's result file, which will lead to a program execution error. The second problem is that the DOS window appears when the ShellExecute function calls an executable file: it is the so-called black screen problem. To solve these two problems, the following solutions are proposed in this paper.

(3) Solve the asynchronous execution problem
The ShellExecute function is established as a process handle, and then the WaitForSingleObject function is used to wait until the shell function stops. After the termination of the shell function, the CloseHandle function will close the process.

This method can effectively solve the problem of the asynchronous execution of the program. The definitions of the above two Windows API functions in VC++ are as follows:

```
DWORD WaitForSingleObject(
HANDLE hHandle,
DWORD dwMilliseconds);
```

Here, hHandle is the handle for the target and dwMilliseconds is the time delay (in milliseconds). Its value can be “0” or “INFINITE.” “0” means not to wait, and execute directly the following statement. “INFINITE” means that indefinite period of time after the execution of the target is completed, and then execute the following statement.

In the optimization system design, we choose “INFINITE” to wait for the implementation of the FORTRAN optimization program:

```
BOOL CloseHandle(
HANDLE hObject);
```

Here, hObject is the handle for the target to be closed.

(4) Solve the “black screen” in the program
The prototype of the shell execution function ShellExecute is defined as follows:

```
HINSTANCE ShellExecute(
HWND hwnd,
LPCTSTR lpOperation,
LPCTSTR lpFile,
```

```
LPCTSTR lpParameters,
LPCTSTR lpDirectory,
INT nShowCmd);
```

Here, hwnd is the parent window handle. In the text it was set to NULL; lpOperation is the operating mode for files. It can be "print," "explore," or "open." Select "open" in the text; lpFile is the name for the operating file which is generated by FORTRAN. For example, the generated executable file name can be "project1.exe".The two parameters of lpParameters and lpDirectory are set to NULL respectively. nShowCmd is the display method for the executable file. When it is set to SW_SHOWNORMAL, the implementation of the FORTRAN will display the standard DOS window. When it was set to SW_HIDE, the DOS window can be hidden while running the FORTRAN program. In this chapter, it was set to SW_HIDE parameter, and the "black screen" problem can be solved by the implementation of the program.

Start the FORTRAN program process, and an executable file will be called; the whole process of the termination code is as follows:

```
void CMyDlg::OnYunxing()
{
HANDLE hProcess;//Building process
hProcess=ShellExecute (NULL, "open", "project1.exe", NULL, NULL, SW_HIDE);
//Set the executable file to process
WaitForSingleObject(hProcess,INFINITE);//Wait for the end of the process execution
CloseHandle(hProcess);//Closing process
}
```

7.2.2 Implementation of the Optimization Design System

In the development of the optimization design system, VC++ was used for the preprocessing and postprocessing interface design, whereas the core optimization program design was used by FORTRAN language. VC++ directly calls executable files generated by FORTRAN. The data transfers from preprocessing and postprocessing between VC++ and FORTRAN by the method of calling disk text files. The optimization system mainly includes four modules: the linear programming module, the two programming module, the dual programming module and the general planning module. Each module has a preprocessing interface for the input of the corresponding mathematical model, and a postprocessing interface for the optimization results. The preprocessing interface mainly includes the number of design variables, the number of constraints, the objective function input form, constraint input form and confirm the input data, run the optimization procedure and so on. The main functions of the postprocessing interface include the display of data results, drawing the design variable iteration curve, drawing the iterative curve of the objective function and the contour of the design variable region. The initial interface of the optimized design system is shown in Figure 7.1. The input interface of a

Figure 7.1: The initial interface of an optimization system (source: http://image.so.com/v?q).

two programming model is shown in Figure 7.2. The optimization results are shown in Figure 7.3. The design variable iteration change curve shown in Figure 7.4.

The design of the optimized system is carried out with the method of mixed programming. The method of VC++ invoking the executable file generated by FORTRAN can easily deal with the a large number of data problem . At the same time, the problem of asynchronous execution and "black screen" in the calling process can be solved. The method makes full use of the powerful function of the VC++ development interface and the powerful scientific computing power of FORTRAN language. Therefore, the optimization design system not only has a friendly preprocessing and postprocessing interface, but also can achieve the desired function of the software. It is more important that the system can be designed by many people in parallel, so the method has the economic value and the reference value.

7.3 Topological Optimization in the Design of a Power Transformer Lifting Lug

Lifting lug is a lifting tool which is welded on the oil tank of the transformer. According to the manufacturing method, the lifting lug can be divided into two types: plate lifting

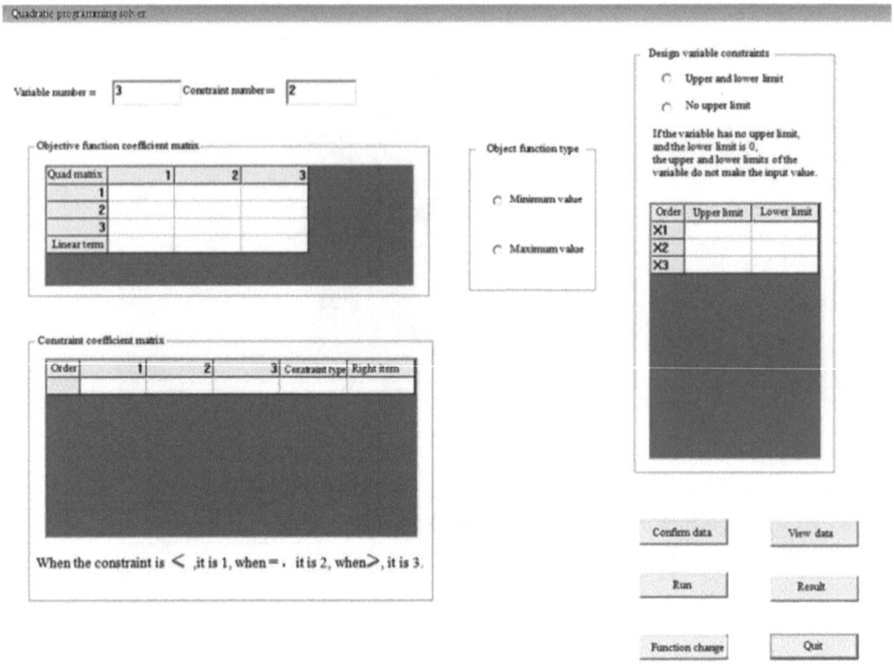

Figure 7.2: The input interface of quadratic programming.

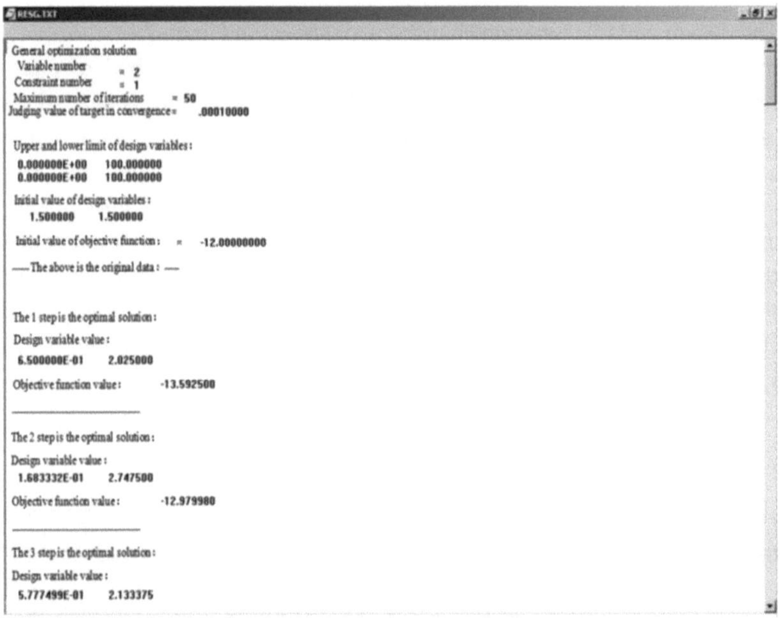

Figure 7.3: Result of optimization.

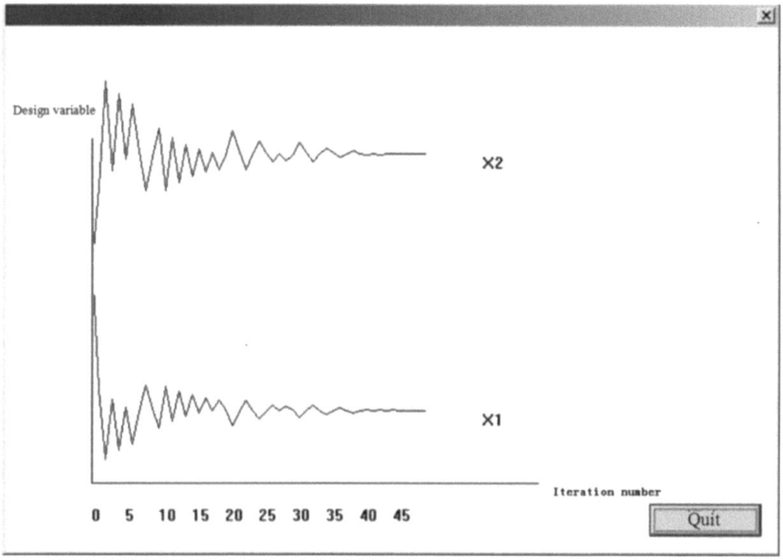

Figure 7.4: Iterative history curve of design variables.

lug and pipe lifting lug. The plate lifting lug is generally used for a barrel-type transformer of small capacity, whereas the pipe lifting lug is generally used for a bell-type transformer of large capacity. In this chapter, we will only consider the plate lifting lug. According to the installation site, the plate lifting lug can be divided into two types: one is installed in the tank cover of the barrel-type transformer. Its function is to lift the box cover quality and the body quality. The other is installed in the barrel-type oil tank along the hanger for lifting the whole quality of the transformer. The lifting lug structure design practice is usually according to requirements, then checks structural calculation and tests the strength requirements. The main disadvantage of this method is that it is difficult to find a reasonable structure, and it is not easy to make an ideal economic and safe design. Obviously, this kind of traditional design method has certain blindness.

In this chapter, the design method of the topology optimization is introduced to design the optimal structure of the lifting lug so that the distribution of the material reaches a reasonable state, and the structure can meet the economy and safety requirements.

7.3.1 Establishment of a Mathematical Model of Topology Optimization and a Finite Element Model of Lifting Lug

(1) Topology optimization mathematical model and base structure of finite element

In order to solve the topology optimization problem, the ICM method is used to establish a continuum topology optimization model with the minimum structure quality as the goal function and the structural stress as the constraint:

$$\begin{cases} \text{find} & \boldsymbol{t} \in E^N \\ \min & W = \sum_{i=1}^{N} f_w(t_i) w_i^0 \\ \text{s.t.} & \sigma_{il} \le \overline{\sigma} \quad (l = 1, \ldots, L) \\ & 0 \le t_i \le 1 \quad (i = 1, \ldots, N) \end{cases} \tag{7.1}$$

Among them, $\boldsymbol{t}$ is topology design variables, W is the total quality of the structure, σ_{il} is i-th element stress under the l-th condition, $\overline{\sigma}$ is the stress constraint limit, L is the total number of conditions, and N is the total number of elements. We use dual programming to solve this optimization model. After several iterations, we get the topological variables of each element at minimum structure quality, and change the topology graph. When the topology variables of the unit are $t_i = 0$, the unit associated with the design variable is deleted and white is displayed in the topology; when $t_i = 1$, the unit associated with the design variable is reserved and the topology is red.

The shape and size of the hanging plate and the lifting lug in [67] are established, and the topology optimization-based structure is established. The base structure of the hanging plate is a 140 × 100 mm rectangular plate, the thickness is 24 mm, opening a Φ 40 mm hole in the center, the elastic modulus of the rectangular plate is $E = 200$ GPa, the Poisson's ratio is $\nu = 0.32$, the density is $\rho = 7800$ kg/m^3, the fixed support is below, and the total force is $F = 40{,}000$ N. The model is divided into 20 × 28 four node quadrilateral elements, the finite element model that is divided into grids is shown in Figure 7.5 (a).

The base structure of the lifting lug is a 100 × 110 mm rectangular plate, the thickness is 16 mm, opening a Φ24 mm hole in the center, the elastic modulus of the rectangular plate is $E = 200$ GPa, the Poisson's ratio is $\nu = 0.32$, the density is $\rho = 7{,}800$ kg/m^3, the fixed support is on the left, and the total force is $F = 10{,}000$ N. The model is divided into 20 × 22 four node quadrilateral elements, the finite element model that is divided into grids is shown in Figure 7.5 (b).

(2) Calculation of boundary distribution load on the hanging hole

The distribution of load on the boundary of the hanging hole is not in accordance with the actual working conditions, so it should be applied to the distribution force. In the actual lifting process, the main part of the contact between the hanging hole and the hook or cable is in the upper part of the round hole, and the size of the load is not evenly distributed. The maximum load is at the top of the lifting hole, the load at the bottom of

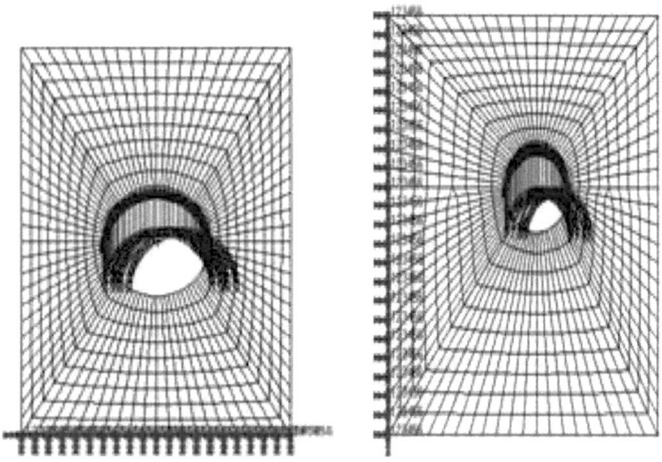
(a) (b)

Figure 7.5: (a) FEM meshing and load distribution diagram of a hanger plate, (b) FEM meshing and load distribution diagram of a lifting lug.

the upper half circle can be considered zero. According to the load features on the role, the force on the boundary of the suspension can be approximated by a sine function to describe its distribution:

$$f = A\sin(\pi x/2r) \tag{7.2}$$

Among them, f is the distributed load, A is the largest acting load, x is the horizontal coordinate value of the node, and r is the radius of the hanging hole.

The integral of the distribution load is equal to the total force acting on the hole:

$$\int_0^{2r} f dx = F \tag{7.3}$$

Formula (7.2) was taken into formula (7.3) to obtain

$$\int_0^{2r} A\sin(\pi x/2r)dx = F \tag{7.4}$$

The solution obtained is as follows:

$$A = F\pi/4r \tag{7.5}$$

Formula (7.5) was taken into formula (7.2), and the distribution load is expressed as

$$f = F\pi/4r\sin(\pi x/2r) \tag{7.6}$$

Because the load can only be applied to a finite number of nodes of the finite element model, the load formula for one node is given as follows:

$$f_n = \int_a^b F\pi/4r\sin(\pi x/2r)dx \tag{7.7}$$

In the above formula, n is the node number of the applied load, a is the lower bound of the distributed load and b is the upper bound of the distributed load. According to formula (7.7), the node load is obtained in order to prevent the load from generating stress concentration on a node; the five nodes add the average load in the integral cell field. Table 7.1 shows the load distribution of the hanger plate. Table 7.2 shows the load distribution of the lifting lug load. The load applied diagram is shown in Figure 7.5 (a), (b).

7.3.2 Topology Optimization Result Analysis and Hanger Plate Design

Finite element static analysis is performed on the MSC.Patran/Nastran platform, and the results of stress analysis are shown in Figure 7.6 (a). The maximum stress on the top of the hanging hole is 89.9 MPa. According to the data in the literature [66], manufacturing materials of the lifting lug is the A3 steel rolling plate, the yield limit is

Table 7.1: Load of hanger plate's nodes.

Field of integration (mm)	0–4	4–8	8–12	12–16	16–24	24–28	28–32	32–36	36–40
Area load (N)	979	2840	4425	5576	12361	5576	4425	2840	979
Node load (N)	196	568	885	1115	2472	1115	885	568	196

Table 7.2: Load of lifting lug's nodes.

Field of integration (mm)	0–3	3–6	6–9	9–15	15–18	18–21	21–24
Area load (N)	381	1084	1623	3828	1623	1084	381
Node load (N)	77	217	325	766	325	217	77

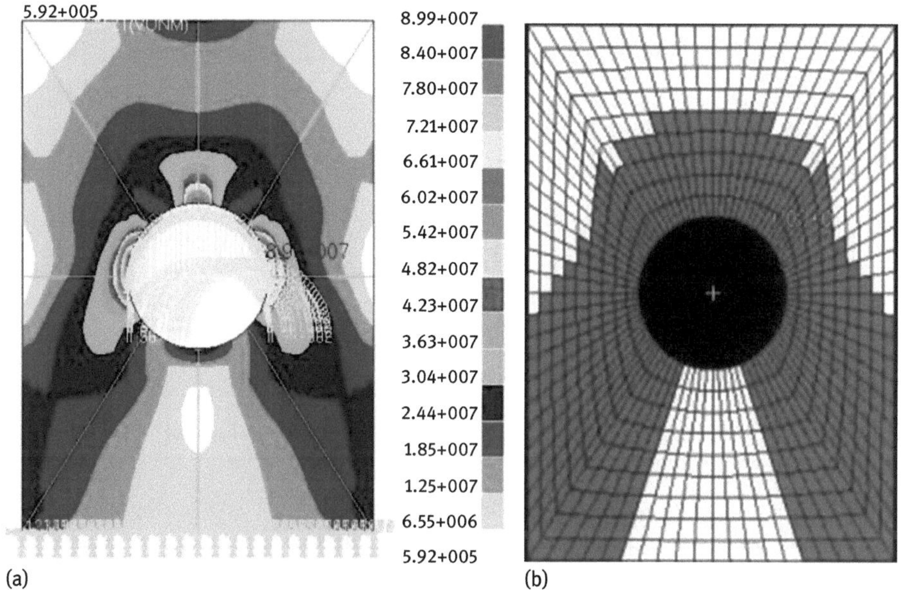

Figure 7.6: (a) Stress distribution of the base structure of a hanger plate, (b) Optimal topology configuration of a hanger plate.

$\sigma_s = 225$ MPa, the safety factor is $n_s = 1.5$, and the allowable stress is $[\sigma] = 150$ MPa. The maximum stress of the lifting hole is less than the allowable stress, so the steel plate meets the requirement of the strength. The maximum displacement is 0.0204 mm, which meets the requirements of stiffness.

The topology optimization program used in the topology optimization of the base structure was developed by the Engineering Numerical Simulation Center of Beijing University of Technology. Due to the allowable stress of the steel $[\sigma] =$ 150 MPa, the allowable stress value is = 135 MPa. After 18 iterations, the topology optimization structure is obtained as shown in Figure 7.6 (b).

In the topology, red is the reserved area which accounts for about 1/3 of the entire region, and white is the deletion of the region. It can save about 2/3 of the steel from the point of view of saving steel. From the point of view of stress, the topological structure can fully meet the requirement of strength under the condition of allowable stress. But in the actual structural design, stiffness, fatigue, reliability and other factors should also be taken into account. According to the results of finite element analysis and topology optimization, the hanger plate can be designed to be the structure as shown in Figure 7.7 (b). Figure 7.7 (a) shows the diagram of a traditional hanger plate structure.

Finite element analysis is carried out on a traditional hanging plate structure and an improved structure, and the results are shown in Figure 7.8 (a), (b). Comparative

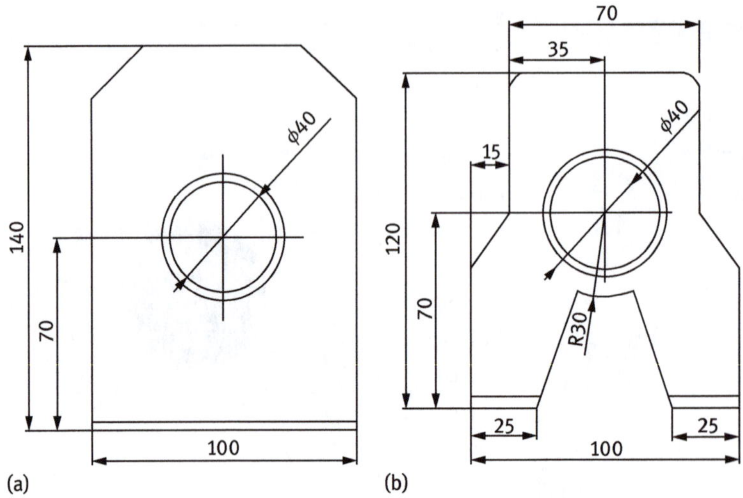

Figure 7.7: (a) Sketch map of a traditional hanger plate, (b) Sketch map of an improvement hanger plate.

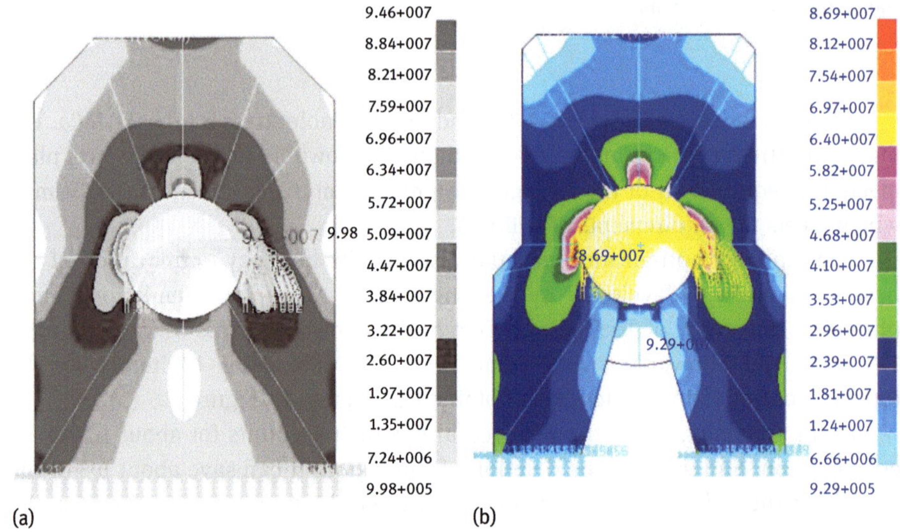

Figure 7.8: (a) Stress distribution of a traditional hanger plate, (b) Stress distribution of an improvement hanger plate.

analysis results in the total mass, maximum stress and maximum displacement of the structure, as shown in Table 7.3.

From Table 7.3, we can see that the material of the improved structure saves about 17% more mass than that of the original structure, the structure of the maximum stress

Table 7.3: Comparison of FEM results of hanger plates.

	Total structural mass (kg)	Maximum stress (MPa)	Maximum displacement (mm)
Hanging plate structure	2.310886	94.6	0.0197
Hanging plate improved structure	1.917287	86.9	0.0233

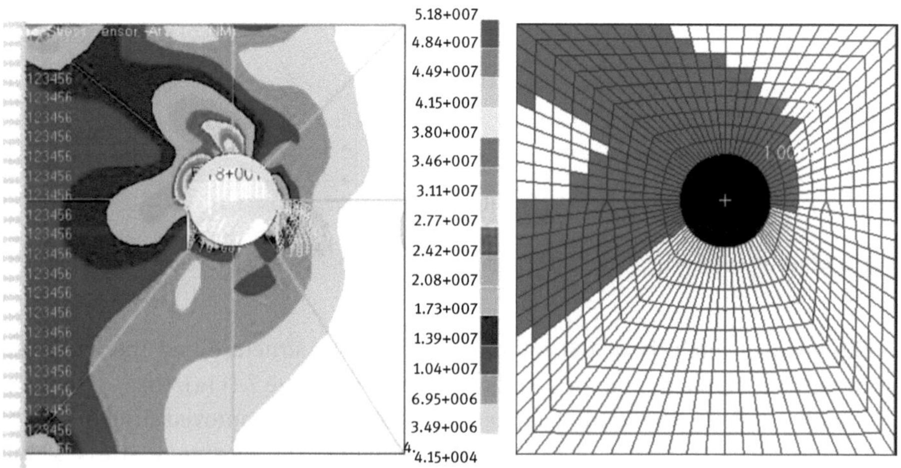

(a) (b)

Figure 7.9: (a) Stress distribution of lifting lug's base structure, (b) Optimal topology configuration of lifting lug.

reduced than that of the original structure, the maximum displacement increases slightly, the maximum stress is less than the allowable stress, and stiffness also meets the stiffness requirements. In the case of mass production, the processing technology cannot be considered complicated, and the improved hanging plate structure is feasible in the actual production. The new structure of the hanging plate was used in the S9-6300/35 transformer, and there was no abnormal situation in lifting and moving.

In the same way, the base structure of a lifting lug is analyzed by finite element method, and the results of stress analysis are shown in Figure 7.9 (a). The maximum stress on the top of the lifting hole is 51.8 MPa. The lifting lug's restraint stress is $[\sigma] =$ 135 MP, in the base structure topology optimization. After 30 times iterative optimization, such as the topology configuration shown in Figure 7.9 (b), reserved area accounted for about a quarter of the entire region.

The improved structure of the lifting lug is shown in Figure 7.10 (b), based on Figure 7.9 (a), (b). Figure 7.10 (a) shows the diagram of a traditional lifting lug structure.

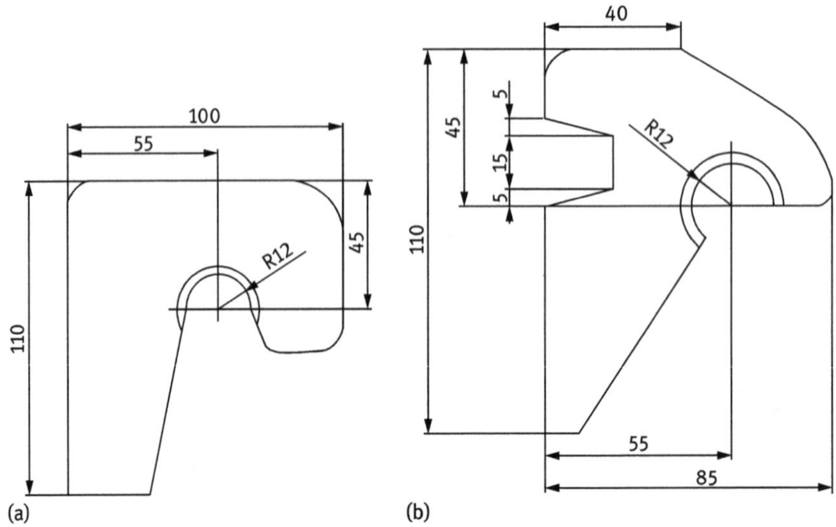

Figure 7.10: (a) Sketch map of traditional lifting lug, (b) Sketch map of improvement lifting lug.

Finite element analysis of the traditional lifting lug structure and the improved structure is performed, and the results are shown in Figure 7.11 (a), (b).

Table 7.4 shows the result data that the material of the improved structure saves 21% more mass than that of the original structure. The maximum stress of the structure increases amplitude, but still meets the requirement of allowable stress; the maximum displacement decreases, but still meets the requirement of stiffness. Analysis of the data shows that the improved structure is feasible, the improved lifting lug is used in the S9-1250/10 transformer, and the actual application results show that the improved structure is safe.

The maximum stress value is far less than that of steel yield limit in numerical analysis. All deformations were within the elastic range; the material will not appear in permanent deformation.

7.4 Finite Element Method and Topological Optimization in a Small-sized Transformer Fuel Tank

Oil tank is one of the main components of the transformer. It is used to protect the body of the transformer and it contains oil. In addition, it is the carrier of the external attachment of the transformer, so the transformer oil tank must have a certain mechanical strength and stiffness to meet the following requirements: one is bearing the transformer body and transformer oil and transformer overall lifting quality; the other is able to carry the quality of all the accessories

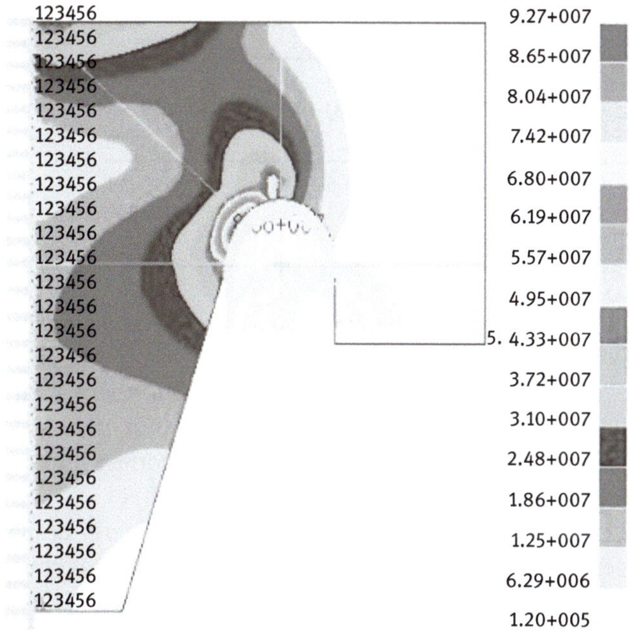

(a)

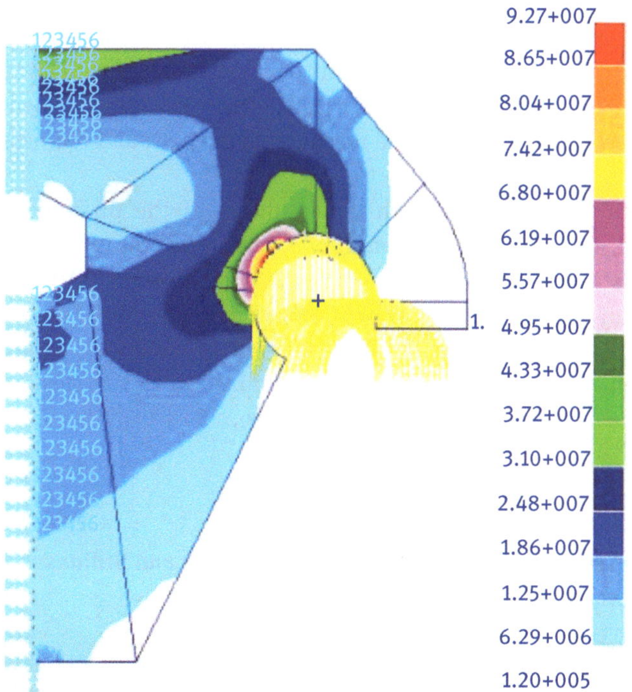

(b)

Figure 7.11: Stress distribution of traditional lifting lug, (b) Stress distribution of improvement lifting lug.

Table 7.4: Comparison of FEM results lifting lugs.

	Total structural mass (kg)	Maximum stress (MPa)	Maximum displacement (mm)
Lifting lug structure	0.8507050	88.6	0.0342
Lifting lug improved structure	0.6715663	92.7	0.0265

of the transformer; the third function is operating impact acceleration during transport under earthquake or wind. The fourth is to withstand the effects of air pressure vacuum oiling, to the large transformer oil tank, but there is no requirement to the small- and medium-sized transformer. The fifth is able to withstand the internal pressure, but also to ensure that the transformer oil tank shall not burst in the internal accident.

In traditional small transformer design, experience design and analogy design are mostly used for the mechanical strength and rigidity design of oil tank. Because the traditional hand calculation method is too cumbersome and the efficiency is low, detailed analysis and calculation of stress and deformation were not used. The mechanical strength and stiffness of the transformer oil tank by the experience design and analog design method usually have no definite design data. In addition, the products were not tested for the strength and rigidity of the oil tank; this will bring security risks in the installation and operation of the transformer. The finite element method can effectively solve the problem of the mechanical strength and stiffness analysis of the transformer.

At present, the research on the strength and stiffness of the oil tank is mostly concentrated on the large transformer oil tank, and there is less research on the analysis and optimization of the small transformer oil tank. The structure of two small transformers' tank was analyzed in the MSC.Patran/Nastran software platform, three-dimensional finite element models were established, and finite element analysis and structural optimization design results were given. All results have certain reference significance.

7.4.1 Test Standards and Test Parameters of Mechanical Strength and Stiffness of the Small Transformer Oil Tank

(1) Test standards and test parameters of mechanical strength

In the first part of the general GB1094.1 power transformer, the mechanical strength of the tank is presented. In 7.2 additional items (in the case of applicable), there are "tank and storage tank vacuum tolerance."

In GB/T6451 and GB/T16274 “three-phase oil-immersed power transformer technical parameters and requirements” 6–10 kV voltage grade product testing project defined: “Transformer oil tank and oil storage tank (if there is) should be the strength (positive pressure) test, after 5 min should be no damage and not allowed to appear the permanent deformation. This test is the type test. The purpose of the test is to verify the safety and reliability of the tank structure design.

The technical parameters of the oil tank are as follows: 6–10 kV small transformer's oil tank is the general structure (noncorrugated tank). All capacity transformer oil tanks are pressurized at 49 kPa, after 5 min should be no damage and without the permanent deformation.

According to the intensity data in the literature [66, 68], tank manufacturing materials commonly used is the Q235-A or Q235-B steel plate, the yield limit is σ_S = 225 MPa, and the maximum stress of the tank is less than the yield strength.

(2) Test parameters of fuel tank stiffness
Test parameters are not clearly defined in the national standard for the deformation of the oil tank. The requirements of the permanent deformation of the oil tank are not the same in each enterprise, but roughly controlled in the following range:
(1) Tank wall 1.5δ (δ is the plate thickness)
(2) Tank cover 0.5δ
(3) Reinforcing iron 1δ
(4) The maximum amount of elastic deformation ≤ 2 times the amount of permanent deformation

7.4.2 Finite Element Analysis and Optimization Design of an Oblong Tank

(1) Finite element model
The small transformer S9-50/10KVA is taken as an example, the oil tank is oblong, and the connection group number of the transformer is Y, yn0. Basic dimensions of the oil tank are as follows: tank width B = 270 mm, total length L = 700 mm, linear part length l = 430 mm, radius of a circular arc R = 135 mm, and tank height H = 640 mm. The tank wall is a 4 mm thick steel plate. The recovery is a 4 mm thick steel plate. The tank cover is a 6 mm thick steel plate. The detailed size of the oil tank is shown in Figure 7.12.

Based on the specific size of the tank, an oil tank model is established. In order to build the model, the following assumptions were made:
(1) Ignore the bottom bracket at the bottom of the tank and replace it with a fixed support.

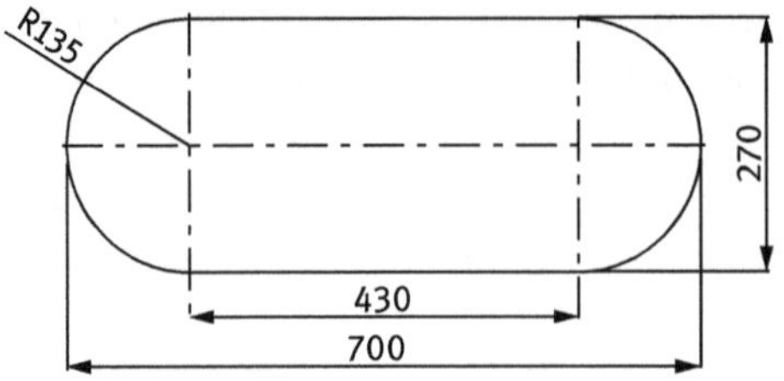

Figure 7.12: Sketch of oval shape tank's cross-sectional size.

(2) The tank cover and the tank wall are considered as a whole by ignoring the connecting bolt of the upper tank along the tank and the tank cover.
(3) The tank cover is a complete steel plate, and the installation hole of the tank cover is ignored.
(4) Ignore the position of the oil tank weld, and assume that the weld and the tank wall have the same strength and stiffness.

A fixed support welding position is at the bottom of the tank which gives limit of six freedom degrees. According to the standard, the applied pressure is 49 kPa, the elastic modulus of the steel plate is E = 200 GPa, the Poisson's ratio is $\mu = 0.3$, and the density is ρ = 7,800 kg/m^3. The entire tank is divided into 14492 node shell elements. The finite element mesh model is shown in Figure 7.13. The direction of applying the pressure is shown in Figure 7.14.

(2) Comparison of finite element analysis and test results
The stress results of finite element analysis are shown in Figure 7.15. Displacement results are shown in Figure 7.16. It can be seen from the figure that the maximum stress is 198 MPa, which is caused by the stress concentration at the tank cover and the tank wall. The maximum stress value is less than the yield limit, and the entire tank structure can meet the standard requirements and not the yield limit. It can be seen from the maximum displacement cloud diagram that the maximum displacement is 6.60mm, which occurs in the middle of the tank wall, and the maximum displacement is less than two times of the tank wall thickness.

The strength of the oil tank test is the type test in the actual production. The test results of numerical analysis data and the oblong tank results are shown in Table 7.5.

As can be seen from the table, the test results and numerical analysis results meet the requirements of the test standards. The maximum elastic deformation of the test results is slightly higher than the numerical results. This is mainly because the steel plate in the numerical analysis is considered as an ideal material, which does not take into account the defects in the actual structure. Another reason is the measurement error caused by the deviation. From the comparison between the finite element analysis and the test results, it can be concluded that the structural design is reasonable.

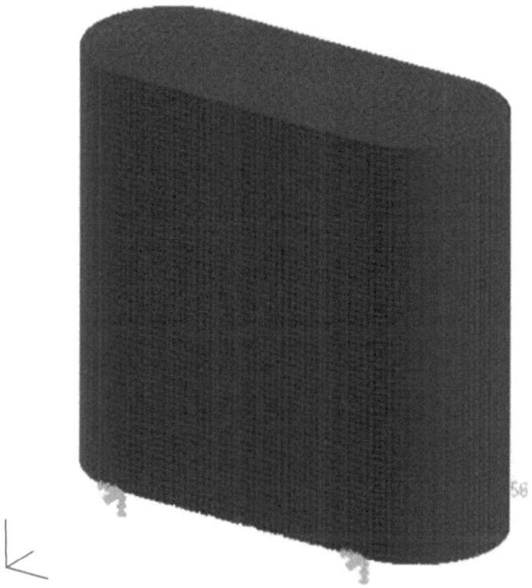

Figure 7.13: Three-dimensional model of an oval shape tank.

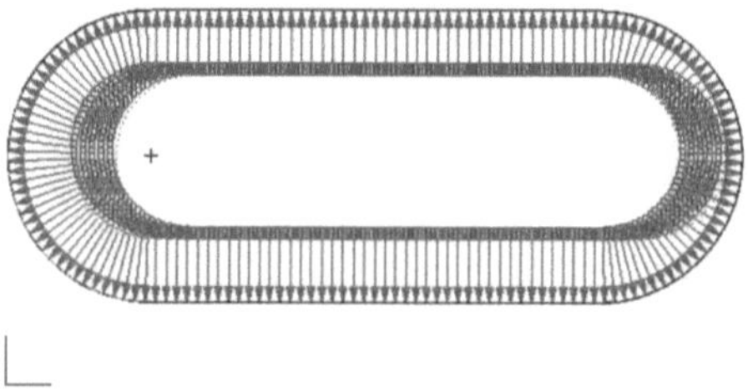

Figure 7.14: Sketch of oval shape tank's cross section with pressure.

(3) Structural optimization design

Before optimizing the structure, the mathematical model of optimization is established, which includes the design variables, objective functions and constraints. According to the specific structure of the fuel tank, the optimization model is set up as follows:

$$\begin{cases} \text{find} \quad td,\ td,\ tg \\ \min \quad M = \sum_{i=1}^{N} m_i^0 \\ \text{s.t.} \quad \underline{\sigma_i} \le \sigma_i \le \overline{\sigma_i} \quad (i = 1, \ldots, N) \\ \qquad \underline{u_j} \le u_j \le \overline{u_j} \quad (j = 1, \ldots, J) \\ \qquad \underline{tb} \le tb \le \overline{tb} \\ \qquad \underline{td} \le td \le \overline{td} \\ \qquad \underline{tg} \le tg \le \overline{tg} \end{cases} \tag{7.8}$$

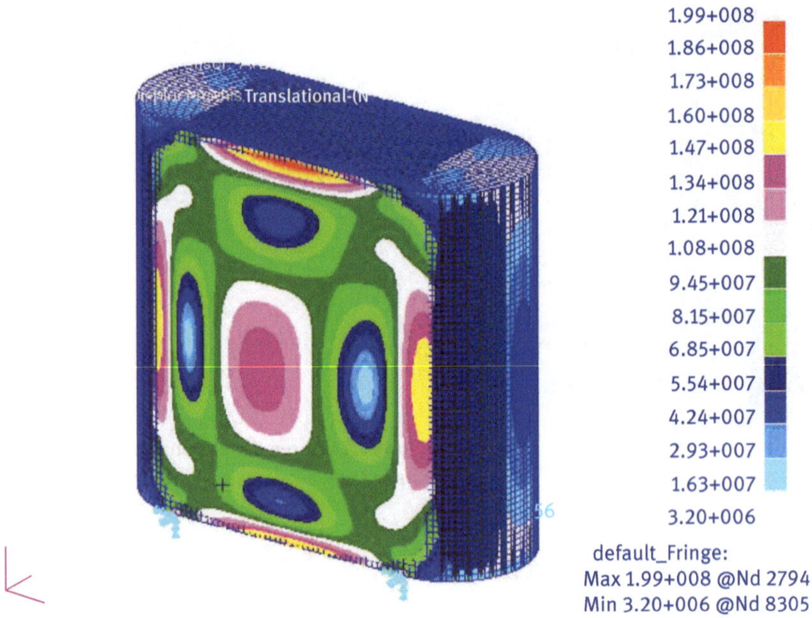

Figure 7.15: Stress analysis result of an oval shape tank.

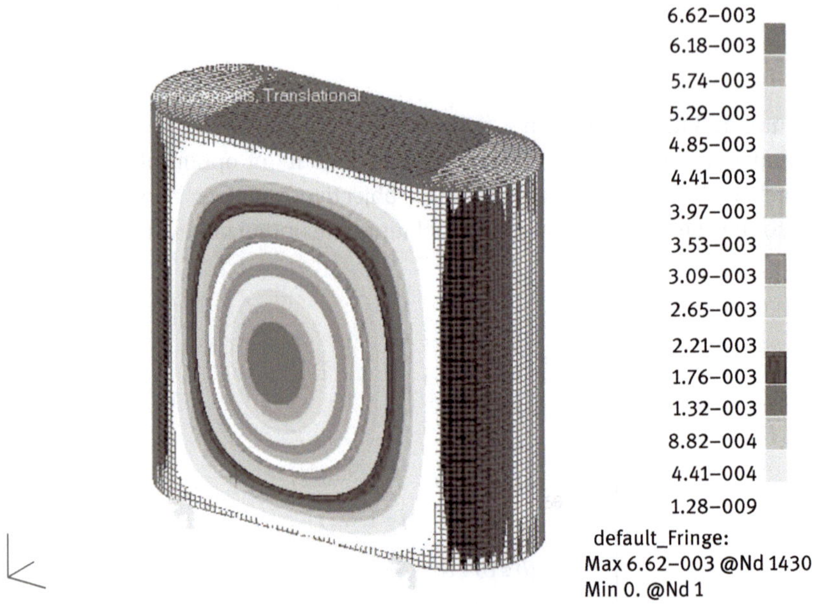

Figure 7.16: Displacement analysis result of an oval shape tank.

Table 7.5: Comparison of oval shape tank's distortion by FEM and test results.

	Maximum elastic deformation (mm)	Maximum permanent deformation (mm)
Test standard requirements	12.00	6.0
Numerical analysis results	6.60	Software unable to calculate temporarily
Positive pressure test result (49 kPa)	7.50	2.65

Here, *tb, td, tg* are design variables; *M* is the total mass of the structure; $\overline{\sigma}_i$ is the allowable stress upper limit; $\underline{\sigma}_i$ is the allowable stress lower limit; $\overline{u}_j$ is the allowable displacement upper limit; $\underline{u}_j$ is the allowable displacement lower limit; *N* is the total number of constraint units; and *J* is the total number of constrained nodes.

The design variables were the tank wall thickness *tb*, the tank bottom thickness *td* and the tank cover thickness *tg*. According to the design principle, the thickness of the common steel plate of the small transformer oil tank is taken as the design variable:

$$3\text{ mm} \le tb \le 12\text{ mm}, 3\text{ mm} \le td \le 12\text{ mm}, 3\text{ mm} \le tg \le 12\text{ mm}.$$

The objective function is the minimum quality of the entire oil tank.

Stress and displacement constraints, respectively, were

$$-199\,\text{MPa} \le \sigma \le 199\,\text{MPa}, \quad -6.6\,\text{mm} \le u \le 6.6\,\text{mm}.$$

After 16 iterations, the final stress optimization results are shown in Figure 7.17, and the maximum stress of the optimized structure is 198 MPa. The optimized displacement results are shown in Figure 7.18, and the maximum displacement of the structure is 6.6 mm. The iterative optimization curve of the design variables is shown in Figure 7.19. The iterative optimization curve of the objective function is shown in Figure 7.20. The optimized design variables and the specific data for the objective function are listed in Table 7.6.

As can be seen from the data in the table 7.6, only the tank cover thickness changes greatly, the tank wall and tank bottom thicknesses changed slightly, and oil tank quality changed little. Because the holes in the tank cover and other accessories are not considered, there will be a change in the optimization quality of the tank cover. After considering the above factors, the tank cover can be thickened to 6 mm. From the above analysis, it can be concluded that the initial design of the oil tank size meets the strength and stiffness requirements and is feasible. In the prescribed intensity and displacement, the different stress and displacement values were input, and different sizes of the tank

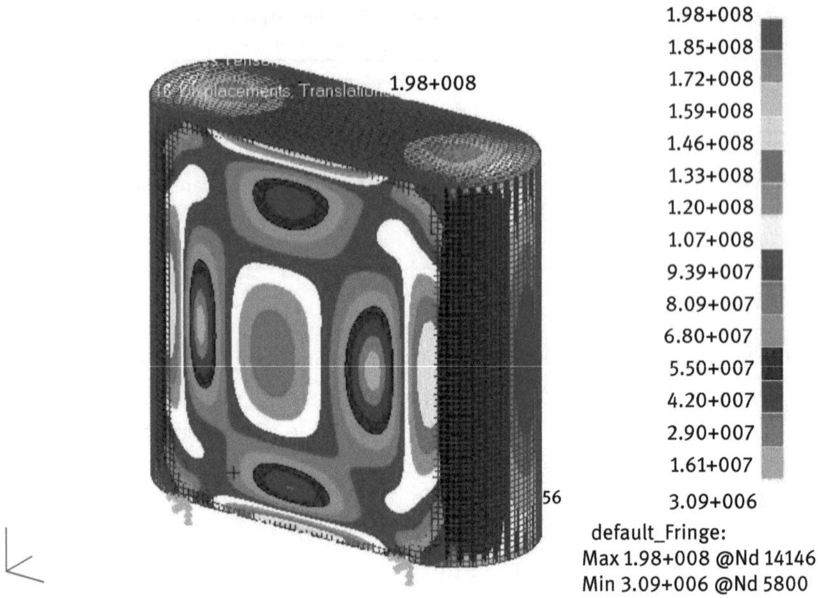

Figure 7.17: Optimal stress result of an oval shape tank.

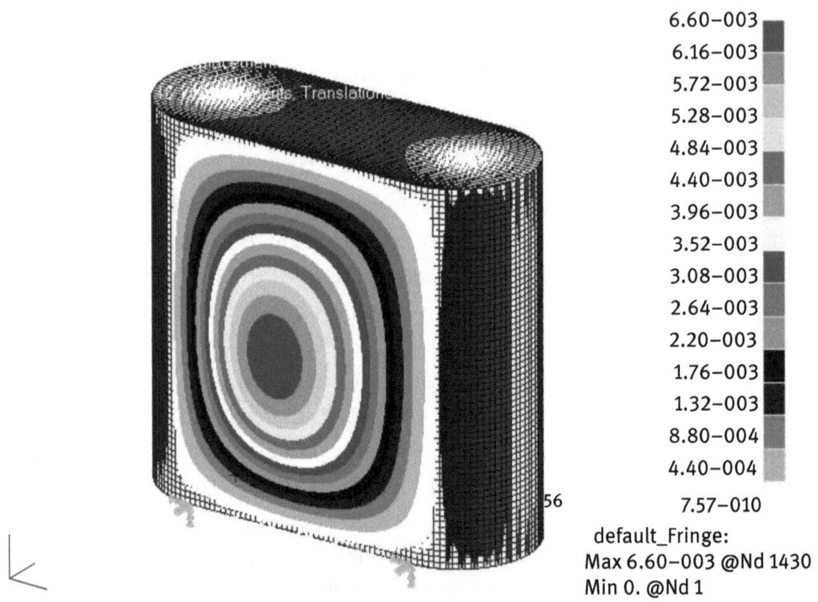

Figure 7.18: Optimal displacement result of an oval shape tank.

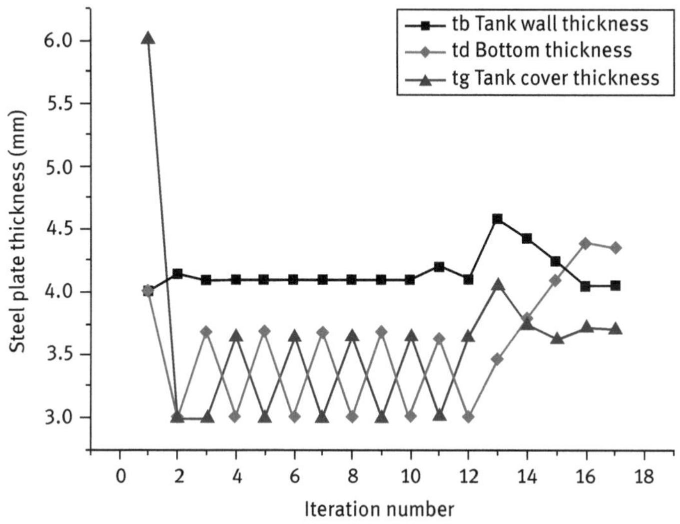

Figure 7.19: Design variable iterative curve of an oval shape tank.

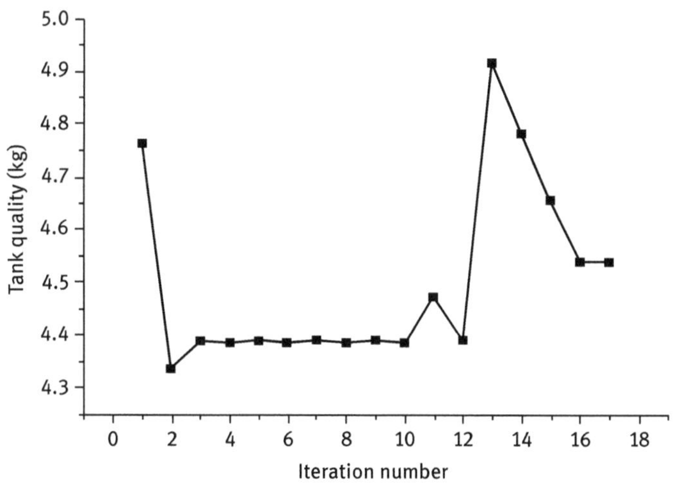

Figure 7.20: Objective function iterative curve of an oval shape tank.

structure can be given. In the optimization design, the design variables are continuous; the thickness of the oil tank is not a standard integer, so the thickness of the oil tank component will be in response to the actual use of plate thickness.

Table 7.6: Comparison of oval shape tank's size before optimization and after optimization.

	Tank wall thickness *tb* (mm)	Tank bottom thickness *td* (mm)	Tank cover thickness *tg* (mm)	Tank quality *W* (kg)
Initial value	4.0	4.0	6.0	47.62358
Optimal value	4.0467	4.3385	3.7031	45.37486

7.4.3 Finite Element Analysis and Optimization Design of a Rectangular Oil Tank

(1) Model building

Taking the small transformer S9-1600/10 kVA as an example, the section of the oil tank is rectangular. The connection group number of the transformer is Y, yn0. The basic dimensions of the oil tank are as follows: width B = 520 mm, length L = 1,340 mm, and height H = 1340 mm. The tank wall is a 6 mm thick steel plate, the tank bottom is an 8 mm thick steel plate, and the tank cover is a 10 mm thick steel plate. The detailed size of the oil tank is shown in Figure 7.21.

The oil tank model was set up according to the size. In the models, the same hypothesis with oblong tank is applied. A finite element model is established for Figure 7.22. Reinforce the iron at the 1/3 and 2/3 points of the model and add the welding edge along the upper part of the tank to increase the strength of the tank, and adopt the same section size in the model. At the bottom of the oil tank welding support position limits six degrees of freedom. Apply the pressure of 49 kPa inside the tank. The elastic modulus of the steel plate is E = 200 GPa, the Poisson's ratio is μ = 0.3, and the density is ρ = 7,800 kg/m^3. The whole tank body is divided into four 16146 node shell elements (shell4). Horizontal reinforcing iron and tank girder element are used with rectangular section (beam). The finite element grid model is shown in Figure 7.22, and the pressure is shown in Figure 7.23.

(2) Comparison of finite element analysis and experimental results

The result of the stress analysis of the rectangular oil tank is shown in Figure 7.24, and the result of the displacement analysis is shown in Figure 7.25. It can be seen from the stress analysis that the maximum stress is of 181 MPa. It also occurs in the tank cover and the tank wall welding because of the stress concentration. The maximum stress value is far less than the yield limit, and the entire tank structure will not yield and meet the stress standard requirements. The maximum displacement is 5.10 mm, which occurs in the middle of the tank wall, and is less than 1.5 times the thickness of the steel plate.

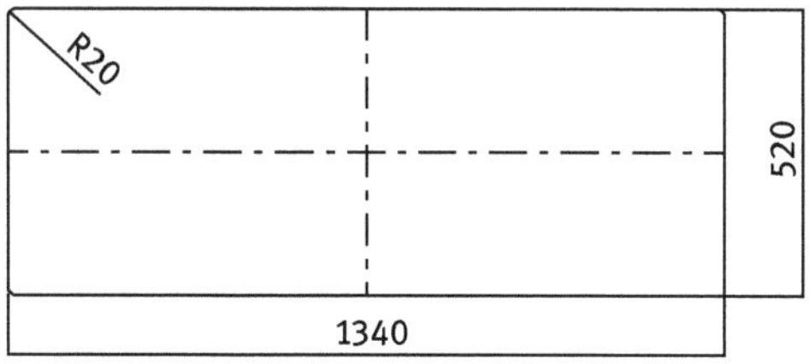

Figure 7.21: Sketch of rectangle tank's cross-sectional size.

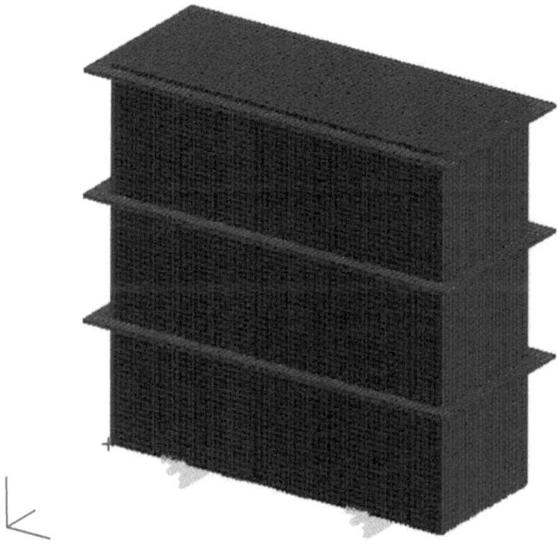

Figure 7.22: Three-dimensional model of a rectangular tank.

Figure 7.23: Sketch of rectangle tank's cross section with pressure.

The comparison of numerical results with experimental results for a rectangular tank is shown in Table 7.7.

It can be concluded from the table that the test results and numerical analysis results can meet the requirements of the test standards. The maximum elastic deformation of the test results is slightly higher than that of numerical analysis. According to the comparison of the results of stress analysis and the deformation in Table 7.7, the design of a rectangular tank structure with reinforced rib is reasonable, which meets the requirements of strength and stiffness.

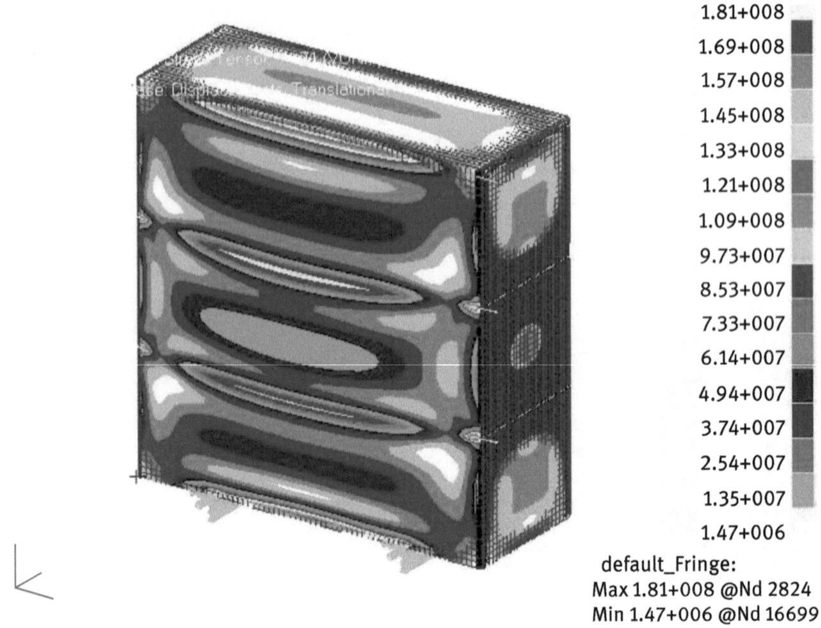

Figure 7.24: Stress analysis result of a rectangular tank.

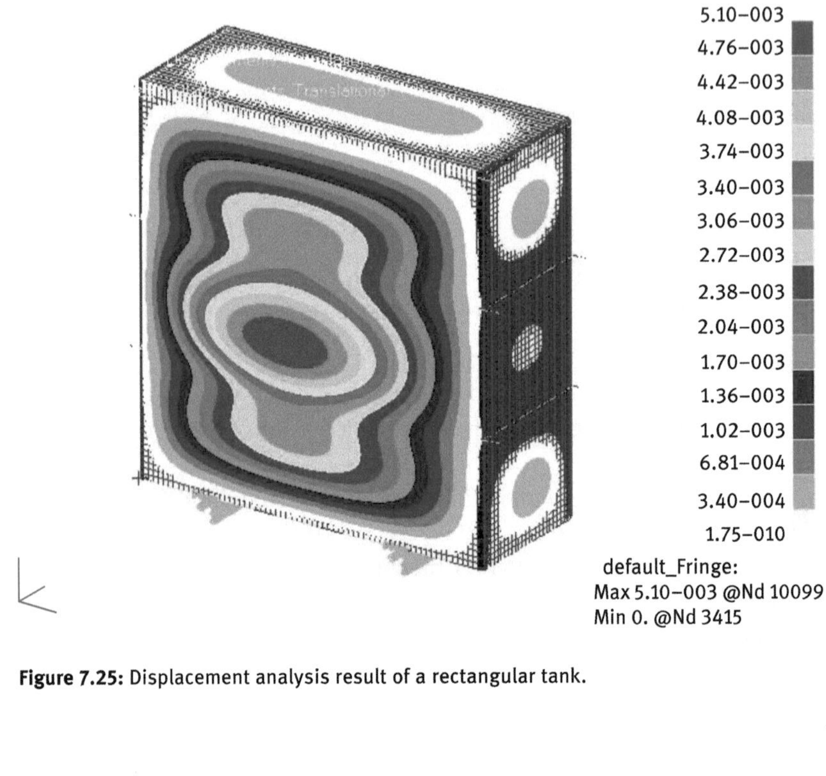

Figure 7.25: Displacement analysis result of a rectangular tank.

Table 7.7: Comparison of rectangle tank's distortion by FEM and test results.

	Maximum elastic deformation (mm)	Maximum permanent deformation (mm)
Test standard requirements	18.00	9.0
Numerical analysis results	5.10	Software unable to calculate temporarily
Positive pressure test result (49 kPa)	6.20	2.15

(3) Structural optimization design
A rectangular tank increases the reinforcing ribs, so the design variables increased stiffener size. An optimization model is established:

$$\begin{cases} \text{find} \quad tb, td, tg, w, h \\ \min \quad M = \sum_{i=1}^{N} m_i^0 \\ \text{s.t.} \quad \underline{\sigma_i} \le \sigma_i \le \overline{\sigma}_i \quad (i = 1, \ldots, N) \\ \qquad \underline{u_j} \le u_j \le \overline{u}_j \quad (j = 1, \ldots, J) \\ \qquad \underline{tb} \le tb \le \overline{tb} \\ \qquad \underline{td} \le td \le \overline{td} \\ \qquad \underline{tg} \le tg \le \overline{tg} \\ \qquad \underline{w} \le w \le \overline{w} \\ \qquad \underline{h} \le h \le \overline{h} \end{cases} \tag{7.9}$$

Here, *tb,td,tg,w,h* are the design variables; M is the total mass of the structure; $\overline{\sigma}_i$ is the allowable stress upper limit; $\underline{\sigma_i}$ is the allowable stress lower limit; $\overline{u}_j$ is the allowable displacement upper limit; $\underline{u_j}$ is the allowable displacement lower limit; N is the total number of constraint units; and J is the total number of constrained nodes.

The design variables were the tank wall thickness *tb*, the bottom thickness *td* and the tank cover thickness *tg*, and strengthen iron and the tank along the thickness W and width h. According to the design principle, the thickness of the steel plate of the small transformer oil tank is commonly used:

$$3\text{ mm} \le tb \le 12\text{ mm}, 3\text{ mm} \le td \le 12\text{ mm}, 3\text{ mm} \le tg \le 12\text{ mm}$$

$$3\text{ mm} \le w \le 16\text{ mm}, 0.1\text{ mm} \le h \le 0.8\text{ mm}.$$

The objective function is the minimum quality of the whole oil tank.

The stress constraint is $-186\,\text{MPa} \le \sigma \le 186\,\text{MPa}$, and the displacement constraint is $-5.1\,\text{mm} \le u \le 5.1\,\text{mm}$.

After 28 iterations, the final stress optimization results are shown in Figure 7.26, and the maximum stress of the optimized structure is 186 MPa. The optimized displacement results are shown in Figure 7.27, and the maximum displacement of the structure is 4.15 mm. The iterative optimization of the design variables is shown in Figure 7.28. The iterative optimization of the objective function is shown in Figure 7.29. Table 7.8 lists the original values and the optimized values of the design variables and objective functions.

As can be seen from the data in Table 7.8, the thickness of the back wall of the tank, bottom thickness and a strengthening rib change a little; the thickness of the tank cover and strengthen rib width change larger; tank quality is greatly reduced. This is also because the tank cover is not considered to dig holes and the other accessories when the model is made, so there will be a significant reduction in the thickness of the tank cover after the optimization. Considering the above factors, the box cover can be thickened. In addition, it can be concluded that by increasing the width of the reinforcement the steel plate thickness can be reduced under the condition of meeting the strength and rigidity of the oil tank, thereby reducing the total quality of the oil tank.

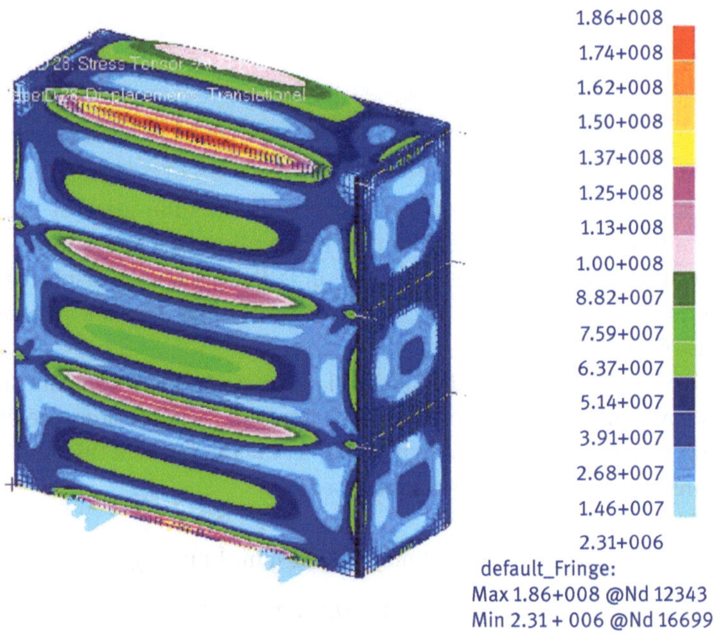

Figure 7.26: Optimal stress result of a rectangular tank.

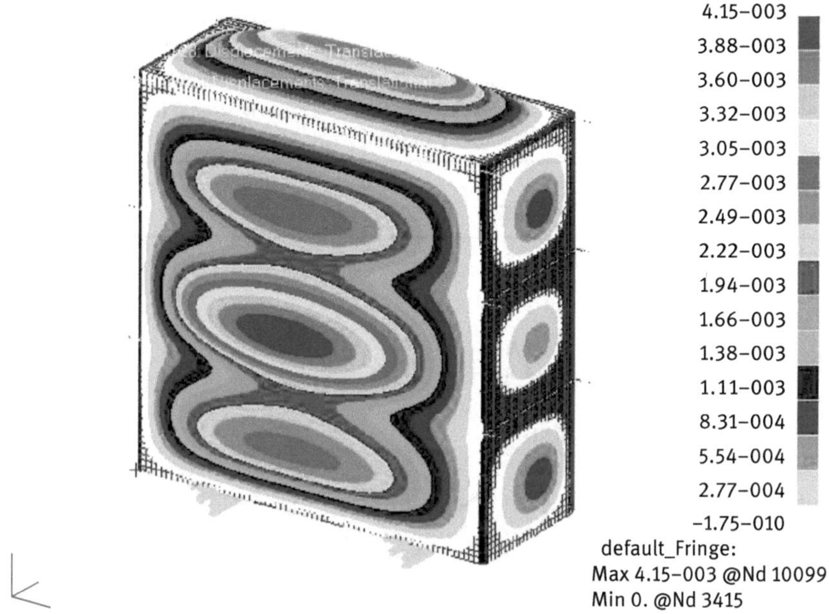

Figure 7.27: Optimal displacement result of a rectangular tank.

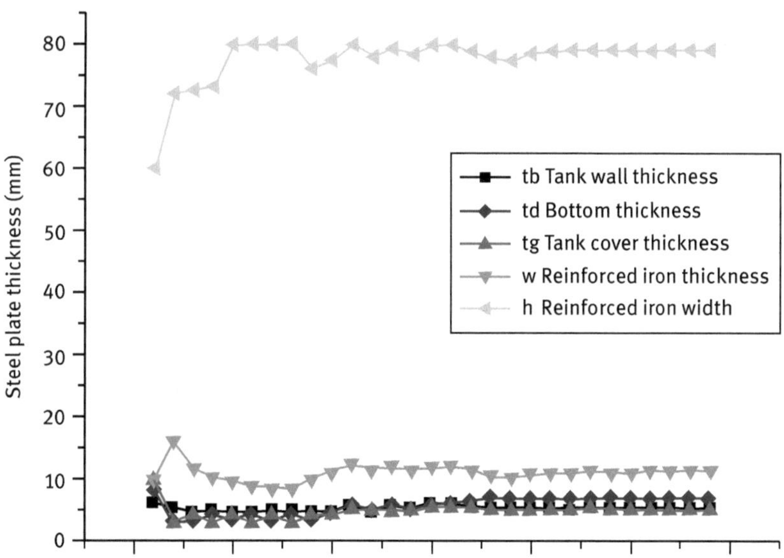

Figure 7.28: Design variable iterative curve of a rectangular tank.

Table 7.8: Comparison of rectangle tank's size before optimization and after optimization.

	Tank wall thickness *tb* (mm)	Bottom thickness *td* (mm)	Tank cover thickness *tg* (mm)	Reinforcing iron *w* (mm)	Reinforcing iron *h* (mm)	Oil tank quality *W* (kg)
Initial value	6.0	8.0	10.0	10.0	60.0	384.2979
Optimal value	5.1881	7.1257	5.3012	11.198	79.156	349.3190

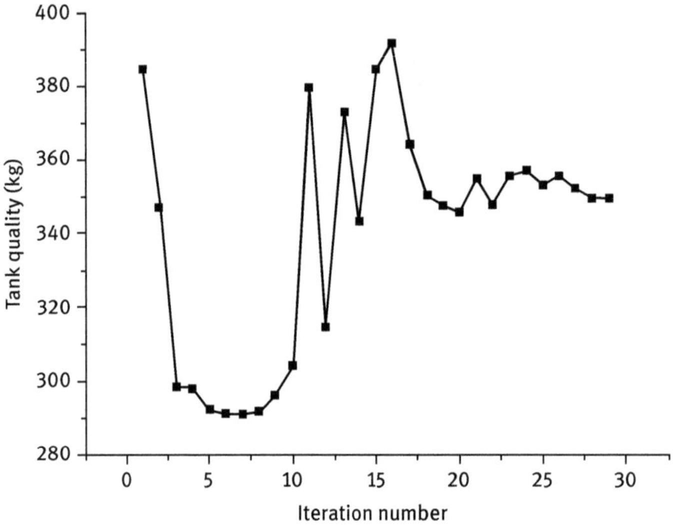

Figure 7.29: Objective function iterative curve of a rectangular tank.

The conclusion can be drawn as follows:

(1) Whether it is a small oblong transformer tank or a rectangle tank, the maximum stress was born in a long-side direction of the tank edge or at the intersection of the tank wall and the bottom. Therefore, we should try to avoid too much stress concentration.

(2) The effect of the reinforcement on the strength and stiffness of the oil tank is very significant. Therefore, the reasonably arranged position, quantity and the section of the reinforcing rib can be effectively improve the strength and rigidity of the oil tank.

7.5 Topological Optimization in the Design of Vacuum Tank

Vacuum tanks for transformers are mainly used arid for 35 kV and below; the capacity of small- and medium-sized transformers is 31,500 kVA. Its temperature can reach

120 degrees and the ultimate residual pressure is 133 pa. After the transformer's body is put into the furnace, the bottom of the vacuum tank and the surrounding heater are heated by steam heating, and the transformer body is heated by the convection and heat radiation of the air in the furnace. In the process of heating the vacuum pumping intermittent and continuous, the water vapors in the tank are taken away to prevent the iron core rust. When the body temperature reaches 110 degrees and the iron core temperature reaches 70–80 degrees, the vacuum pump is started to carry out continuous vacuum drying, and the time required for the process maintained. The temperature, vacuum degree and insulation resistance are measured in the whole process. Therefore, the strength and rigidity of the vacuum tank is very important.

The finite element analysis and the optimization design of the tank are mainly studied in the chapter. Yumin Song used software ANSYS to analyze the stress and deformation of the shell of the 210 T iron tank of the converter and evaluate the strength and rigidity. Zhiliang Zhang designed the large spherical tanks using finite element method and assessment and checked strength on the dangerous section of the stress. At the same time, the pillar height, the plate thickness and the cylinder wall thickness were optimized. Dongping Yang built a finite element model of a blasting tank using the parametric design and finite element analysis in ANSYS software parametric design language (APDL). The optimization model of a blasting tank was given, and the shape and size of a blasting tank were determined. Guo Jiang gave model of the aerodynamic ash car in minimum quality and built the Matlab program using the Matlab optimization function. According to the characteristics of the use of 120 t types of kinds of iron tank, the three kinds of molten iron tank finite element analysis of the stress field were carried out, which provide the theoretical basis for the analysis and design of the molten iron tank. The finite element model of the horizontal gray car body was established, and the displacement field and stress field of the tank structure were calculated, the weak link of the structure was found, the improvement suggestions were put forward, and the dynamic characteristic analysis and calculation were carried out.

In the traditional structure design, the design of the mechanical strength and stiffness of the vacuum tank mainly adopts the experience design and the analogy design, but the detailed analysis and calculation of the stress and deformation does not do. This is mainly because the hand calculation method is too tedious and the efficiency is low. The mechanical strength and stiffness of the vacuum tank obtained by the traditional design method usually do not have the exact design data. This will bring security risks in the installation and operation of the vacuum tank. The finite element method can be effectively used to solve the above problems in the structure design of the vacuum tank.

The three-dimensional finite element model of the octagonal-shaped horizontal vacuum tank structure was established in the Solidworks/COSMOS 3D modeling and finite element analysis software platform, and finite element analysis and structural optimization design were given. The results had certain reference significance for actual production.

7.5.1 Finite Element Analysis of a Vacuum Tank

Take the capacity of the 31,500 kVA power transformer as an example. The basic dimensions of the eight-sided section of the vacuum tank body are as follows: its total width is $B = 3{,}000$ mm, the length hemline is 1,500 mm, the length of a side vertical edge is 1,400 mm, the angle between the side bevel edge and the bottom edge is 120 degrees, the total length is $L = 4{,}500$ mm, the total height is $H = 3{,}500$ mm and tank walls are of a 10 mm thick steel plate. The tank is reinforced with I-beam half in accordance with T-shaped cross section welding on the outer surface of the vacuum tank, the initial vacuum tank model of rib is made of half of the 22 A-type I-steel, and its height is 110 mm. Between the surface adjacent I-beam spacing is 900 mm in the long axis, and between adjacent I-beam spacing in the octagonal tank body section is 700 mm; relative surface were at the same interval arrangement. The three-dimensional model of the vacuum tank was established in the SolidWorks platform. The model of the vacuum tank is shown in Figure 7.30.

To facilitate the establishment of the model, the following assumptions were made:

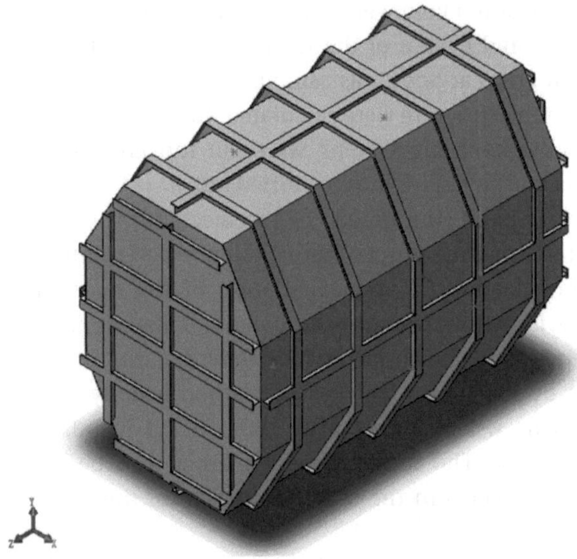

Figure 7.30: Three-dimensional model of a vacuum tank.

(1) Ignore the bottom bracket of the vacuum tank and replace it with the fixed support.
(2) Ignore the connection bolt of the upper tank and the one end of the vacuum tank flange, and the tank door and body can be considered as a whole.
(3) The vacuum tank is taken as a complete steel plate, and the tank installation holes are ignored.
(4) The weld seams of the tank are ignored and the joints of the tank are assumed to have the same strength and stiffness.

According to the intensity data in the literature [68], vacuum tank body manufacturing materials commonly used is the Q235-A or Q235-B steel plate, the yield limit is σ_S = 225MPa, and the tank maximum stress is less than the yield limit.

For the deformation of the tank, the national standard has not been clearly defined. The requirements of the permanent deformation in various enterprises are not the same, but generally controlled in the following range:
(1) The maximum permanent deformation is 0.5δ (δ is the plate thickness).
(2) The maximum amount of elastic deformation ≤2 times the amount of permanent deformation.

At the bottom of the vacuum tank plus a fixed support, limit the freedom of the six directions. Apply the pressure of 10167 Pa, the elastic modulus of the steel plate is E = 210 GPa, the Poisson's ratio is μ = 0.3, and the density is ρ = 7,800 kg/m^3. The whole tank is divided into eight-node body unit.

Based on the three-dimensional model of the vacuum tank built on the SolidWorks platform, the finite element analysis of the model is carried out by using COSMOS. The stress results of finite element analysis are shown in Figure 7.31, and the displacement results are shown in Figure 7.32. It can be seen from the figure that the maximum stress is 280.9 MPa, which occurs at the welding point of the tank cover and the tank wall. The maximum stress value is far greater than the yield limit, and the whole tank structure will yield, which does not meet the standard requirements.

It can be seen from the maximum displacement figure that the maximum displacement is 9.517 mm, which occurs in the middle part of the eight-sided section of the tank body; the maximum displacement is greater than 0.5 times the thickness of the tank wall: it cannot meet the requirements of the tank.

Through the analysis of strength and stiffness, the following conclusions can be drawn. The eight-sided horizontal vacuum tank which uses 10 mm thick steel plate and 22a I type steel 1/2 as the stiffener the stiffness and strength cannot meets the requirements, so the existing structure must be strengthen. There are a variety of options to improve the program. It is found that the distance and height of stiffeners on the side wall of the long axis and the side

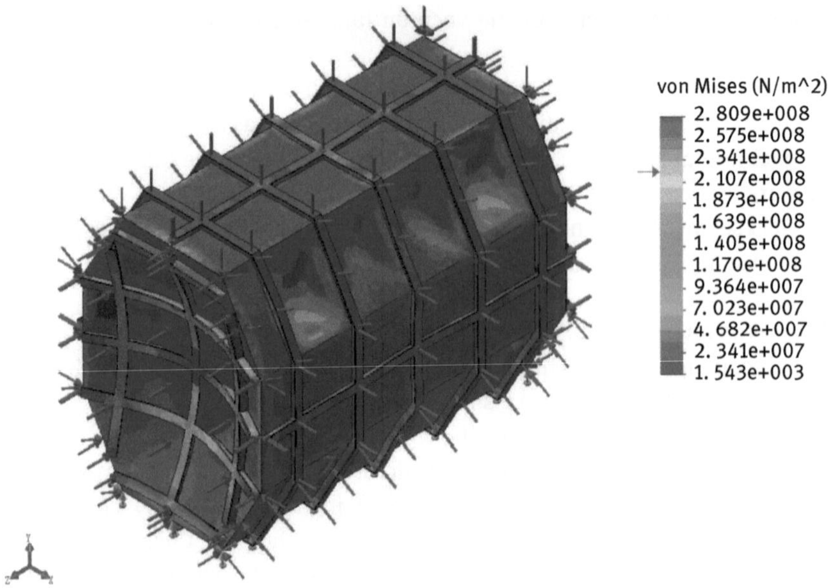

Figure 7.31: Stress analysis result of the original model.

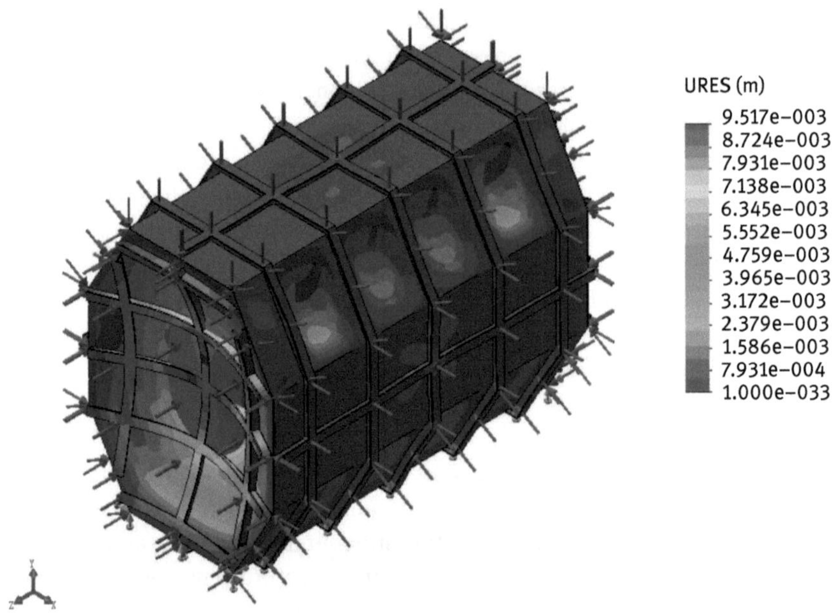

Figure 7.32: Displacement analysis result of the original model.

wall of the eight sides are the most effective to improve the strength of the tank body. It is beneficial to reduce the quality of the tank and save steel under the condition of meeting the strength and stiffness.

7.5.2 Optimization Design of a Vacuum Tank

The spacing of the side wall reinforcement is constant, and changing the height and thickness of the reinforcing rib is the most effective method to improve the strength and rigidity of the tank body. If the structure of the whole tank is optimized, it needs to be divided into a large number of finite element mesh, which needs a lot of memory and computation time. Sometimes, due to the optimization of the number of iterations, the memory capacity is not enough for the termination of the iteration optimize design, and the result is not ideal convergence. At the same time, due to the symmetry of the structure of the tank body, all the reinforced ribs have the same section, so simplifying the initial structure of the tank is feasible.

The two eight-sided walls of the vacuum tank is the largest and the most easily yielding and deforming. Therefore, the simplified model only takes one side, which is covered with reinforcing ribs, all sides of the plate are fixed and supported, and the vertical pressure of a large air pressure is outside. The simplified model of the applied load and the support is shown in Figure 7.33.

The first task is to establish a mathematical model of optimization before structural optimization. Three elements of the structural optimization mathematical model include design variables, objective functions and constraints. According to the concrete structure of the vacuum tank, the thickness of the steel plate is constant, the height and thickness of the reinforcement are used as the design variables, the total mass of the tank body of the vacuum tank is taken as the objective function, and the constraints are, respectively, the stress and displacement as the constraint function. The optimization model is as follows:

$$\begin{cases} \text{find} & tb, td \\ \min & M = \sum_{i=1}^{N} m_i^0 \\ \text{s.t.} & \underline{\sigma_i} \le \sigma_i \le \overline{\sigma_i} \quad (i = 1, \ldots, N) \\ & \underline{u_j} \le u_j \le \overline{u_j} \quad (j = 1, \ldots, J) \\ & \underline{tb} \le tb \le \overline{tb} \\ & \underline{td} \le td \le \overline{td} \end{cases} \tag{7.10}$$

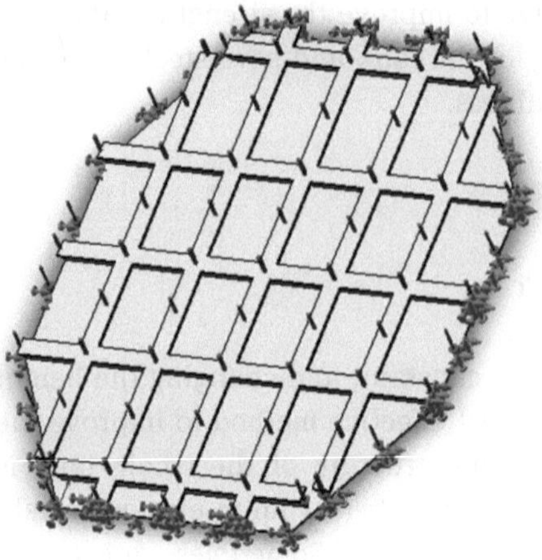

Figure 7.33: Simplified model of a vacuum tank.

Here, *tb*, *td* are the design variables, respectively, for the height and thickness of the reinforcement; *M* is the total mass of the structure; $\overline{\sigma}_i$ is the allowable stress upper limit; $\underline{\sigma}_i$ is the allowable stress lower limit; $\overline{u}_j$ is the allowable displacement upper limit; $\underline{u}_j$ is the allowable displacement lower limit; *N* is the total number of constrained units, and *J* is the total number of constrained nodes.

Design variables are taken to be *tb*, the height of reinforcement, and *td*, the thickness of reinforcement. According to the thickness of the steel plate used in the vacuum tank, the value of the design variable is

$$50\ \text{mm} \le tb \le 300\ \text{mm}, 1\ \text{mm} \le td \le 50\ \text{mm}.$$

The objective function is the minimum quality of the whole tank.

The stress constraint is $-180\ \text{MPa} \le \sigma \le 180\ \text{MPa}$, and the displacement constraint is $-5\ \text{mm} \le u \le 5\ \text{mm}$.

The results of the design variables before optimization are shown in Figure 7.34, in which the height of the reinforcing bar is 110 mm and the thickness is 3.75 mm. After 11 iterations, the final optimization results are shown in Figure 7.35, in which the height of the optimized reinforcement is 213.53 mm and the thickness is 1.88 mm.

From the above data, it can be seen that only the height of the reinforcement increases, the thickness of the reinforcement was in the small trend, and finally the side wall of the tank meets the stress and displacement of the constraint requirements. Since the whole tank structure is not considered when the optimization model is established, it is necessary to verify the stress

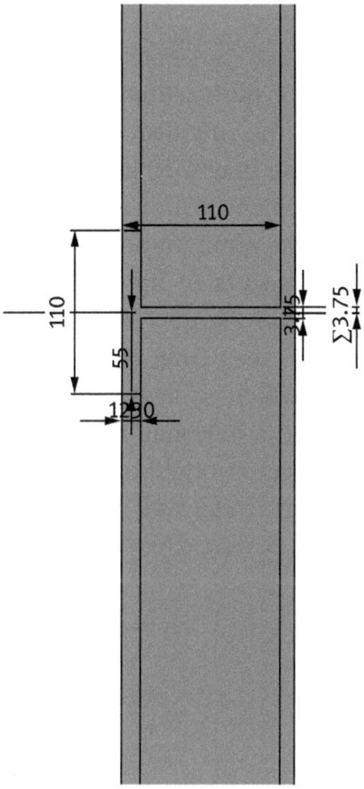

Figure 7.34: Original size of stiffeners.

and displacement of the whole vacuum tank after the height and thickness of reinforcement are rounded.

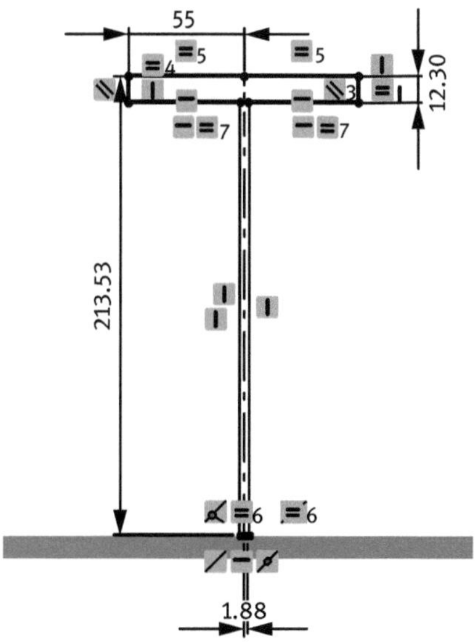

Figure 7.35: Optimal size of stiffeners.

7.5.3 Modify and Validate the Design of a Vacuum Tank

According to the optimization design of the side wall of the vacuum tank, the height and thickness of the reinforcement bars need to be rounded. According to [68], half 32A type I-beam is the reinforcing rib whose height is 160 mm and thickness is 9.5 mm. In the case of keeping the reinforcement distance constant, the section size of all reinforced bars is modified, and the strength and stiffness of the modified vacuum tank model are verified. The results of finite element stress analysis of the modified vacuum tank are shown in Figure 7.36, and the displacement results are shown in Figure 7.37. It can be seen from the analysis results that the maximum yield stress is 163.8 Mpa, which is lower than the yield stress of carbon steel. The maximum displacement is 4.96 mm, which is 5 mm less than the permanent deformation; the entire vacuum tank body mass is 10.3 tons. According to the method, it can be seen that it can quickly get the safe and optimal structure of the vacuum tank.

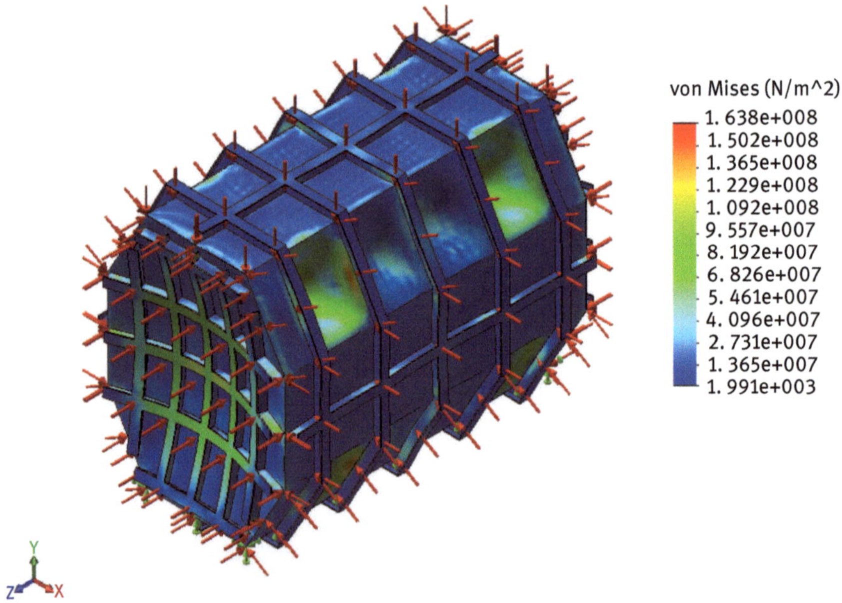

Figure 7.36: Stress result of a modified vacuum tank.

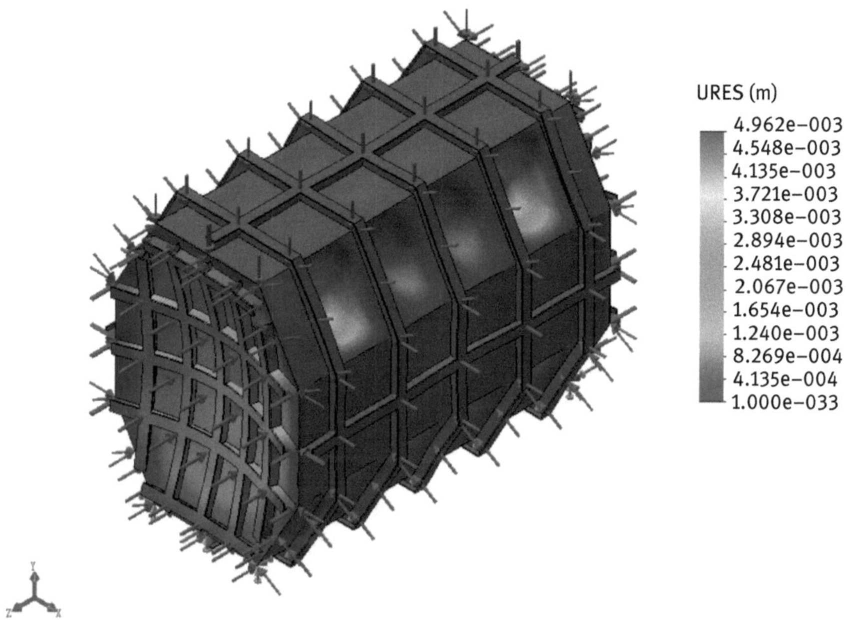

Figure 7.37: Displacement result of a modified vacuum tank.

7.5.4 Conclusion

(1) Vacuum tank is important equipment in the manufacturing process of the transformer. The design of the vacuum tank is mainly focused on the strength and stiffness of the tank.

(2) Vacuum tank is important equipment in the manufacturing process ofThe effect of reinforcement on the strength and stiffness of the tank body is very significant. The position of the reinforcing rib is reasonably arranged, and the number and the section of the reinforcing rib can effectively improve the strength and rigidity of the tank.

7.6 Summary

The design of an optimization system is carried out, based on the method of VC++ calling FORTRAN mixed programming. Directly call the executable approach easily dealing with large amounts of data, and solve the "black screen" problem appearing in the process of the asynchronous call. Make full use of the powerful function of the VC++ development interface and the powerful scientific computing power of the FORTRAN language. The developed optimization design system not only has a friendly and preprocessing and postprocessing interface, but also can achieve the function of the expected structure optimization design of the software.

The topology optimization design and the structure design are combined in design of the hoist. Two new structural schemes are proposed based on the finite element static analysis results and the topology optimization results, and the finite element analysis of the strength and stiffness of the new scheme is made. It is the first time that the topology optimization method is applied in the structure design of a transformer spreader.

The maximum deformation and the maximum stress site of the oil tank were given, and the comparison of measured data in engineering was given according to the finite element analysis of the small transformers' rectangular and oblong tank. The optimum size of the oil tank and the section size of the reinforcement bar are obtained by the method of size optimization. This method can effectively improve the design quality of a transformer oil tank, reduce the blindness of the strength and rigidity design, and improve the accuracy of product design, safety and economy.

By using the method of finite element analysis and the structural optimization method in the design of a vacuum tank body, it can effectively improve the quality of the design, reduce the blindness in the tank design, reduce body quality ensuring the strength and stiffness requirements, and improve the economic efficiency of product design, accuracy, and safety.

8 Prospect

At present, the research on topology optimization mainly focuses on stress, displacement, frequency constraint types, and fewer buckling constraints. Topology optimization of continuum structures with respect to size and shape optimization is a higher-level optimization, and because their mathematical model is difficult to describe, it is considered to be the most challenging subject in the field of topology optimization. Therefore, topology optimization of continuum structures considering buckling will be more challenging. In this book, independent continuous mapping (ICM) i.e., independent continuous mapping) topology optimization method considering buckling is used for the topology optimization problem of continuum structures.

At present, the ICM method is applied to truss, rigid frame, shape and topology optimization of the continuum structure. The constraints include stress, displacement and frequency response, and satisfactory results are obtained. This book confirms that the ICM method can effectively deal with the topology optimization of continuum structures with buckling constraints, buckling and displacement constraints, buckling and stress constraints and a combination of all the three constraints.

1. The continuum topology optimization model is established.

Four kinds of optimal models with the object function of weight, buckling constraints, displacement constraints, stress constraints and all static constraints combination were constructed and solved via the ICM method. The continuous independent topological variables are used in this problem. The element weight, element stiffness matrix and element geometric stiffness matrix are introduced by different filter functions.

On the basis of the ICM method and homogenization theories, filter function expressions of the unit weight, element stiffness matrix and filter function expression of the element geometric stiffness matrix of the strain energy were derived. By numerical calculation, the filter function exponent of element stiffness and element geometric stiffness matrix was fit with the least square method.

2. Approximate treatment of topology optimization model.

The objective function is nonlinear. The objective function is approximately expressed as a second-order expressions and buckling constraints were explicitly expressed by structure's mode strain energy and mode geometric strain energy. Displacement constraints were approximately expressed as an explicit function by using virtual unit-load method. Using the globalization method of stress constraints and the von Mises's yield criterion in mechanics of materials, the local stress constraints were translated into the whole structure strain-energy constraint. Finally, optimization models were translated into a dual program and solved by the sequence

https://doi.org/10.1515/9783110462272-008

second-order programming method. The number of the variable is reduced and the model's scale is diminished.

3. Selection criterion of constrained value.

A suitable set of curves about "Ratio of Quality" and "Ratio of Mechanics Property" of continuum structures topological optimization were drawn and fitted. It is convenient to create the optimal model. The concepts of "Ratio of Quality," "Ratio of Buckling Constraints," "Ratio of Displacement Constraints" and "Ratio of Stress Constraints" were defined. The distribution curve of Quality ratio and buckling constraint ratio, the distribution curve of Quality ratio and displacement constraint ratio, and the distribution curve of Quality ratio and stress constraint ratio were given by one numerical example. Using the least square method, the curve of Quality ratio and buckling constraint ratio, the curve of Quality ratio and displacement constraint ratio, and the curve of Quality ratio and stress constraint ratio were fit and expressed as a function. Finally, numerical examples were given to verify the functions of the curves.

All topology optimization algorithms are developed on the MSC.Patran/Nastran software platform, and an independent topology optimization program module is established. The topological optimization program has a user-friendly interface result output function and the help document can be read expediently by the user. The topology optimization program extends the research scope of continuum structure topology optimization. The optimal arithmetic model can not only deal with the topology optimization of two-dimensional continuum structure and three-dimensional continuum body but also deal with the cylinder shell. All algorithms can calculate the topology optimization problems of multistage buckling modes under single working condition and multi working condition of continuum problem.

A few conclusions are given above. In addition, there are still some problems to be further studied in the future.

1. In this book, only the topology optimization of continuum structures under the static state was considered, and the frequency and dynamic response of the structure are to be further studied.
2. The present structural topology optimization research focuses on the truss, frame, film, plate, three-dimensional entity and so on, in the single structure type. In this book, owing to the buckling constraints, the topology of the structure concentrated more on the plate, shell and three-dimensional structure. In practical engineering, the combined structure of a variety of structures has a large proportion, so future research should focus on the combination of topology optimization algorithm and topology optimization of composite structure.
3. In this book, the buckling topology optimization is based on the linear buckling analysis. The loading history and the initial imperfection of the structure are not considered in the linear buckling phase, so the buckling critical load has a great error. The results considering nonlinear buckling will be more accurate.

4. The arithmetic development language is Patran command language. It is an interpretation and implementation language, so the execution speed is slow. Moreover, it is in the text format; therefore, software copyright protection may not be conducive. Further software development work has to concentrate on the integration of the algorithms and the pre- and postprocessing procedures.

References

[1] Han Qiang. Advances in research for static and dynamic buckling of structures. Advances in Mechanics, Vol. 28, No. 3, 1998, pp. 349–360.

[2] Feng Lei. Strength analysis of typical structure based on the buckling theory. Master degree thesis. Chang'an University, 2003.

[3] Sun Huanchun, Chai Shan, Wang Yuefang. Structural optimization with discrete variables. Dalian University of Technology Press, 1995.

[4] Qian Lingxi. Optimization design of engineering structure. Water Resources and Electric Power Press, 1983.

[5] Cheng Gengdong. Introduction to optimum design of engineering Structures. Water Resources and Electric Power Press, 1984.

[6] Sui Yunkang. Modeling, transformation and optimization-New advances in structural synthesis method. Dalian University of Technology Press, 1996.

[7] Michell A G M. The limits of economy of materials in frame structures. Philosophical Magazine, Series6, Vol. 8, No. 47, 1904, pp. 589–597.

[8] Cheng K T, Olhoff N. An investigation concerning optimal design of solid elastic plates. International Journal of Solids and Structures, Vol. 17, No. 3, 1981, pp. 305–323.

[9] Bendsoe M P, Kikuchi N. Generating optimal topology in structural design using a homogenization method. Computer Methods in Applied Mechanics and Engineering, Vol. 303, No. 69, 1998, pp. 197–224.

[10] Chen Tieyun, Shen Huishen. Buckling of the structure. Shanghai Science and Technology Literature Press, 1993.

[11] Kalyanaranan V, Pekoz T, Winter G. Unstiffened compression elements. Journal of Structural Division, ASCE, Vol. 103, No. 9, 1977, pp. 1013–1019.

[12] Rajaskaran S, Murray D W. Coupled local buckling in Wide-flange Beam-columns. Journal of Structural Division, ASCE, Vol. 99, No. 6, 1973, pp. 1003–1023.

[13] Johnson C P, Will K M. Beam buckling by finite element procedure. Journal of Structural Division, ASCE, Vol. 100, No. 3, 1974, pp. 669–685.

[14] Hancock G J. Local distortional, and lateral buckling of I Beams. Journal of Structural Division, ASCE, Vol. 104, No. 11, 1978, pp. 1787–1798.

[15] Koiter W T. On the stability of elastic equilibrium. Holland, Delft, 1945.

[16] Van der Neut A. The interaction of local buckling and column failure of thin walled compression members. Proceedings 12th International Congress on Applied Mechanics, Stanford University, 1969, pp. 389–399.

[17] Guo Yanlin. Cold-formed steel column local and global buckling. Journal of Xi'an Institute of Metallurgical Construction, Vol. 21, No. 2, 1989, pp. 75–81.

[18] Zheng Hong. Development of steel structure related to buckling theory. Journal of Northwest Institute of Architectural Engineering, June 1999, pp. 1–5.

[19] Prager W, Rozvany G I N. Optimization of structural geometry in dynamical systems. Bednark A R, Cesari L, Eds. Dynamical systems, Academic, New York, 1997, pp. 265–294.

[20] Rozvany G I N. Exact least-weight truss layouts for rectangular domains with various support conditions. Structural Optimization, Vol. 6, No. 1, 1993, pp. 63–69.

[21] Dorn W, Gomory R, Greenberg H. Automatic design of optimal structures. Journal de Mechanique, Vol. 3, No. 1, 1964, pp. 25–52.

[22] Ringertz U. On topology optimization of trusses. Engineering Optimization, Vol. 9, No. 3, 1985, pp. 209–218.

[23] Kirsch U. Optimal topologies of truss structures. Computer Methods in Applied Mechanics and Engineering, Vol. 72, No. 1, 1989, pp. 15–28.

https://doi.org/10.1515/9783110462272-009

[24] Duan Baoyan, Ye Shanghui. Topology optimization of truss structure considering state constraints under multiple load cases. Acta Mechanica Sinica, Vol. 24, No. 1, 1992, pp. 59–69.

[25] Tan Zhongfu, Sun Huanchun. The modified simplex method for topology optimization of space truss with multiple loading conditions. Acta Mechanica Sinica, Vol. 26, No. 1, 1994, pp. 90–97.

[26] Sheu CY, Schmit LA. Minimum weight design of elastic redundant truss under multiple static loading conditions, AIAA Journal, Vol. 10, No.1, 1972, February, pp. 155–162.

[27] Ringertz U. A branch and bound algorithm for topology optimization of truss structures. Engineering Optimization, Vol. 10, No. 2, 1986, pp. 111–124.

[28] Wang Yuefang, Sun Huanchun. Topology optimization design of truss with discrete size variables subject to multiple constraints and loading cases. Acta Mechanica Sinica, Vol. 27, No. 3, 1995, pp. 366–369.

[29] Chai Shan, Lianshuan Shi, Sun Huanchun. Topology optimization of truss structures with discrete variables including two kinds of variables. Acta Mechanica Sinica, Vol. 31, No. 5, September 1999, pp. 574–583.

[30] Kirsch U. Optimal topologies for structures. Applied Mechanics Reviews, Vol. 42, No. 8, 1989, pp. 223–239.

[31] Willam Keith Rule. Automatic truss design by optimized growth. J Struct Eng, Vol. 120, No. 10, 1994, pp. 3063–3070.

[32] McKeown J J. Growing optimal pin-jointed frames. Structural Optimization, Vol. 15, 1998, pp. 92–100.

[33] Liu J S, Parks G T, Clarkson P J. Metamorphic development: A new topology optimization method for continuum structures. Structural and Multidisciplinary Optimization, Vol. 20, pp. 288–300.

[34] Liu J S, Parks G T, Clarkson P J. Topology shape optimisation of axisymmetric continuum structures – A metamorphic development approach. Structural and Multidisciplinary Optimization. DOI 10.1007/s00158-004-0445-5

[35] Luo Ying, Duan Baoyan. Topology optimization design method of structure evolution of triangular mesh. Chinese Journal of Applied Mechanics, Vol. 19, No. 3, 2002, pp. 50–53.

[36] Zhou M, Rozvany G I N. Difficulties in truss topology optimization with stress local buckling constraints. Structural Optimization, Vol. 11, No. 3, 1996, pp. 134–136.

[37] Sved G, Ginos, Z. Structural optimization under multiple topology optimizations. Structural Optimization, Vol. 10, No. 10, 1968, pp. 803–805.

[38] Cheng G D, Jiang Z. Study on topology optimization with stress constraints. Engineering Optimization, 1992, Vol. 20, No. 3, pp. 129–148.

[39] Cheng Geng Dong, GuoXu. ε-relaxed approach in structural topology optimization. Structural Optimization, Vol. 13, No. 4, 1997, pp. 258–266.

[40] Guo X, Cheng G.D, Yamazaki K. A new approach for the solution of singular optima in truss topology optimization with stress and local buckling constraints. Structural and Multidisciplinary Optimization, Vol. 22, No. 5, 2001, pp. 364–372.

[41] Zou Wensheng, Zuo Zhengxing. Truss topology optimization considering structural stability. Journal of Beijing Institute of Technology, Vol. 19, No. 1, 1999, pp. 29–33.

[42] Liu Wei. Topology optimization design of space truss buckling constraints. Master degree thesis. Dalian University of Technology, 2002.

[43] Xu Suqiang, Xia Renwei. Truss topology optimization with genetic algorithm. Computational structural mechanics and applications, Vol. 11, No. 4, 1994, pp. 436–446.

[44] May S A, Balling R J. A filtered simulated annealing strategy for discrete optimization of 3D frameworks. Structural Optimisation, Vol. 4, No. 2, 1992, pp. 142–148.

[45] Cai Wenxue, Cheng Gengdong. Truss topology optimization with simulated annealing method design. Journal of South China University of Technology, Vol. 26, No. 9, 1998, pp. 78–84.
[46] Zhao Guozhong, Chen Biaosong, Kang Zhan, Gu Yuanxian. Buckling and stiffness design optimization of beams of various section. Engineering Mechanics. Vol. 19, No. 3, 2002, pp. 44–49.
[47] Liu Xiaolin, Pang Baojun, Li Tao, Ji Jin. Optimization and stability analysis of large depressed center flat chassis structure. Railway vehicles, Vol. 35, No. 11, 1997, pp. 35–38.
[48] Sui Yunkang, Yang Deqing, Sun Huanchun. Truss geometry optimization based on two-step method linked with sensitivity of optimal objective. Acta Mechanica Sinica, Vol. 29, No. 4, 1997, pp. 470–475.
[49] Sui Yunkang, Yang Deqing. Two kinds model of topology optimization design of truss structure variable density. Mechanics in Engineering, Vol. 19, No. 5, 1997, pp. 61–62.
[50] Sui Yunkang, Yang Deqing. A new method for structural topological optimization based on the concept of independent continuous variables and smooth model. Acta Mechanica Sinica (English Edition), Vol. 18, No. 2, 1998, pp. 179–185.
[51] Sui Yunkang, Yang Deqing, Sun Huanchun. Uniform ICM theory and methods on optimization of structural topology for skeletal and continuum structures. Chinese Journal of Computational Mechanics, Vol. 17, No. 1, 2000, pp. 28–33.
[52] Sui Yunkang, Yu xin Ye Baorui. The Uniform Model Based on the Exist-Null Combination for the Truss and Membrane Topological Optimization with Stress Constraint. Acta Mechanica Solidasinica, Vol. 22, No. 1, 2001, pp. 15–22.
[53] Sui Yunkang, Yu Xin. The topological optimization for truss structures with stress constraints based on the exist-null combined model. Acta Mechanica Sinica, Vol. 14, No. 4, 1998, pp. 363–370.
[54] Yun Kang Sui. ICM method of topological optimization for truss, frame, continuum structure, WCSMO-4, Dalian, June, 2001, pp. 4–8.
[55] Yun Kang Sui. Modeling and optimization of topology design for skeleton and continuum structure. ICOTA2001, December, 2001, pp. 16–18.
[56] Yun Kang Sui, Lian Chun Long, Xu Chun Ren. Topological optimization of frame structures with stress and displacement constraints under multi-loading cases. WCSMO-4, Dalian, June, 2001, pp. 4–8.
[57] Yang Deqing. Study on continuous variables of structural shape and topology optimization design theory and method. Doctoral dissertation, Dalian University of Technology, 1998.
[58] Yu Xin. Topology optimization design of structures with and without complex method and program. Doctoral dissertation. Dalian University of Technology, 1999.
[59] Du Jiazheng. Frame structure section optimization and topology optimization method and software development. Doctoral dissertation. Beijing University of Technology, 2004.
[60] Ren Xuchun. Frame structure topology optimization and inclined centralizer position optimization. Master thesis. Beijing University of Technology, 2000.
[61] Sui Yunkang, Peng Xirong. Topology optimization method for improving ICM. Acta Mechanica Sinica, Vol. 37, No. 2, 2005, pp. 190–198.
[62] Peng Xirong. The static and dynamic topology optimization of continuum structures.Doctoral dissertation, Beijing University of Technology, 2004.
[63] Ye Hongling. Method and software development of static topological optimization of continuum structures. PhD thesis, Beijing University of Technology, 2005.
[64] Guan S, Grant P, Xie Y M. Evolutionary structural optimization incorporating tension and compression materials. Advances in Structural Engineering, Vol. 2, No. 4, 1999, pp. 273–288.

[65] João Carlos Arantes Costa Jr, Marcelo Krajnc Alves. Layout optimization with h-adaptivity of Structures. International Journal for Numerical Methods in Engineering, Vol. 58, No. 17, 2003, pp. 83–102.
[66] Xie Yucheng. Manual of power transformer. Mechanical Industry Press, Beijing, 2003.
[67] Magazine editorial department of transformer. Manual of power transformer. Liaoning science and Technology Press, Shenyang, 1989
[68] The mechanical properties of Engineering Materials Data Handbook. Mechanical Engineering Materials Data Handbook. Mechanical Industry Press, Beijing, 1994.

Disclaimer for the software

There are high-performance computing workstations, servers and high-performance computers, and other ancillary in the Engineering Numerical Simulation Center of Beijing University of Technology.

At the same time, the center has many famous CAE software, such as MSC/PATRAN, the structure analysis software NASTRAN, dynamics analysis software DYTRAN and STRAND 7, nonlinear analysis software MSC/MARC, the fatigue analysis software MSC/FATIGUE. The “American MSCCAE product technical training center” was established by the center with the cooperation of the MSC company, which is responsible for the MSC/PATRAN and MSC/NASTRAN software training in the northern district.

Dr. Bingchuan Bian graduated from the Engineering Numerical Simulation Center of Beijing University of Technology, and his tutor is Professor Sui Yun Kang, a famous mechanics expert. After graduation, Dr. Bian has maintained close research cooperation with the engineering numerical simulation center of Beijing University of Technology. The programs in text were completed using MSC/PATRAN-Nastran with the cooperation of the Engineering Numerical Simulation Center of the Beijing University of Technology.

Tablet of the American MSC/CAE product technical training center

https://doi.org/10.1515/9783110462272-010

Index

https://doi.org/10.1515/9783110462272-011